T0201997

Essays in the Philosophy of Chemistry

Essays in the Philosophy of Chemistry

Edited by ERIC SCERRI AND GRANT FISHER

UNIVERSITY PRESS

Oxford University Press is a department of the University of Oxford.
It furthers the University's objective of excellence in research, scholarship,
and education by publishing worldwide. Oxford is a registered trade mark of
Oxford University Press in the UK and certain other countries.

Published in the United States of America by Oxford University Press
198 Madison Avenue, New York, NY 10016, United States of America

© Oxford University Press 2016

All rights reserved. No part of this publication may be reproduced,
stored in a retrieval system, or transmitted, in any form or by any means,
without the prior permission in writing of Oxford University Press,
or as expressly permitted by law, by license, or under terms agreed with
the appropriate reproduction rights organization. Inquiries concerning
reproduction outside the scope of the above should be sent to the
Rights Department, Oxford University Press, at the address above.

You must not circulate this work in any other form
and you must impose this same condition on any acquirer.

Library of Congress Cataloging-in-Publication Data
Names: Scerri, Eric R., editor. | Fisher, Grant (Grant Andrew), editor.
Title: Essays in the philosophy of chemistry / edited by Eric Scerri and
Grant Fisher.
Description: Oxford : Oxford University Press, [2015] | Includes
bibliographical references and index.
Identifiers: LCCN 2015031319 | ISBN 978-0-19-049459-9
Subjects: LCSH: Chemistry—Philosophy.
Classification: LCC QD6 .E87 2015 | DDC 540.1—dc23 LC record available at
http://lccn.loc.gov/2015031319

9 8 7 6 5 4 3 2 1
Printed by Sheridan, USA

Eric Scerri would like to dedicate this volume to his late mother, Ines, who passed away in 2015 and who spent a good deal of her life in nurturing and encouraging his academic interests.

Grant Fisher dedicates this volume to his late mother, Jeannette. The editors together also dedicate the volume to the Estonian philosopher of science, Rein Vihalemm, one of our contributors to this volume, who passed away in 2015.

CONTENTS

CONTRIBUTORS

Marina Paola Banchetti-Robino, Department of Philosophy, Florida Atlantic University.

Nicholas W. Best, Department of History and Philosophy of Science, Indiana University.

Alan Chalmers, Unit for History and Philosophy of Science, University of Sydney.

Hasok Chang, Department of History and Philosophy of Science, University of Cambridge.

Joseph E. Earley, Sr., Department of Chemistry, Georgetown University.

Grant Fisher, Graduate School of Science and Technology Policy, Korea Advanced Institute of Science and Technology.

Kostas Gavroglu, Department of History and Philosophy of Science, University of Athens.

Rom Harré, Department of Psychology, Georgetown University and Linacre College, University of Oxford.

Robin Findlay Hendry, Department of Philosophy, Durham University.

Hinne Hettema, Department of Philosophy, The University of Auckland.

Noretta Koertge, Department of History and Philosophy of Science, Indiana University.

Jean-Pierre Noël Llored, Linacre College, University of Oxford and SPHERE, Paris Diderot University.

Lee McIntyre, Center for Philosophy and History of Science, Boston University.

Richard M. Pagni, Department of Chemistry, University of Tennessee.

Guillermo Restrepo, Laboratorio de Química Teórica, Universidad de Pamplona.

Eric R. Scerri, Department of Chemistry and Biochemistry, University of California Los Angeles.

Ana Simões, Centro Interuniversitário de História das Ciências e Tecnologia, Faculdade de Ciências, Universidade de Lisboa.

Rein Vihalemm, Institute of Philosophy and Semiotics, University of Tartu.

Essays in the Philosophy of Chemistry

Introduction

ERIC SCERRI AND GRANT FISHER

THE PHILOSOPHY OF CHEMISTRY emerged in Europe during the 1990s—or to be more precise, in the year 1994.[1] Since that time, the field has grown in stature and importance, offering a unique perspective on chemistry and its place within the natural sciences. For example, the International Society for the Philosophy of Chemistry (ISPC) was formally established in 1997, following some earlier gatherings among the early enthusiasts of the field, and has held meetings every year since. The journal of the society, *Foundations of Chemistry*, began appearing in 1999 and is now in its sixteenth year of publication, with full recognition from the Science Citation Index.[2] But the emergence of the philosophy of chemistry has hardly been an easy process. As Joachim Schummer points out in his editorial for the journal *Hyle*, the other journal dedicated to the philosophy of chemistry, the philosophy of chemistry had been mostly ignored as a field, in contrast to that of physics and, later, biology.[3] This seems to have been due to a rather conservative, and at times implicitly reductionist, philosophy of physics whose voice seemed to speak for the general philosophy of science. It has taken an enormous effort by dedicated scholars around the globe to get beyond the idea that chemistry merely provides case studies for established metaphysical and epistemological doctrines in the philosophy of physics. These efforts have resulted in both definitive declarations of the philosophy of chemistry to be an autonomous field of inquiry and a number of edited volumes and monographs.[4]

---

[1] A number of articles have examined the state and growth of the discipline, including Van Brakel, J., Vermeeren, H., "On the philosophy of chemistry," *Philosophy Research Archives*, 7, 1405–56, 1981; Scerri, E.R. & McIntyre, L. (1997) "The case for the philosophy of chemistry," *Synthese* 111: 213–232, 1997; Van Brakel, J., "Philosophy of science and philosophy of chemistry," *Hyle*, 20, 11–57, 2014.

[2] Foundations of Chemistry, http://www.springer.com/philosophy/epistemology+and+philosophy+of+science/journal/10698.

[3] Schummer, J., Editorial for the Special Issue on "'General Lessons from the Philosophy of Chemistry' on the Occasion of the Twentieth Anniversary of *Hyle*," *Hyle* 20: 1–10, 2014.

[4] We have gathered below what we hope is a complete bibliography on all monographs and edited collections on the philosophy of chemistry and closely related topics.

Philosophy of chemistry, like any other field of inquiry, has a historical and social context. But from a broad conceptual perspective, its birth pains seem difficult to reconcile with a rather obvious property of chemistry: Within the natural sciences, chemistry's domain borders both physics and biology. In this regard, philosophy of chemistry is potentially unrivalled in its philosophical importance within the philosophy of natural sciences. Since it shares it boundaries with both physics and biology, no other discipline has the capacity to do more to edify the complex interactions between the life and physical sciences. Philosophy of chemistry is quite literally at the heart of the philosophies of natural sciences. It is for this reason, among numerous others, that the present volume collectively expresses issues, topics, and problems in a field that has its own voice, its own agenda, and its own projects. But engaging with philosophy of chemistry is an opportunity for a deeper understanding of chemistry and the natural sciences generally. This volume explores the ways in which this might be achieved.

The philosophy of chemistry is still a young field, yet it has rapidly established a recognized infrastructure for the dissemination of research through its journals, societies, and meetings.[5] This is a measure of both the interest in the field and a commitment to its long-term development. The current volume represent a stage in the development of the philosophy of chemistry, providing a repository for some of the best current thought in the field, and a guide for both the initiated and uninitiated. It serves to map out some of the distinctive features of the philosophy of chemistry, to establish recent developments in the field as chemistry evolves as a set of dynamic practices and conceptual frameworks. The boundaries that distinguish philosophy of chemistry from the philosophies of the other natural sciences are a matter of continuing evolution. This is just as it should be. Philosophies of the natural sciences must continue to evolve along with the sciences constituting the object of inquiry. And internal developments do not happen in a vacuum. As well as the historical development of chemistry itself, there are developments in metaphysics, epistemology, and methodology that are both informed by and shape the philosophy of chemistry. *Essays in the Philosophy of Chemistry* recognizes these connections and the importance of integrating history of chemistry with the philosophy of chemistry. And it demonstrates how some parts of contemporary philosophy are all the poorer for its neglect of chemistry.

Philosophy of chemistry is notable in that working chemists of philosophical caliber contribute to the field in addition to professional philosophers. As Kostas Gavroglu and Ana Simões stress in their contribution to this volume (see chap. 3), chemists have been keen to engage in philosophical problems—unlike many of their contemporaries in physics, who may well have been aware of these problems but rarely intervened. This dynamic is crucial. Not

---

[5] A list of meetings and their locations, has been published in a recent editorial of *Foundations of Chemistry*. E.R. Scerri, Editorial 46, *Foundations of Chemistry*, 16, 1–2, 2014. Since that editorial appeared the 2014 meeting was held at the London School of Economics and the 2015 meeting in Rio de Janeiro, Brazil.

only do chemists bring their expertise to the field, they can also identify its distinctive projects. Philosophy of chemistry is, in a sense, a "naturalized" philosophy of the natural sciences because taking account of actual scientific practice is not just an aim but a genuine achievement. Bringing the distinctive objects, practices, models, theories, technologies, experimental techniques and divisions of labor of chemistry to philosophical investigation has undercut some old objections to philosophy of chemistry. For example, an argument against the possibility of a philosophy of chemistry was that, while physics had foundational metaphysical problems associated with quantum mechanics and general relativity, and biology had foundational projects concerning evolutionary theory, chemistry lacked comparable philosophical counterparts. However, the concept of molecular structure and the status of the periodic law are just two philosophical counterparts to foundational problems in physics and biology. The philosophical complexities and importance of these issues rival anything in the philosophies of physics and biology.

One of the themes of this collection is how philosophy of chemistry can make a contribution to problems of philosophy more generally. For example, can chemistry and quantum chemistry contribute to philosophy of mind? In chapter 4, Marina Paola Banchetti-Robino and Jean-Pierre Noël Llored argue that the "hard problem" of consciousness in the philosophy of mind (explaining the relationship between physical states like brain states and experience such as mental states) should be tackled in a manner that parallels philosophy of chemistry. They propose an extended mereology for philosophy of mind, drawing on the nonstandard mereology developed by Rom Harré and Llored[6] for quantum chemistry. Contributions to the philosophy of mind from the philosophy of chemistry arise because of the central place of reduction as a topic in both fields. It is not surprising, then, that attempts are being made to connect philosophy of chemistry with not only the philosophy of mind, but also with metaphysics and epistemology more broadly. Lee McIntyre points out the irony in philosophy of chemistry's "discovery" of antireductionism, emergence, and supervenience (chap. 5), while philosophers of mind appear to be abandoning them. But McIntyre argues that news of supervenience's demise is premature; not only can supervenience be defended in chemistry, if one can cash out the idea that chemical properties are supervenient then why not chemical *laws*? Supervenient chemical laws are epistemologically emergent yet materially dependent on physical relationships. Thereby supervenience can provide a supporting role for the autonomy of explanation in chemistry. On a contrasting note, Eric Scerri (chap. 6) offers an illuminatingly reflective account of his changing ideas on the reduction of chemistry to physics, from his past caution about claims of reduction to one of caution with respect to claims about the failure of reduction. Assessing the past successes and failures of the reduction of the periodic systems, it is now not so clearcut that the periodic system *hasn't*

---

[6] Harré, R., and Llored, J.-P. "Mereologies as the grammars of chemical discourses," *Foundations of Chemistry* 13, 63–76, 2011.

been reduced, and Scerri highlights how important it is to keep a close eye on scientific developments and to seek more accurate experimental data.

Other contributions shed light on old as well as enduring problems in the philosophy of science. Hinne Hettema argues in chapter 13 that engaging with chemistry can shed new light on the concept of a theoretical term. He tackles Hempel's "theoretician's dilemma": Theoretical terms either serve a purpose in establishing definite connections between observables or they do not. If they serve their purpose they can be dispensed with, and if they don't they should be dispensed with. While Hempel's response was to argue that theoretical terms do more than provide links in derivational chains, motivating a realist constructive account of theoretical terms in chemistry is difficult because many of its theoretical terms are inconsistent with those having counterparts in physics. Hettema illustrates these difficulties by drawing on Fukui's frontier orbitals and Paneth's classification of an element, offering a resolution to Hempel's dilemma based on a specific proposal for the ontology of chemical objects based on existence and grounding claims.

While Hettema resurrects an old and perhaps by now a somewhat neglected topic, Hasok Chang (chap. 11) confronts one of the most central issues in contemporary philosophy of science—scientific realism. Chang describes how the debate about scientific realism is distinctive in chemistry and how that distinctiveness might inform the debate as it applies across the sciences. The history of chemistry reveals the complexity of epistemic attitudes to unobservable entities like atoms in the nineteenth century. Chang argues that there was no realism *tout court* in chemistry. Realism is partial and selective. Empirical success confers warrant only on parts of theories. Empirical success can result from a very partial knowledge of the systems of interest, as in nineteenth-century structural chemistry. Chang goes on to critique preservative realism because success need not correlate with those parts of theories which get preserved in theory change. He defends "conservationist pluralism," a position which accepts that empirical and interventionist success has been achieved by many theoretical routes and that we should retain them and add new ones with their own specific domains of application.

The contributions to this volume can enlighten the philosophies of the natural sciences because the neglect of chemistry can often end up polarizing metaphysical debates. In chapter 9, "Causality in Chemistry: Regularities and Agencies," Rom Harré surveys the intellectual terrain comprising the metaphysics of causation. He highlights the neglect of the active, agentive powers of substances in the philosophies of the natural sciences. Causality in chemistry, Harré argues, is of a hybrid structure, residing in the observable regularities between events and powerful particulars capable of bringing about both chemical change and the maintenance of chemical stability. Conversely, it could hardly be claimed that "chemistry" is neglected in discussions concerning the status of natural kinds. Gold and water, after all, have a central place in these debates. But Robin Hendry argues (chap. 12) that a priori requirements for natural kinds are undermined by chemistry. There are various ways of individuating substances but Hendry argues this is done in chemistry at the level of microstructures. He defends this view on the grounds that it captures the interest-dependent classificatory practices of chemistry and because macroscopic individuation fails. Hendry draws on a causal argument for microstructural essentialism to support this claim and concludes that chemical kinds are natural kinds.

The chapter by Joseph E. Earley Sr. (chap. 10) can only be described as a tour de force in every sense of the term. Earley's writing, perhaps more than that of any other currently active author, displays a complete mastery of both the philosophical and the chemical literature. Nothing that could be said briefly here would do justice to his essay, which ranges across a diverse set of topics including the nature of substance, properties and relations, bundles, molecular properties, onticity, wholes and parts and much more.

The neglect of chemistry can diminish philosophical appreciation of central aspects of scientific practice barely touched on by philosophy of science. Richard M. Pagni argues in chapter 8 that philosophical interest in reaction mechanisms in organic chemistry has been somewhat minimal and yet its richness should render it of significant import. He outlines how philosophical conceptions of mechanisms might apply to chemistry; how reaction mechanisms attain their explanatory power; their status as causal mechanisms; and how causation connects with explanatory power. But that does not mean chemists and philosophers of chemistry are, as it were, "off-the-hook" when it comes to neglecting issues of central philosophical importance, or that philosophers of chemistry couldn't learn from other sciences. In "Contingencies in Chemical Explanation" (chap. 7) Noretta Koertge argues that chemists and philosophers of chemistry would benefit from learning about the importance of contingency in biology. In spite of the coherence of chemistry's explanatory structure based on the periodic table and quantum mechanics, some chemical materials confound chemists' attempts at integration. Materials such as allotropes, clathrates, and alloys depend on the contingent environmental conditions in which they are formed. Koertge considers the implications of these contingencies, especially for education in chemistry and philosophy of science.

While this volume is dedicated to philosophy of chemistry, it appreciates the indispensability of history of chemistry to philosophical analysis. Without appreciating historical context there is little hope of understanding the cohesiveness or dissent within and between communities of practitioners, the nature of their practices, or the development of their conceptual frameworks. Some of the contributions to this volume offer important contributions to the history and historiography of chemistry. Nicholas W. Best (chap. 2) defends the revolutionary status of the Chemical Revolution and Lavoisier's place within it. According to Best, the Chemical Revolution was a multifaceted scientific revolution encompassing theoretical, linguistic, and methodological change, and a shift in the philosophical theory of meaning. Best argues that a natural-kind—based theory of reference was tacitly introduced, encouraging the idea that there had been a continuity of reference when the same chemical names were used in spite of profound theoretical upheavals. Alan Chalmers (chap. 1) makes an important contribution to a lively debate about the impact of Robert Boyle's version of the corpuscular theory on the rise of modern chemistry. Chalmers challenges the extent to which Robert Boyle's attempts to wed chemistry and the mechanical philosophy in the seventeenth century advanced chemistry because Boyle's version of corpuscular theory was not sufficiently connected to observation and experiment.

Gavroglu and Simões (chap. 3) discuss a number of philosophical themes in quantum chemistry. They present a historiographical proposal for writing the histories of "in-between" sciences such as quantum chemistry, among others. The role of theory in chemistry, especially the ontological status of theoretical entities, is considered alongside reductionism as it has been received by the practitioners of quantum chemistry. Gavroglu and Simões argue that the reductionism debate depends on the perspectives of specific communities of practitioners, as do other issues in the philosophy of chemistry. The histories of (sub)disciplines are crucial to the framing of the philosophical issues as well as specific demands and expectations practitioners have concerning the role of theory compared to those of practitioners within physics.

Disciplinary identity, a theme closely connected with the Gavroglu and Simões contribution, is pursued by Guillermo Restrepo in "Mathematical Chemistry, A New Discipline" (chap. 15). Restrepo explores the relationship between mathematics and chemistry, and argues that mathematical chemistry satisfies Mary Jo Nye's (1993) conditions of discipline formation.[7] One important implication of the forging of this new identity is that mathematical chemistry presents opportunities for historical and philosophical study in its own right. There is, of course, much to be said for analyses that contribute to our understanding of the science of chemistry at its most general level. Our understanding of disciplinary identities can be forged in a number of ways, not only historically and socioculturally, but also philosophically. One crucial philosophical perspective is Rein Vihalemm's longstanding contribution to the conception of chemistry as a science. In chapter 16 of this volume, Vihalemm builds on his idea of the dual character of chemistry. Vihalemm argues that chemistry as a science should not be thought to possess a specific character; rather chemistry has a hybrid character: a constructive-hypothetico-deductive ("φ-Science") inquiry and a classifying-historico-descriptive inquiry.

One other theme explored by Gavroglu and Simões that connects well with the other contributions to this volume is the impact of computers on quantum chemistry: how postwar development of computers resulted in a reconceptualization of the idea of an "experiment." Digital computers are of crucial historical importance to quantum chemistry and have attracted some attention from philosophers of science for a number of years. The methodological implica-

---

[7] Nye, M.J., *From Chemical Philosophy to Theoretical Chemistry—Dynamics of Matter and Dynamic of Disciplines*, 1800–1950, University of California Press, Berkeley, 1993.

tions of computing in late twentieth century organic chemistry are pursued in Grant Fisher's contribution "Divergence, Diagnostics, and a Dichotomy of Methods" (chap. 14). Fisher draws on the "dichotomy of methods"—disagreements between communities of quantum chemists concerning the veracity of approximation methods used to study pericyclic reaction mechanisms. He argues that divergence in semi-empirical and ab initio predictions of transition state geometries and energies for the Diels-Alder reaction saw quantum chemists attempt to probe the consequences of pragmatic trade-offs between computational manageability and predictive accuracy by what he calls "diagnostics."

This volume reflects current developments in a growing and maturing discipline. We hope the reader will find in it some of the ways in which philosophy of chemistry raises novel problems and projects, how history and philosophy can contribute to broadening our understanding of chemistry, how new insights into old philosophical problems in philosophy broadly might be gained, how neglected problems in philosophy of science come to regain importance when viewed through the lens of chemistry, and how chemistry can help us grapple with enduring debates in epistemology and metaphysics. Perhaps the reader will even find how philosophy of chemistry might contribute to complementary philosophical projects at the boundaries between physics and biology, even influencing future philosophical cooperation between autonomous philosophical fields, cooperation that is often reflected in current scientific practice.

Finally, we wish to end with two cautionary notes. First, the philosophy of chemistry as a distinct academic discipline is now about twenty years old. (The discipline was founded in 1994.) There have been several books published on the subject but the number remains rather small, a fact that we wish to exploit in order to justify another collection of essays on the subject. We also take the opportunity to provide a full bibliography of the books (in several languages) that have appeared up to this point in the hope of further stimulating scholarship in this area.

Second, many of the previous books and articles on the philosophy of chemistry begin in a self-congratulatory fashion by describing how quickly the field has grown and become fully established. We wish to resist fully embracing such an approach in this volume, since we believe that matters are still very much in an early stage of development. For example, it may be the case that sessions on the philosophy of chemistry frequently appear on the biennial program of the prestigious Philosophy of Science Association. However these sessions largely consist of speakers talking among each other or preaching to the converted. These encounters generally lack the presence of experts in "real chemistry" who might provide a more critical edge to the proceedings. As a result one has to wonder about the extent to which such activities really serve the reputation of the field. We would like to propose that a new goal for the field might be to attempt to hold occasional, or even regular sessions, on the philosophy of chemistry at meetings of the American Chemical Society or the Royal Society of Chemistry.[8] One might even envisage the eventual forma-tion of divisions for the philosophy of chemistry within these societies, in much the same way that the history of chemistry has its own divisions in the world's major scientific societies.

References

Bibliography of previous books on the philosophy of chemistry and related subjects

Baird, D., Scerri, E.R., McIntyre, L. (eds.). (2006). *Philosophy of Chemistry, Synthesis of a New Discipline* (Boston Studies in Philosophy of Science, vol. 242). Berlin: Springer.

Baird, D., Scerri, E.R., McIntyre, L. (eds.). (2014). *Filosofía de la química. Síntesis de una nueva disciplina* (Spanish Edition) Mexico DF: Fondo de Cultura Económica.

Bensaude-Vincent, B. (2005). *Faut-il avoir peur de la chimie?* Paris: Empêcheurs de Penser en Rond, Paris.

Bensaude-Vincent, B. (2008). *Matière à Penser, Essais d'hisoire et de philosophie de la chimie.* Paris: Presse Universitaire de Paris Ouest.

Bensaude-Vincent, B., and Simon, J. (2008). *Chemistry—The Impure Science.* London: Imperial College Press.

Berson, J.A. (2003). *Chemical Discovery and the Logician's Program.*, Weinheim: Wiley-VCH.

Bhushan, N., Rosenfeld, S. (eds.). (2000). *Of Minds and Molecules.* New York: Oxford University Press.

Chamizo, J.A. (ed.) *Historia y Philosophia de la Química,* siglo XXI editors, Mexico DF.

Earley, J. (ed.). (2003). *Chemical Explanation.* New York: New York Academy of Sciences.

Hettema, H. (2012). *Reducing Chemistry to Physics.* Groningen, Netherlands: Rijksuniversitat.

Janich, P. (ed.). *Philosophische Perspektiven der Chemie.* Mannheim: Bibliographisches Institut.

Janich, P., Psarros, N. (eds.). *Die Sprache der Chemie.* Würtzburg: Koningshausen and Neumann.

Janich, P., Psarros, N. (eds.). (1998). *The Autonomy of Chemistry.* Würtzburg: Koningshausen and Neumann.

Klein, U. (ed.). (2001). *Tools and Modes of Representation in the Laboratory Sciences.* Dordrecht: Kluwer.

Nye, M.-J. (1993). *From Chemical Philosophy to Theoretical Chemistry.* Berkeley: University of California Press.

Primas, H. (1983). *Chemistry, Quantum Mechanics and Reductionism.* Berlin: Springer.

---

[8] As a matter of fact, one such session, which was spread over two days, was arranged by one of us (E.S.) at a general meeting of the American Chemical Society in 1997. *Yes, Chemistry and Philosophy Are Miscible,* Session on Chemistry & Philosophy, American Chemical Society, Annual Meeting, San Francisco, April, 1997. Although the session raised a great deal of interest and enthusiasm among society members, such an exercise has never been repeated since.

Psarros, N., Ruthenberg, K., Schummer, J. (eds.). (1997). *Philosophie der Chemie.* Würzburg: Bestandsaufnahme und Ausblick.

Psarros, N., Gavroglu, K. (eds.). Ars Mutandi, *Issues in Philosophy and History of Chemistry*, Leipzig: Leipsizer Universitätsverlag.

Psarros, N, Ruthenberg, K., and Schummer, J. (eds.). (1996). *Philosophie der Chemie— Bestandsaufnahme und Ausblick.* Würzburg: Königshausen and Neumann.

Rocke, A.J. (1984). *Chemical Atomism in the Nineteenth Century: From Dalton to Cannizzaro.* Columbus: Ohio State University Press.

Rocke, A.J. (2010). *Image & Reality.* Chicago: University of Chicago Press.

Ruthenberg, K., Van Brakel, J. (eds.). (2008). *Stuff, the Nature of Chemical Substances.* Würtzburg: Konigshausen and Neumann.

Scerri, E.R., McIntyre, L. (eds.). (2015). *Philosophy of Chemistry, Growth of a New Discipline* (Boston Studies in Philosophy and History of Science, vol. 306). Berlin: Springer.

Scerri, E.R. (2009). *Selected Papers on the Periodic Table.* London: Imperial College Press.

Scerri, E.R. (2008). *Collected Papers on Philosophy of Chemistry.* London: Imperial College Press.

Scerri, E.R. (2007). *The Periodic Table, Its Story and Its Significance.* New York: Oxford University Press.

Sobczynska, D., Zeidler, P., Zielonaka-Lis, E. (eds.). (2004). *Chemistry in the Philosophical Melting Pot.* Frankfurt am Main: Peter Lang.

Schummer, J. (1996). *Realismus und Chemie: Philosophischen Unteruchungen der Wissenschaft von der Stoffen.* Würtzburg: Konigshausen and Neumann.

Van Brakel, J. (2000). *Philosophy of Chemistry.* Leuven: Leuven University Press.

Woody, A., Hendry, R.F., Needham, P. (2012). *Philosophy of Chemistry* (Handbook of the Philosophy of Science Series). Amsterdam: Elsevier.

History and Philosophy
of Chemistry

| Robert Boyle's Corpuscular
Chemistry

Atomism before Its Time

ALAN CHALMERS

1 *Introduction*

In her important and pioneering work on Robert Boyle's contributions to chemistry Marie Boas Hall (Boas 1958; and Hall 1965, 81–93) portrayed Boyle's advances
as being tied up with and facilitated by his adoption of the new world view, the
mechanical or corpuscular philosophy, as opposed to Aristotelian or Paracelsian
philosophies or world views. In recent decades such a reading has been challenged.
Historians of chemistry such as Frederic L. Holmes (1989), Ursula Klein (1994,
1995, 1996) and Mi Gyung Kim (2003) have portrayed modern chemistry as
emerging in the seventeenth century by way of a path closely tied to technological
and experimental practice and relatively independent of overarching philosophies
or world views. Such a perspective raises questions about how productive Boyle's
attempts to wed chemistry and the mechanical philosopher were as far as the emergence of modern chemistry is concerned. This is the issue I will investigate.

In recent work on Boyle's chemistry William Newman (2006) has also taken
issue with what he calls the "traditional accounts," especially that of Hall.
Newman's quarrel with the traditional accounts is the extent to which they read
Boyle's corpuscular chemistry as emerging out of the atomism of Democritus
and Lucretius and its reincarnations in the hands of early mechanical philosophers such as Descartes and Gassendi, neglecting a corpuscular tradition that
has its origins in Aristotle's *Meteorology*. In a range of detailed and pioneering
studies Newman (1991, 1996, 2001, 2006) has documented the elaboration of
the latter tradition in the works of the thirteenth century author known as Geber
and its passage to Boyle, especially via Daniel Sennert, a Wittenburg professor
of medicine in the early seventeenth century. While Newman's work has led to
a substantial and significant re-evaluation of the sources of Boyle's corpuscular
chemistry there is a sense in which he does not break from the "traditional"
view insofar as he reads the revolutionary aspects of Boyle's chemistry in terms

of a change from an Aristotelian to a mechanical matter theory. Newman (2006, 2–3) invokes "the great disjunction between the common view of matter before and after the mid-seventeenth century" and wishes to stress the role of chemistry in bringing about that change, culminating in Boyle's removal of the Aristotelian forms that were presumed to animate the corpuscles of Geber and Sennert resulting in a corpuscular theory that was thoroughly mechanized. It was "Boyle's ceaseless war on hylomorphism and his reduction of the sensible world to mechanical causes" that marks that aspect of his work that warrants the name of "revolution" according to Newman (Newman 2006, 225).

There is no doubt that Boyle was centrally concerned with articulating a mechanical world view as an alternative to the Aristotelian one and that he sought both to support that philosophy and to employ it to improve chemistry by formulating chemistry in mechanical terms. Those, like Klein and Kim, who wish to locate the origins of modern chemistry closer to technological and experimental practice than philosophical matter theory do not need to question the detailed historical account that Newman has given of Boyle's contribution to that latter change. Rather, they can ask, rhetorically, what did the change from Aristotelian to mechanical matter theory have to do with the emergence of modern chemistry? That is an issue on which Newman has locked horns with critics, especially Klein and myself, in recent years.[1]

As a contribution to the debate outlined above I focus on Boyle's version of corpuscular chemistry. I question the extent to which it was productive as far as the advancement of chemistry was concerned. I argue that Boyle's incorporation of chemistry into a mechanical version of the corpuscular theory was too remote from experimental practice to usefully inform it. Insofar as Boyle was able to portray chemical knowledge in mechanical terms this was achieved by imposing contrived corpuscular mechanisms onto chemistry known by other means. Boyle's corpuscular chemistry was accommodated to rather than supported by the experimental evidence, a distinction that Boyle came close to making himself as we shall see in section 5.

While casting doubt on the productiveness of an incorporation of the mechanical philosophy into chemistry I do not wish to imply that a science, such as chemistry, should dispense with theory. I do not wish my position to be interpreted as proposing the "positivist manifesto" that Newman (2010, 204) interprets me as advocating. The introduction of Paracelsian principles or Aristotelian elements might well be read as attempts to bring some theoretical order and guidance into chemistry. However, Boyle's detailed critique of those attempts, especially in *The Skeptical Chemist*, is sufficient to reveal the serious shortcomings of those attempts. If I am right that Boyle's attempt to rectify the problem by introducing mechanical corpuscles was a failure, then how was an adequate theoretical dimension to be introduced? Here I side with Klein

---

[1] Klein (2007) has criticized Newman in an essay review of Newman (2006), and Newman (2008) has returned the compliment in an essay review of Klein and Lefèvre (2007) and Kim (2003). Newman (2010) is a response to Chalmers (2010) on which I comment in Chalmers (2011).

(1994), who has argued that a crucial turning point lay in the introduction of the notions of chemical compound and chemical combination articulated with the aid of affinity tables, the first of which was published by Etiènne-François Geoffroy in 1718. Like pressure in Boyle's hydrostatics and gravity in Newton's physics, affinities were experimentally accessible causes and they were peculiar to a specific science (chemistry) and not general, ultimate causes of the kind that mechanical philosophers hoped to locate in the shapes, sizes and motions of corpuscles. They conform to Boyle's conception of intermediate causes, the details of which I address in the next section.

2 Boyle's Distinction between Intermediate and Ultimate Causes

The emergence of science in the seventeenth century can be instructively viewed from the point of view of a distinction, introduced by Boyle himself, between intermediate causes and explanations, on the one hand, and ultimate explanations and causes, on the other. Ultimate explanations appeal to ultimate causes, causes that are not themselves susceptible to, nor in need of, explanation at a deeper level. Aristotelian elements, Paracelsian principles, and the atoms of mechanical philosophers were typically treated as ultimate causes. Ultimate causes serve to explain phenomena that can be observed and experimented on but are not themselves directly accessible to observation or experiment. Intermediate causes are like ultimate ones insofar as they are explanatory, but differ from ultimate ones, first insofar as they are susceptible to explanation at a deeper level and second insofar as they can be explored experimentally. Weight and elasticity are ready examples of intermediate causes.

Boyle introduced a distinction between levels of causes early in his life as an "experimental philosopher." He had taken up serious experimenting in the second half of the 1650s, while resident in Oxford. It was during the same period that he first articulated his version of the mechanical philosophy. The latter philosophy traced all phenomena of the material world back to the shapes, sizes, arrangements and motions of atomic corpuscles that were understood to be the sole components of that material world. The atoms, each possessing an unchanging shape and size, were ultimate causes and characterized as such by Boyle. It was not until the early 1660s that Boyle's experimental work and his articulation of the mechanical philosophy began to be published. The first publication was his "New experiments physicomechanical touching the spring and weight of the air and its effect" containing a detailed account of his pneumatics, defended by experiments with the air pump (Boyle 2000 vol. 1, 141–300). The weight and spring of the air that Boyle invoked as causes of pneumatic phenomena were not ultimate, corpuscular causes. The status of the noncorpuscular causes figuring here and in Boyle's experimental work generally was an issue that he explicitly took up in his next publication.

In 1661 Boyle (2000, vol. 2, 5–203) published "Certain physiological essays" containing accounts of his experimental work in areas other than pneumatics. Most of those essays were concerned with his experimental work in chemistry. The collection began with a "proemial essay" in which Boyle described and explained the character of the "experimental learning" in which he was engaged. It was in this essay that Boyle introduced the notion of intermediate causes and explanations as distinct from those involving mechanical atoms. In a discussion that takes up four pages of his essay, Boyle (2000, 21–24) explicitly makes the following points: There are degrees of explanation ranging from those that appeal to the "primitive affections," the shapes, sizes and motions of atoms of universal matter, to those involving "more familiar" qualities such as weight and fluidity. Explanations that appeal to intermediate causes such as gravity are not ultimate explanations but they are explanations for all that. There is plenty of legitimate work to be done involving the identification and experimental investigation of intermediate causes that falls short of the identification of ultimate causes.[2] The experimenter may "without ascending to the top of the series of causes, perform things of great moment."

Boyle's distinction between intermediate and ultimate causes makes immediate sense of, and was presumably influenced by, Boyle's pneumatics. In "New experiments physico-mechanical, touching the spring of the air and its effects," Boyle experimentally identifies and explores the weight, pressure, and spring of the air and employs those notions to explain a range of phenomena, some of them familiar and others novel phenomena produced using the air pump. Boyle is quite clear in this work and elsewhere, first that the weight and spring of the air, being intermediate causes, are in need of explanation at a deeper level and second that he, in common with his fellow mechanical philosophers, is unable to supply those deeper, ultimate or atomic, explanations.

The characteristic of Boyle's pneumatics that I have highlighted features even more clearly in his essay "Hydrostatical paradoxes made out by new experiments," published in 1666, which involved an exposition and extension of Pascal's hydrostatics. Much of the work is devoted to a detailed description of experiments devised and conducted by Boyle to illustrate and support the principles of hydrostatics. Boyle (2000, vol. 5, 193–194) was concerned to defend this experimental work as a legitimate part of *philosophy*.

> Hydrostatics is a part of philosophy, which I confess I look upon as one of the ingeniousest [sic] doctrines that belong to it. Theorems and problems of this art, being most of them pure and handsome productions of reason duly exercised on attentively considered subjects and making in them such discoveries as are not only pleasing but divers of them surprising, and such as would make one at first wonder by what ratiocination men came to attain the knowledge of such unobvious truths. Nor are the delightfulness and subtlety of the hydrostatics the only things for which

---

[2] Boyle used phrases such as "the most general causes," "first principles," "atomic principles," and "the primitive affections of the smallest parts of matter" to refer to those fundamental causes not themselves in need of explanation. I have used the term "ultimate causes' to encapsulate all of this.

we may commend them: For there are many, as well as of the more familiar, as of the more abstruse phenomena of nature, that will never be thoroughly understood nor clearly explicated by those that are strangers to the hydrostatics.

Boyle (2000, vol. 5, 207 and 236) insisted that in his hydrostatics he was intent on explaining why various propositions describing the phenomena are true as well as establishing that they are true, and this was to be done by identifying their "true causes" (Boyle 2000, vol. 5, 213 and 214).

For Boyle, then, hydrostatics developed with the aid of detailed experimentation was a legitimate branch of philosophy because it explained both familiar and novel ("unobvious") phenomena by identifying their causes. But those causes were not the ultimate causes figuring in the mechanical philosophy. Boyle's essay on hydrostatics makes no reference whatsoever to the shapes, sizes, and motions of corpuscles. Rather, the causes invoked by Boyle (as by Pascal before him) were weight and pressure.[3] For Boyle, hydrostatics and pneumatics were a legitimate part of "experimental learning," and, indeed, of "philosophy" by virtue of the fact that they explained phenomena by identifying intermediate causes accessed by experiment.

It should be noted here that appropriate notions of intermediate causes cannot simply be read off of nature, nor is experimental contact with them straightforward. As Boyle notes in the passage cited above, the necessary notions are "pure and handsome productions of reason exercised on attentively considered subjects." Boyle could help himself to a concept of weight that already had a long history and could build on the work of Archimedes, Stevin, and Pascal to develop a conception of pressure. The notion of the spring of the air was largely of Boyle's own devising, gradually fashioned in the course of his own experimental endeavors involving the integration of work with head and hand. This kind of work transcended that of a sooty empiric, drudge, or mere artisan.

In the light of Boyle's understanding of experimental knowledge as the identification and exploration of intermediate causes, well and straightforwardly exemplified in his hydrostatics and pneumatics, we now turn our attention back to chemistry. The discussion of this section suggests the following question: Did Boyle's experimental chemistry involve and invoke intermediate causes on a par with the pressure and weight that informed his pneumatics? As is discussed in the following section, the answer is that it did not.

3 Boyle's Experimental Chemistry and Its Relation to Philosophy

I noted above that most of the works appearing in "Certain physiological essays" were concerned with chemistry. One of those essays was explicitly concerned with the relation between experimental chemistry and the corpuscular

---

[3] The appeal to intermediate, rather than ultimate, causes in the hydrostatics of Pascal and Boyle is discussed in more detail in Chalmers (2012b).

philosophy. In the opening paragraph of that essay, Boyle (2000, vol. 2, 85) took issue with the idea that experimental chemistry is appropriately practiced by illiterate, "sooty empiricks" but has no place in the deliberations of philosophers. His defense of chemistry went beyond the recognition that its productions are of use in various trades and especially in medicine. Boyle was intent on showing that experimental chemistry can be "very assistant even to the speculative naturalist of his contemplations and enquiries."

The details of Boyle's discussion make it clear that, insofar as he wished to argue for a productive link between experimental chemistry and the contemplations and enquiries of speculative naturalists, it is naturalists that advocate a version of the corpuscular philosophy that he has in mind. This is already apparent from the title of his essay, "Some specimens of an attempt to make chymical experiments useful to illustrate the notions of the corpuscular philosophy," closely followed by Boyle's advocacy of "the desirableness of a good intelligence between corpuscularian philosophers and the chymists."[4]

We do not need to speculate about what Boyle means by the "corpuscular philosophy" here because he spells out what he has in mind in the course of the essay. He explains that he does not wish to align himself with the details of a specific version of the corpuscular philosophy, such as that spelled out by the ancient atomists or by Descartes. Rather, he is content with the assumption, shared by all corpuscularians, that the phenomena of nature are to be explained by appeal to "little bodies variously figured and moved" (Boyle 2000, vol. 2, 87), the bodies themselves differing "but in the magnitude, figure, motion or rest, and situation of their component particles" (Boyle 2000, vol. 2, 91). That is, by the "corpuscular philosophy" Boyle means the mechanical philosophy that he articulated repeatedly and in detail in a number of works. Phenomena are to be explained by appeal to the motions and arrangements of corpuscles rather than "inexplicable forms, real qualities, the four peripatetic elements or so much as the three chymical principles" (Boyle, 2000, vol. 2, 88).

Toward the end of the essay outlining his "attempt to make chymical experiments useful to illustrate the notions of the corpuscular philosophy," Boyle reiterates the nature of the project he is advocating. He stresses that he is keen to overcome, on the one hand, the view that experimental chemistry is the work of illiterate practitioners which at best has practical benefits but is of no relevance to the deliberations of an advocate of the corpuscular philosophy and, on the other hand, the view that the corpuscular philosophy is too general and speculative to be of use to an experimental chemist. Boyle (2000, vol. 2, 92) concludes his essay by insisting, on the contrary, that "as many chymical experiments may be happily explicated by corpuscular notions, so many of the

---

[4] Newman and Principe (1998) advocate the adoption of Boyle's term "chymistry" to refer to the practice he was engaged in to distinguish it from modern chemistry. While allowing Boyle his own spelling I have used "chemistry" to apply to both Boyle's practice and the modern version. In doing so I do not, of course, wish to deny key differences in the two practices; rather, I wish to identify and explore them.

corpuscular notions may be commodiously either illustrated or confirmed by chymical experiments."

The extent to which Boyle's contributions to the corpuscular philosophy and chemistry can be seen as helping to fulfil these ambitions is an issue that will be dealt with later. Here I wish to stress a contrast between the way in which Boyle strove to capture a philosophical dimension of chemistry, on the one hand, and of pneumatics and hydrostatics on the other. Both were considered to be philosophical insofar as they were explanatory. But the causes involved were of different kinds. While Boyle attempted to trace chemical phenomena to corpuscular causes, he was content with the intermediate causes of weight, pressure and elasticity in pneumatics and hydrostatics. Boyle's discussion of the philosophical credentials of chemistry presupposes a contrast between empirical knowledge of artisans that is non-explanatory and philosophical knowledge that explains by way of the identification of corpuscular causes. Intermediate causes that served Boyle so well in his characterization of pneumatics and hydrostatics do not figure in his chemistry.

Boyle's stand on the philosophical credentials of chemistry that I have highlighted is well brought out by the experiment involving the decomposition and recomposition of saltpeter or niter, an experiment that Boyle regarded as providing particularly strong support for his views.[5] An account of it was included by Boyle in "Certain physiological essays" and was repeatedly invoked thereafter in his chemical and philosophical texts. Boyle appeals to this experiment to cast doubt on some assumptions about the role of substantial forms and principles that he attributes to Aristotelians and Paracelsians, respectively. He contrasts the problems faced by his opponents with a demonstration of the compatibility of the experimental results with the corpuscular philosophy. It is the mechanical or corpuscular explanation of the experiment that makes Boyle's account "philosophical" rather than merely empirical, in Boyle's view.

In his experiment, Boyle first purified some saltpeter by dissolving it in water, allowing it to crystallize and then filtering off the crystals. These crystals were then fused in a crucible and burning coals added, one by one, until the resulting fulmination ceased. Heating the crucible continued for a further quarter of an hour to ensure that none of the volatile substance driven off in the process remained. A substance bearing the properties of saltpeter, and weighing only a little less than the saltpeter with which the experiment began, could now be reconstituted by adding "spirit of nitre" to the "fixed nitre" in the crucible.

Boyle's experiment runs counter to the idea that naturally occurring substances (such as saltpeter), as opposed to artefacts, owe their nature to the possession of a substantial form, which, in Boyle's derogatory terminology "gives it its being and denomination and from whence all its qualities are in the vulgar philosophy, by I know not what inexplicable ways, supposed to flow"

---

[5] The substance Boyle refers to as "saltpeter" in the title of his publication is referred to as "saltpeter" or "nitre" more or less interchangeably in the body of the text. From a modern point of view they are both potassium nitrate.

(Boyle 2000, vol. 2, 108). The reconstituted niter bears all the properties of the original saltpeter and is yet clearly an artefact arising from the addition of spirit of niter to fixed niter.

Boyle stresses the extent to which the properties of the reconstituted niter differs from the properties of the spirit of niter and the fixed niter from which it is prepared and uses this to argue against those "chymists" who attribute the properties of substances to the presence of some principle or other responsible for them. For instance, saltpeter is "not only inflammable, but burns very fiercely and violently," whereas neither fixed niter nor spirit of niter are inflammable, the former "having already suffered the loss of all that fire could deprive it of" and the latter "being observed to be more apt to quench than foment fire" (Boyle 2000, vol. 2, 102).

Boyle complements his use of the redintegration of niter to criticize his opponents with his own explanation of the phenomenon. That explanation appealed to the shapes, sizes, and motions of corpuscles. The status and role of Boyle's corpuscular explanations of chemical phenomena, both here and in his chemistry generally, is an issue that is discussed in detail below.

4 Boyle's Mode of Defense of the Corpuscular Philosophy

Boyle articulated and defended his version of the corpuscular or mechanical philosophy in detail and on numerous occasions. There were two key ways in which Boyle argued for the fundamental matter theory that was involved. One way appealed to the "intelligibility" of the mechanical philosophy. The other appealed to the extent to which phenomena known through observation and experiment were compatible with that philosophy. A clarification of what those two modes of support for the mechanical philosophy involved, from Boyle's point of view, will set the scene for a discussion of how chemistry contributed to it.

When Boyle insisted that the mechanical philosophy was intelligible and those of his Aristotelian and "chymical" opponents unintelligible he meant something quite specific. In both of the central texts in which Boyle articulated and defended the mechanical philosophy, *The Origin of Forms and Qualities* and *The Excellency and Grounds for the Mechanical Philosophy*, intelligibility is discussed in the context of the ultimate nature of material reality and the properties that can intelligibly be attributed to matter at that ultimate level. Boyle insisted that a portion of matter must of necessity possess some definitive shape and size and relative orientation, together with some degree of motion or rest, by virtue of being such. In *The Excellency of the Mechanical Philosophy*, Boyle (2000, vol. 8, 105–106) insisted that "men do easily understand one another's meaning" when they characterize matter in this way, and such characterizations are "simple" in the sense that they are not "resolvable into any things, whereof it may be truly, or so much as tolerably, said to be compounded" In *Origin of Forms* (2000, vol. 5, 309) Boyle raised the question of the status of the forms and principles that he saw his opponents as introducing as ontological primitives. "Nor could I ever find it intelligibly made out what these real qualities may be that they [the scholastics] deny to be either matter, or modes of matter, or immaterial substances." Boyle's talk of intelligibility was concerned with the properties that can be attributed to material reality ultimately or primitively. A portion of matter must possess some shape, size, relative orientation, and degree of motion or rest by virtue of being such and it is perfectly clear what is involved in such characterizations. By contrast, the status of Aristotelian forms or Paracelsian principles is unclear from an ontological point of view, insofar as such entities are presumed to qualify matter and be other than matter.

The recognition that Boyle's notion of intelligibility applies only at the ultimate ontological level is important lest his own appeal to intermediate causes be branded as unintelligible. As was stressed above, Boyle's hydrostatics and pneumatics involved ascribing weight, pressure, and elasticity to portions of fluids. Since those notions are not included among those Boyle lists as being intelligibly ascribable to matter, it might appear that his hydrostatics and pneumatics must be classified as unintelligible. But this would be a mistake. Secondary causes, such as weight, pressure, and elasticity, are used by Boyle to characterize fluids, but in doing so, he does not imply that fluids possess such properties primitively. On the contrary, Boyle insists that such notions are, in principle, reducible to the motions and arrangements of portions of matter possessing shape and size only.

Let us now turn to the second way in which Boyle sought to defend his mechanical philosophy, involving its compatibility with the phenomena. When it comes to the question of empirical support for his various knowledge claims, it is not difficult to identify tensions, if not contradictions, in Boyle's utterances. For instance, Boyle (2000, vol. 8, 166) criticizes the appeal by "chymists" to their principles, mercury, salt, and sulfur, and insists that his criticism would hold even if the explanations of his opponents were "to come home to the phenomena." This is because explanations by appeal to the three principles "are not primary, and, if I may so speak, fontal enough." Yet, as we have seen, this criticism can be applied to Boyle's own pneumatics, since the explanations he offers there involve appeal to weight, pressure and spring of the air that are not primitive or fontal. When, in The Excellency of the Mechanical Philosophy, Boyle (2000, vol. 8, 114) claimed empirical support for the mechanical philosophy, he listed hydrostatics as one area in which that philosophy had been successfully applied, notwithstanding the fact that elsewhere, and repeatedly, he insisted that hydrostatics did not provide explanations that were fontal in the sense required by the mechanical philosophy because of its appeal to weight and pressure.

Elsewhere I have attempted to resolve these tensions in Boyle's writings to the best of my ability.[6] Here I summarize my case, which involves a distinction between confirmation by empirical evidence and a weaker notion, accommodation to such evidence. Knowledge of intermediate causes is amenable to

---

[6] See, especially, Chalmers (2009, 110–114).

empirical confirmation whereas fundamental matter theories like the mechanical philosophy, involving ultimate causes, are supportable only by accommodating them to the evidence. Boyle did not explicitly make this distinction in as many words, but it is implicit in much of what he did and wrote.

Boyle's essay on hydrostatics provides as good an example as any of knowledge of intermediate causes confirmed by experiment. A fluid presses on a surface to a degree that is proportional to the area of the surface and by an amount that is determined by the depth below the fluid surface. Pressure, so understood, when added to weight, can be appealed to in order to straightforwardly explain a wide range of experimental evidence, to some extent involving previously known phenomena but also novel, or what Boyle called unobvious, phenomena. In would be a remarkable coincidence if the principles of hydrostatics could naturally explain such a wide range of phenomena and be false nevertheless, just as one can be confident that one has found the correct cipher to a code if that cipher straightforwardly yields readable messages in a wide variety of cases including novel ones.[7]

Accommodation of claims to the evidence is a mode of support distinct from, and less demanding than, confirmation. The mechanical or corpuscular philosophy can be accommodated to the phenomena by devising corpuscular mechanisms capable of reproducing those phenomena. For instance, the mechanical philosophy can be accommodated to the fact that observable materials differ in density by attributing that density to the degree to which corpuscles of universal matter are closely packed. Accommodation differs from confirmation insofar as the accommodations are adapted to phenomena otherwise known and there is no independent evidence to support the claim that the contrived explanation is the correct one. The argument from coincidence illustrated by the code analogy does not apply to accommodation insofar as the accommodations are contrived on a case-by-case basis in light of the phenomena to be accommodated. It is no coincidence that the mechanical philosophy can be accommodated to a wide range of phenomena if there are no constraints on one's freedom to devise a fresh mechanism for each phenomenon to be accommodated.

The extent to which the mechanical philosophy could be accommodated to the phenomena was invoked by Boyle in response to the charge that a world composed only of particles possessing shape, size, and motion is much too stark to account for the vast range of properties and happenings in it. That response invoked the degree to which variety could be traced back to the vast numbers of shapes, sizes, and motions capable of being possessed by corpuscles and the great number of relations they could bear to one another.[8] The

---

[7] Boyle (2000, vol. 8, 115) himself employed the analogy with code cracking, but his application of it in the context of support for the mechanical philosophy was unwarranted, for reasons that will become apparent.

[8] Boyle repeatedly stressed this flexibility of the corpuscular philosophy. Two key passages in which it is discussed at length are in "A history of particular qualities" (Boyle 2000, vol. 6, 276—277) and in "The excellency of the mechanical hypothesis" (Boyle, 2000, vol. 8, 112–114).

variety of phenomena exhibited in the world results from the motions and arrangements of corpuscles of various shapes and sizes just as works of litera- ture arise from arrangements of letters of the alphabet. On account of this va- riety, the mechanical philosophy is "so general and pregnant that among things corporeal there is nothing *real*—that may not be derived from, or brought to a subordination to, such comprehensive principles" (Boyle, 2000, vol. 8, 114). This point did constitute a response to a key objection of his opponents and so did constitute an argument for the general philosophy. However, it involved accommodation to rather than confirmation by, the phenomena, and was a much weaker kind of support than that involved in the new experimental sci- ence, including some of Boyle's own contributions to it.

5 Boyle's Chemistry and Support for the Mechanical Philosophy

There is no doubt that Boyle looked to chemistry for the strongest support for the corpuscular philosophy. In "Certain physiological essays" Boyle (2000, vol. 2, 91) declared that "there are scarce any experiments that may better accommo- date the [corpuscular philosophy] than those that may be borrowed from the laboratories of the chymists." He went on to explain why: A study of chemistry promises to reveal information about the deep structure of matter insofar as it involves significant qualitative changes in matter that take place fairly rapidly. The experimental chemist can practically intervene to simplify the situations in which such changes take place and shield them from extraneous influences. "Bodies employed by the chymists being for the most part active ones, the progress of nature in an experiment, and the series of successive alterations through which the matter passes from first to last, is wont to be made more nimbly, and consequently becomes the more easy to take notice of and com- prehend" (Boyle 2000, vol. 2, 91).

In the first of the quotations from Boyle in the preceding paragraph Boyle himself describes his goal as that of "accommodating" the corpuscular philos- ophy to chemistry rather than anything stronger. However, I wish to maintain that Boyle had only limited success at contributing to that goal. In crucial places Boyle does not provide the corpuscular mechanisms that his accommo- dations require and in others he devises mechanisms that are highly contrived and not fully up to the task of reproducing the phenomena they are devised to accommodate. Let me substantiate this negative assessment.

The experiment on saltpeter that figured so prominently in Boyle's corpus- cular treatment of chemistry is a case where Boyle simply does not specify the corpuscular mechanisms that he invokes, and which he declares to lie behind and be responsible for the change from saltpeter to salt of tartar and back again. Spinoza, in an interchange with Boyle conducted via Oldenburg as interme- diary, criticized Boyle for not providing a corpuscular account of the changes involved and attempted to repair the deficiency. In response, Boyle declined to specify a corpuscular mechanism. Rather, he retreated to an insistence on the negative point that his experiment undermined an understanding of chemical change by appeal to substantial forms, a reply that did not impress Spinoza since he took the rejection of substantial forms for granted.[9]

Boyle's most direct attempt to supply corpuscular mechanisms for chemical changes occurs in his essay "The mechanical origin of qualities." There he devises mechanisms for explaining such things as the action of acids and precipitation. As a sample of what he has to offer, I quote at length his account of the transformation of highly corrosive sublimate (mercuric oxide) into mild-tasting *mercuris dolcis* (mercurous oxide) by adding additional mercury to the former and subliming the mixture.

> For most part of the salts, that made the sublimate so corrosive, abide in the *mercuris dulcis*; but by being compounded with more quicksilver they are diluted by it, and (which is more considerable) acquire a new texture, which renders them unfit to operate, as they did before, when the fretting salts were not joined with a sufficient quantity of the mercury to inhibit their corrosive activity. It may perhaps somewhat help us to conceive, how this change may be made, if we imagine, that a company of mere knife-blades be first fitted with hafts, which will in some regards lessen their wounding power by covering or casing them at the end which is designed for the handle;—and that each of them be afterwards sheathed (which is, as it were, a hafting of the blades too;) for then they become unfit to cut or stab, as before, though the blades be not destroyed: Or else we may conceive these blades without hafts or sheaths to be tied up in bundles, or as it were in little faggots with pieces of wood somewhat longer than themselves, opportunely placed between them. For neither in this new constitution would they be able to cut and stab as before.—But, whether these or any other like changes of disposition be fancied, it may by mechanical illustrations become intelligible how the corrosive salts of common sublimate may lose their efficacy, when they are united with a sufficient quantity of quicksilver in *mercuris dulcis*.
>
> (BOYLE 2000, vol. 8, 470)

Boyle here assumes that sweetness and bitterness are due to the sharpness of the corpuscles of the substances possessing those tastes. Even if we grant him that, he has merely offered two rough possibilities of how mercury could mechanically transform bitter sublimate into sweet *mercuris dulcis*. Boyle does not pretend otherwise. In the introduction to his essay on the mechanical origin of qualities he makes it clear that his proposed corpuscular mechanisms are designed to show no more than that mechanical, and therefore, for Boyle, intelligible, explanations of a variety of phenomena are possible, thereby establishing that "substantial forms, sympathy, antipathy etc" are not necessary (Boyle 2000, vol. 8, 326). Boyle draws attention to the fact that he often offers

---

[9] Details of the Spinoza/Boyle interchange can be found in Hall and Hall (1965, 458–470). It has been discussed by Antonio Clericuzio (1990, 561–589) in a way that gels with my interpretation.

more than one possible explanation for the same phenomenon, as is the case in the example involving mercury sublimate. Boyle is content with "mechanical accounts of particular qualities themselves, as are intelligible and possible, and are agreeable to the phenomena whereto they are applied" Boyle, 2000, vol. 8., 327). There is no suggestion by Boyle that he is capable of establishing the correct corpuscular explanation of chemical, or any other, phenomena. This stands in stark contrast to Boyle's attitude to the explanations, for example of the workings of the mercury barometer, in his pneumatics. In response to doubts concerning those explanations raised by critics, such as Linus and Hobbes, Boyle modified his experiments and conducted new ones to further defend his position. There is no doubt that, in his pneumatics, Boyle sought to establish the right explanations, not merely possible ones.[10]

One of the chapters in Boyle's tract on the mechanical origins of qualities deals with precipitation. It reveals the same general pattern as the one I have noted earlier. Mechanisms, involving the relationships between particle shapes that enable them to interlock with or dislodge, each other are contrived as possible explanations of why one substance can displace and sometimes replace another in a compound leading to the precipitation of the latter. But here there is a difficulty that stood in the way of accommodating precipitation to the corpuscular philosophy, one that was explicitly raised during Boyle's lifetime. The difficulty is posed by series of precipitations in which substances that give rise to precipitation are themselves precipitated. Silver dissolved in nitric acid is precipitated by the addition of copper. But the copper is itself precipitated by the addition of iron, and the iron in turn precipitated by the addition of zinc. The phenomenon was clearly described by Christopher Glaser (1667) in a book published in 1663, and translated into English in 1667, a year or two after the publication of Boyle's "mechanical origin of qualities."

> The silver dissolved in the aqua-fortis, and poured into the vessel of water, precipitates, and separates itself from the dissolvent, by putting a plate of copper into it—the silver is found at the bottom. It must be washed, dried, and kept (if you please) in the form of a calx, or else reduced into an ingot in a crucible, with a little salt of tartar. But if into this second water, which is properly a solution of copper, you put a body more earthy and porous than copper, as iron is, the copper precipitates, and the corrosive spirits of the aqua-fortis fasten to the substance of iron; which may likewise be precipitated by some mineral more earthy and porous than iron, as lapis calminaris and zinc.[11]

The problem for the corpuscularist is to explain how particles of a substance can have shapes and sizes that enable them to dislodge other particles from an association with particles of nitric acid, and combine with them in their stead,

---

[10] For a detailed analysis of Boyle's defense of his pneumatics see Shapin and Schaffer (1985). I discuss the status of Boyle's hydrostatics and pneumatics in more detail in Chalmers (2012b) and explore the evolution of Boyle's concept of pressure in detail in Chalmers (2015).
[11] Christopher Glaser (1667) as cited in Klein (1994, 189).

and at the same time, by virtue of those same shapes and sizes, be subject to being dislodged themselves by the particles of a further additive. The speed and selectivity of chemical combinations inevitably posed major challenges to a mechanical or corpuscular philosopher intent on explaining them by recourse to the shapes, sizes, and motions of particles and nothing else.

6 The Mechanical Philosophy as a Guide for Chemists

Boyle (2000, vol. 2, 91) ended his piece on "the desirableness of a good intelligence betwixt the corpuscularian philosophers and the chymists" in a way that suggests there could be some mutual benefit to both.

> I hope it may conduce to the advancement of natural philosophy, if, as I said, I be so happy as, by any endeavours of mine, to possess both chymists and corpuscularians of the advantages that may redound to each party by the confederacy I am mediating between them, and excite them both to enquire more into one anothers philosophy, by manifesting, that as many chymical experiments may be happily explicated by corpuscularian notions, so many of the corpuscularian notions may be commodiously either illustrated or confirmed by chymical experiments.

What is here characterized as a two-way advantage is not really that at all. Explaining chemical experiments by appealing to corpuscular mechanism and looking to chemical experiments to confirm or illustrate the corpuscular philosophy amount to the same thing, since the experiments confirm the corpuscular philosophy only to the extent that they can be interpreted in corpuscular terms. What is missing from the alleged mutual benefits to be gained from bringing "chymistry" and corpuscular philosophy together is some useful guidance that the corpuscular philosophy might give the experimenter. The burden of this section is to show that the corpuscular philosophy had little to offer in that respect.

In our previous discussion we have seen that the corpuscular mechanisms Boyle invoked in the context of chemistry were contrived, and were possible rather than actual mechanisms. We have also documented the extent to which Boyle stressed that it was the flexibility of the corpuscular philosophy, stemming from the freedom to attribute various shapes, sizes, and motions to corpuscles, which made it possible to adapt the corpuscular philosophy to the phenomena. It was precisely because of these features that the corpuscular philosophy could offer little by way of guidance to the experimenter. The corpuscular mechanisms were contrived to fit chemical phenomena known by other means while the flexibility meant that the search for experimental knowledge of chemical phenomena could not be guided since, as Boyle (2000, vol. 8, 113) noted, the mechanical philosophy can be accommodated to any phenomena because "there can be no ingredient assigned, that has a real existence in nature, that may not be derived immediately, or by a row of decompositions, from the universal matter modified by its mechanical affections." Boyle goes on to remind us that this flexibility arises because the shapes and sizes of corpuscles have among them "as great a variety as need be wished for, and indeed a greater than can easily be so much as imagined"!

Boyle (2000, vol. 8, 276) again stressed, in "the mechanical origin of qualities," that mechanical principles "being so general and pregnant that among things corporeal there is nothing real—that may not be derived from, or brought to a subordination to, such comprehensive principles." As a matter of fact, the flexibility of the mechanical philosophy notwithstanding, there were phenomena that were not obviously reducible to mechanical principles. Included among them, at Boyle's own admission, were weight and elasticity, and we noted in the previous section that there were chemical phenomena that posed difficulties too. But insofar as Boyle's claim about the adaptability of the mechanical philosophy is correct, it indicates that adherence to the mechanical philosophy does little by way of anticipating what the phenomena might be like, and so offers little guidance to the experimenter.

Boyle tended to portray the adaptability of the mechanical philosophy as a strength rather than a weakness. Certain changes ruled out as impossible by those who maintain the immutability of elements or substantial forms are permitted from the point of view of the flexible mechanical philosophy.

> [W]hereas the Schools generally declare the transmutation of one species into another, and particularly that of baser metals into gold, to be against nature, and physically impossible; the corpuscular doctrine rejecting the substantial forms of the schools, and making bodies to differ but in the magnitude, figure, motion or rest, and situation of their component particles, which may be almost infinitely varied, seems much more favourable to the chymical doctrine of the possibility of working wonderful changes, and even transmutations in mixed bodies.
>
> (BOYLE, 2000, vol. 2, 91)

This theme, from "Certain Physiological Essays" is reiterated by Boyle (2000, vol. 5, 332) in his more mature work "The Origin of Forms."

> So that I would not say, that anything can immediately be made of everything —as a gold ring of a wedge of gold, or oil or fire of water; yet, since bodies, having but one common matter, can be differenced but by accidents, which seem all of them to be the effects and consequences of local motion, I see not, why it should be so absurd to think, that (at least among inanimate bodies) by the intervention of some very small addition or substraction of matter (which yet in most cases will scarce be needed), and of an orderly series of alterations, disposing by degrees the matter to be transmuted, of almost any thing may at length be made any thing.

The flexibility of the mechanical philosophy enables Boyle to endorse and entertain the ambitions of the most adventurous alchemical adept.

Boyle's corpuscular chemistry, leaving completely unconstrained the devising of corpuscular mechanisms, was quite incapable of putting chemists on the track of possible, as distinct from impossible, chemical phenomena. Boyle simply did not have a chemical theory at the corpuscular level, a level that, in any case, was too remote from what was experimentally accessible to offer useful guidance to the experimenter.

7 The Introduction of Intermediate Causes into Chemistry

With their emphasis on experimental exploration and practical applications, seventeenth-century chemists faced the task of making their discipline philosophically respectable and distinguishing it from the work of mere artisans or what Boyle referred to as "vulgar empirics." The chemists to whom Boyle was opposed accomplished this task by invoking Aristotelian elements or, more commonly, some modification or extension of Paracelsian principles as providing the underlying "philosophical" causes of chemical phenomena. The textbook tradition that was a marked feature of seventeenth-century chemistry in France involved books that typically began with a relatively short chapter that described the theoretical or philosophical underpinnings of chemistry by invoking elements and principles and then proceeded, in subsequent chapters, to describe the experimental dimension of chemistry and its practical applications especially as related to medicine. Hall (1965, 82—83) has drawn attention to the fact that the connection between the theory and the practice was far from tight. Kim (2003) devotes the first chapter of her book to an exploration of the need perceived by seventeenth-century chemists to portray their discipline as philosophically respectable and is sensitive to the fact that that problem took a peculiar form for Boyle insofar as he rejected the elements and principles deployed by his opponents for that purpose. On Kim's reading Boyle's introduction of corpuscles into chemistry was motivated by the desire to provide causes for chemical phenomena that filled the gap left by his rejection of elements and principles. This postulation of an ideological justification for Boyle's corpuscular chemistry is perfectly compatible with my insistence that it lacked an epistemological one.

Boyle's predicament can revealingly be read in terms of his own distinction between intermediate and ultimate, corpuscular causes. We saw in section 2 how Boyle saw himself as rendering hydrostatics philosophically respectable by invoking the newly forged concept of pressure along with the already familiar concept of weight as causes of hydrostatic phenomena. Those causes fitted Boyle's category of intermediate causes and were cited by him as examples of the latter. Once Boyle had rejected elements and principles in chemistry he was faced with the problem of identifying appropriate intermediate causes in that area that could be both explanatory and subject to experimental exploration in the way that pressure was in hydrostatics. The state of development of the chemistry of the day belied an adequate solution to the problem and, as we have seen, Boyle resorted to invoking systems of corpuscles that lay beyond what could fruitfully guide and be tested against experiment. By the early eighteenth century chemistry had developed to a stage where the beginnings of a solution to Boyle's problem became possible. The notion of a chemical compound as resulting from the combination of the substances ordered in the table published by Geoffroy in 1718 can be read as the introduction of intermediate causes into mineral chemistry. In this section I bring my interpretation of Boyle's chemistry into relief by comparing it to Geoffroy's chemistry of affinity, drawing heavily on Klein's account of the latter.

In 1718 Geoffroy (1996) published what can be seen as the first of a series of affinity tables that productively informed eighteenth-century chemistry, together with a commentary on it. Substances were arranged in columns, the parent substance at the head of a column combining with all the substances beneath it to form compounds. The ordering of substances in a column was designed to indicate the facility with which they combined with the parent substance. A substance added to a compound of the parent substance with some other substance lower in the column would displace the latter substance and combine with the parent substance in its stead. Klein (1994, 1995) has persuasively argued that Geoffroy here gives the first systematic account of the modern notions of chemical compound and chemical combination.[12] Substances can combine to form compounds that may have properties qualitatively different from those of either of its components. The components of a compound remain in the compound as its components insofar as they generate the properties of the compound through their action and insofar as they can be recovered from the compound.

The substances represented in Geoffroy's table, together with their relative affinities, or "*rapports*" as Geoffroy referred to them, qualify as intermediate causes. The properties of compounds can be explained by reference to their composition and the displacement of one substance in a compound by another can be explained by appeal to the affinities represented in the table. This explanatory dimension is illustrated by Geoffroy's treatment, in the commentary on his table, of the preparation of mercury sublimate. Geoffroy identifies and describes all the various ways of preparing mercury sublimate known to and employed by the artisans of his day and then explains how and why the methods work as they do by appeal to the affinities represented in his table. Geoffroy's chemical substances certainly act as causes, but they are intermediate, not ultimate ones. Geoffroy presents relative affinities as experimentally determinable facts but he makes no attempt to explain those affinities. The affinities have a status similar to that of gravity in the mechanics of Newton's *Principia*. Newton acknowledged that gravity was in need of an explanation but nevertheless insisted that the validity of his use of the inverse square law of attraction did not depend on or require knowledge of that explanation. Geoffroy's affinities and Newton's gravity qualified as intermediate causes insofar as they were non-ultimate causes that were explanatory and accessible to experimental investigation.

---

[12] I offer a sympathetic critique of Klein's account of the origin of the concept of chemical compound in Chalmers (2012a).

The chemistry encapsulated in Geoffroy's table is concerned with chemical substances and their combination, and not with underlying corpuscular mechanisms that might explain such phenomena. This is clear from the opening sentences of Geoffroy's paper.

> In chemistry one observes different relationships (*rapports*) between different bodies, which act such that they unite easily with each other. These relationships have their degrees and their laws.

In my (and Boyle's) terminology, Geoffroy's chemistry here involves intermediate, experimentally accessible, rather than ultimate, corpuscular, causes and explanations. It, nevertheless, constituted a research program that could be, and was, extended to include more and more chemical substances and their compounds, at least in the realm of mineral, as opposed to vegetable and animal chemistry. It constituted an important part of the "investigative practice" that Larry Holmes (1989) has identified as the eighteenth-century endeavor that set the scene for Lavoisier.

The symbols arranged in Geoffroy's table represent chemical substances that persist through chemical change. They are precisely those substances that are capable of combining by virtue of the rapport existing between them and other substances. Chemical substances have a range of properties, such as color and smell, which might well be used to help with their identification. However, such properties do not figure in the table. Chemical substances combine by virtue of the affinity between them, not by virtue of their color or smell. The abstraction involved here is comparable to that at work in Boyle's hydrostatics. There, the liquids involved have a range of properties, including chemical properties, that get no mention because they have no bearing on the hydrostatic phenomena under investigation. Geoffroy's chemistry involved an experimental investigation of the manifestations of the intermediate causes of affinity just as Boyle's hydrostatics involved an investigation of the results of the action of the intermediate causes, weight and pressure.

It is important to put the scope of the chemistry informed by affinities in perspective. Insofar as it was concerned solely with substances that could be analyzed into and built up from their components, it constituted only a small part of the chemistry of the time. The majority of the substances manipulated by chemists in their laboratories in the seventeenth and eighteenth centuries were plant and animal materials the transformations of which were irreversible. Most substances in this realm could not be synthesized from their components. The chemistry captured in Geoffroy's table was no more the whole of chemistry than Boyle's pneumatics was the whole of physics.

Boyle's chemistry, in contrast to the chemistry informed by affinity, invoked corpuscular causes rather than intermediate ones. Boyle did not possess the notion of a *chemical* substance that is at work in Geoffroy's chemistry. He characterized substances by way of nominal definitions that specified their properties. Gold, for instance, is designated as a substance "that is extremely ponderous, very malleable and ductile, fusible and yet fixed in the fire, and of a yellowish colour" and can "resist aqua fortis" (Boyle 2000, vol. 5, "Origin of forms," 322–323). The notion of affinity (or *rapport*) that supplied Geoffroy with what Boyle lacked in this respect was ruled out by Boyle on the grounds that it was inappropriately anthropomorphic and, in any case, had no place in the ontology of the corpuscular philosophy.[13] Lacking any notions of intermediate causes capable of ordering the experimental investigation of chemical phenomena, and subordinating chemistry to an all-embracing corpuscular hypothesis, Boyle was not able to set in place a practice distinctive of *chemistry* at all.[14]

It is important to appreciate that my comparison of the chemistry of Boyle and Geoffroy is not intended as an admonishment of Boyle.[15] There are good historical reasons why Geoffroy was able to initiate a program in chemistry in a way that Boyle could not. A precondition for the construction of Geoffroy's table was experimental knowledge of the analysis and synthesis of the compounds implicit in it. Some such knowledge of the kind was already involved in the practices of sixteenth-century metallurgists and seventeenth century pharmacists, but it was not until the early eighteenth century that there was available a critical mass of knowledge of reversible reactions sufficient to form the basis of an affinity table.[16]

Concluding Remarks

Boyle's conception of a material world made up of corpuscles of universal matter possessing only shape, size, and a degree of motion or rest was too far removed from what was accessible to observation and experiment in the seventeenth century for it to be of any assistance in the advancement of chemistry, notwithstanding Boyle's affirmations to the contrary. The flexibility of the corpuscular philosophy, which Boyle invoked as a key merit of it, insofar as it facilitated the contrivance of mechanisms capable of reproducing the phenomena, was a sign of its impotence as far as giving guidance to an experimenter is concerned. Boyle had no theory of chemistry at the corpuscular level.

The innovations of Geoffroy, with which I have contrasted Boyle's "chymistry," grew out of work in Paris in the context of the French Academy and the *Jardin du roi*. The chemistry involved was not subservient to an ultimate matter theory to the extent that Boyle's was. Paracelsian notions that had been prominent in French chemistry earlier in the seventeenth century continued to play a role, but so did corpuscular analogies borrowed from the mechanical philosophers.

---

[13] See, for instance, Boyle (2000, vol. 8, 416 and 490–491).

[14] Symptomatic of this is the fact that the two treatises designed by Boyle to promote the desirability of interactions between corpuscular philosophers and chymists include, not only the essay on the redintegration of saltpeter but also "The History of Fluidity and Firmness."

[15] The color indicator tests introduced by Boyle to distinguish between acids, alkalis, and neutral substances were of lasting value and played an important classificatory function appropriately highlighted by Hall (1965, 87–90).

[16] The path of knowledge of reversible reactions from sixteenth-century metallurgy to Geoffroy's table is detailed in Klein (1994).

This was true of the work of the Academicians Nicholas Lemery and William Homberg on some of whose work Geoffroy was able to build. Homberg's conception of a "middle salt," understood as the result of the action of an acid on alkalis, alkaline earths, and metals, caught on precisely because, in the words of Holmes (1989, 38), "it served as a powerful organising principle for experimental chemistry." It was this work that Geoffroy was able to adapt and extend in the construction of his table. The mechanical or corpuscular analogies adopted by the French chemists were as of little aid to their chemistry as they had been to Boyle's.[17]

In my discussion I have construed the "rapports" assumed in Geoffroy's table as examples of what Boyle referred to as intermediate causes, lying behind and giving rise to observable phenomena and explaining them. I contend that the transition from pre- to modern chemistry can be seen as involving a struggle to formulate appropriate intermediate causes accessible to experiment. Boyle's attempts to wed his chemistry to the mechanical philosophy, and thereby furnish chemistry with the ultimate causes that were the concern of the latter, did not contribute to the task of constructing intermediate causes needed for an experimental chemistry. The disengagement of experimental science from philosophical matter theories was arguably one of the key features that gives the term "the Scientific Revolution" its warrant. Boyle's hydrostatics and pneumatics serve as classic examples of this fruitful disengagement. Boyle did not match this achievement in chemistry, which is indicative, not of his deficiencies, but of just how demanding is the task of identifying intermediate causes capable of forming the basis of a science.

References

Boas, Marie. (1958). *Robert Boyle and Seventeenth-Century Chemistry*. Cambridge: Cambridge University Press.

Boyle, Robert. (2000). *The Works of Robert Boyle*. Ed. Michael Hunter and Edward E, Davis. London: Pickering and Chatto.

Chalmers, Alan. (2009). *The Scientist's Atom and the Philosopher's Stone: How Science Succeeded and Philosophy Failed to Gain Knowledge of Atoms*. Dordrecht: Springer.

Chalmers, Alan. (2010). "Boyle and the origins of modern chemistry: Newman tried in the fire," *Studies in History and Philosophy of Science*, 41, 1–10.

Chalmers, Alan. (2011). "Understanding science through its history: A response to Newman," *Studies in History and Philosophy of Science*, 42, 150–153.

Chalmers, Alan. (2012a). "Klein on the origin of the concept of chemical compound," *Foundations of Chemistry*, 14, 37–53.

---

[17] Principe (2007, 4–5) argues that the usefulness of corpuscular models in Lemery's chemistry has been overstated.

Chalmers, Alan. (2012b). "Intermediate causes and explanations: The key to understanding the scientific revolution," *Studies in History and Philosophy of Science*, 43, 551–562.

Chalmers, Alan. (2015). "Robert Boyle's mechanical account of hydrostatics: Fluidity, the spring of the air and their relationship to the concept of pressure," *Archive for History of Exact Science*, 69, 429–454.

Clericuzio, Antonio. (1990). "A redefinition of Boyle's chemistry and corpuscular philosophy," *Annals of Science*, 47, 561–589.

Geoffroy, Etienne François. (1996). "Table of the different relations observed in chemistry between different substances: 27 August, 1718," *Science in Context*, 9, 313–320.

Glaser, Christopher. (1667). *The Compleat Chemist*. London: John Starkey.

Hall, A. Rupert and Hall, Marie Boas. (1965). *The Correspondence of Henry Oldenburg. Vol. 1, 1641–1662*. Madison: University of Wisconsin Press.

Hall, Marie Boas. (1965). *Robert Boyle on Natural Philosophy*. Bloomington: Indiana University Press.

Holmes, F. L. (1989). *Eighteenth-Century Chemistry as an Investigative Enterprise*. Berkeley: University of California Press.

Kim, Mi Gyung. (2003). *Affinity, That Elusive Dream: A Genealogy of the Chemical Revolution*. Cambridge, MA: MIT Press.

Klein, Ursula. (1994). "Origin of the concept of chemical compound," *Science in Context*, 7, 163–204.

Klein, Ursula. (1995). "E. F. Geoffroy's table of different 'rapports' between different chemical substances-A reinterpretation," *Ambix*, 42, 79–100.

Klein, Ursula. (1996). "The chemical workshop tradition and the chemical practice: Discontinuities within continuities," *Science in Context*, 9, 251–287.

Klein, Ursula. (2007). "Styles of experimentation and alchemical matter theory in the scientific revolution. Essay review of Newman (2006)," *Metascience*, 16, 247–256.

Klein, U. and Lefèvre, W. (2007). *Materials in Eighteenth-century Science: A Historical Ontology*. Cambridge, MA: MIT Press.

Newman, William. (1991). *The Summa Perfectionis of Pseudo-Geber*. Leiden: Brill.

Newman, William. (1996). "The alchemical sources of Robert Boyle's corpuscular philosophy," *Annals of Science*, 53, 567–585.

Newman, William. (2001). "Experimental corpuscular theory in Aristotelian alchemy: From Geber to Sennert." In *Late Medieval and Early Matter Theories* (C. Luthy, J. Murdoch and W. Newman, eds.). Leiden: Brill, 291–329.

Newman, William. (2006). *Atoms and Alchemy: Chymistry and the Experimental Origins of the Scientific Revolution*. Chicago: University of Chicago Press.

Newman, William. (2008). "'The chemical revolution and its chymical antecedents.' Essay review of Klein and Lefèvre (2007) and Kim (2003)," *Early Science and Medicine*, 13, 171–19.

Newman, William. (2010). "How not to integrate the history and philosophy of science: A reply to Chalmers," *Studies in History and Philosophy of Science*, 41, 203–213.

Newman, William and Principe, Lawrence. (1998). "Alchemy vs chemistry: The etymological origins of a historiographic mistake," *Early Science and Medicine*. 3, 32–56.

Principe, Lawrence. (2007). "A revolution that nobody noticed: Changes in early eighteenth-century chemistry." In *New Narratives in Eighteenth-Century Chemistry* (L. Principe, ed.). Dordrecht: Springer, 1–22.

Shapin, Steven and Schaffer, Simon. (1985). *Leviathan and the Air Pump*. Princeton, NJ: Princeton University Press.

CHAPTER 2

What Was Revolutionary about the Chemical Revolution?

NICHOLAS W. BEST

THE CHANGES TO CHEMICAL theory and practice that took place in late eighteenth-century France were truly revolutionary because of the radical nature of the theoretical and methodological changes that occurred, because they were deliberately so, and because that was the start of a tradition in the philosophy of chemistry.

What makes the Chemical Revolution unique among scientific revolutions is that it was anticipated by both philosophers and scientists before it occurred. This meant that the chemists who effected those changes were aware of the subversive nature of their reforms and carried out the revolution in a deliberate fashion.

Three major shifts in the science of chemistry coincided in late eighteenth-century France to make the Chemical Revolution *the* turning point in the history of chemistry: Oxygen chemistry overthrew the reigning phlogiston theory; a cadre of prominently political chemists reformed chemical terminology, providing a new system of names based on oxygen theory; and an empirico-pragmatic conception of elements as simple substances replaced a waning belief in hypostatical chemical principles. This last shift (although itself gradual) ensured that the revolutionary changes in theory and nomenclature would be the last truly radical reforms chemistry would ever need.

Furthermore, the Chemical Revolution was itself a revolution in the philosophy of chemistry as it forced a change in tacit assumptions about the nature of both matter and scientific knowledge. Moreover, studies of this revolution have long shaped general philosophy of science and continue to do so.

1 A Brief Historiography

Cherry-picking the history of science for examples to fit an a priori philosophical theory should be even less acceptable in philosophy of the special sciences than in other branches of philosophy. If philosophers of science are to learn from history, it should be by analyzing changes within periods that historians recognize as revolutionary and giving a philosophical account. Hence the Chemical Revolution is a crucial point for even the most minimally naturalistic philosophy of chemistry.

For some time now, historians of science have understood that the chemistry practiced before the 1770s cannot be dismissed as prescientific mysticism, as was once supposed.[1] Just a few generations ago the doyen of British historians, Herbert Butterfield, famously singled out chemistry for having been retarded in its progress toward the level of sophistication that other sciences—particularly mechanics and astronomy—had enjoyed since the seventeenth century. Butterfield made his name as a historiographer by denouncing simplistic historical narratives that interpret the actions of prominent predecessors as intended to lead toward whatever institutions dominate the period in which that historian is writing.[2] Yet when it comes to science he assumes that all disciplines must take a similar course and describes the occurrence of these dramatic changes in chemistry as a "postponed revolution."[3] By characterizing the timing of this revolution as anomalous, Butterfield implies that there exists some sort of natural progression (albeit more subtle than the kind he denounced), which in turn suggests that the revolution was in some way inevitable. On the contrary, we shall see that it was an event that human agents were deliberately trying to create and only inevitable to the extent that all prophecies can be seen as self-fulfilling.

More recently, Hasok Chang has claimed that historians of science have never, strictly speaking, been whiggish about the Chemical Revolution. Rather, he argues, older histories of this period exemplify "a crude triumphalism, which would celebrate anybody who won (at the time), regardless of whether he was right (by today's standards)."[4] Chang champions a very appealing pluralist historiography but toward the end of this chapter I shall argue that a more sophisticated form of triumphalism is entirely appropriate for philosophers of chemistry, since the way we conceptualize chemical kinds does descend from the victors of the Chemical Revolution.

In most sciences, it is safe to assume that the agents of change during the seventeenth century were largely unaware that their contributions to natural philosophy were part of a larger shift in worldview that only later came to be

---

[1] For an introduction to the historiography of that period, see McEvoy (2010); Russell (1988); Debus (1998); Principe and Newman (2001). The denigration of alchemy is very much a twentieth-century phenomenon. For example, one of the earliest works dedicated to the history of chemistry (Thomson 1830) puts Paracelsus on the same footing with Lavoisier as a chemical revolutionary (my thanks to an anonymous reviewer for pointing this out). Similarly, Marcellin Berthelot (1893), originator of the term "Chemical Revolution," also believed that truly scientific chemistry was practiced in the middle ages.

[2] Butterfield (1931).

[3] Butterfield (1957, ch. 11).

[4] Chang (2009, 251).

known as the Scientific Revolution. But this was not the case for the Chemical Revolution of the eighteenth century. Bernard Cohen's landmark *Revolution in Science*[5] argues that the most remarkable feature of the Chemical Revolution was that it was predicted and carried out by the same man—Antoine-Laurent Lavoisier (1743–1794). This is not the whole story; a revolution in chemistry was in fact anticipated a generation before Lavoisier realized one. Even though the changes he undertook differed from those anticipated by earlier chemists and philosophers, Lavoisier was well aware that he was effecting a revolution. With a small group of like-minded chemists he united separate strands of revolutionary activity and together they formed the new chemical regime.

A number of scholars have sought to diminish the significance of these late eighteenth-century upheavals by emphasizing Lavoisier's debt to earlier generations. Ursula Klein, Wolfgang Lefèvre, and Mi Gyung Kim have challenged the novelty of Lavoisier's theoretical contributions by demonstrating the extent to which he was indebted to the affinity chemists of the early eighteenth century, especially Étienne-François Geoffroy.[6] Conversely, Maurice Crosland has claimed that the changes in chemistry effected by Lavoisier were "more than a revolution."[7] He argues for a narrow Chemical Revolution because he sees linguistic change in chemistry as a gradual process that began long before Lavoisier.[8] The events of the 1770s and 1780s were more than a revolution to the same extent that the Scientific Revolution was so much more than its subset the Copernican Revolution. Copernicus' displacement of the sun (while retaining most of the same mechanisms as Ptolemy) could have been an abortive revolution were it not for a host of other changes in natural philosophy over subsequent generations. Likewise, Lavoisier's refutation of phlogiston theory could have been a mere reform of the model of combustion.[9] That is what Lavoisier believed he was doing at first but the revolution he triggered evolved into something more when a team of prominent chemists joined his cause over the following decade.

The intersection of such broad historical trends makes a narrow period revolutionary. The Chemical Revolution was neither a set of merely theoretical changes,[10] nor an entirely metaphysical shift in belief about which entities exist

---

[5] Cohen (1985, 230).

[6] Klein and Lefèvre (2007); Kim (2003). This line of research dates back to Frederic Lawrence Holmes (1989). His aim in downplaying phlogiston in that work was to provide a curative counterbalance to earlier scholarship that depicted French chemistry from the 1700s to the 1770s as dominated by phlogiston theory because historians understood that paradigm in terms of what it was not—oxygen theory. But Klein and Kim go much further than Holmes intended, implying that the true revolution occurred in the early eighteenth century. William Newman (2008) refutes this line of argument by showing that the techniques and ideas that Klein and Kim depict as novel around 1700 actually date from much earlier.

[7] Crosland (2009).

[8] Crosland (1962). Compare Anderson (1984, 1989).

[9] Indeed, Jerry Gough (1988) has argued that this is the correct interpretation.

[10] Although analyses of the changes in entirely theoretical terms have been attempted—e.g. Andrew Pyle (2001); Paul Thagard (1990).

and how their existence is maintained. It was an array of philosophical changes that coincided with scientific discoveries and tied them to theoretical innovation.

Historians who concentrate on a single thread of the Chemical Revolution—whether the overthrow of phlogiston, nomenclature reform, or quantification—can find predecessors that will allow them to argue for successively earlier revolutions or for continuity. This approach does not do justice to either the nature of scientific research in this period or the clear fact that so many of these disparate innovations came from the same few scientific and philosophical minds: Lavoisier, Louis-Bernard Guyton de Morveau (1737–1816), Antoine-François Fourcroy (1755–1809) and Claude-Louis Berthollet (1748–1822).[11] This team worked together to ensure that two radical changes—oxygen theory and the new nomenclature—dovetailed with more gradual shifts toward quantification and away from a transcendental view of chemical principles.

2 A Much-Anticipated Revolution

The use of the term "revolution" to describe a reform in chemistry dates back to Diderot and d'Alembert's *Encyclopédie,* the pinnacle of Enlightenment natural philosophy. In the main article on chemistry, Gabriel-François Venel (1723–1775) predicts a revolution in chemistry; he believes that it will go from being a mere technical craft to being a true science—"the revolution that places chemistry in the rank that it deserves...will place it at least beside mathematical physics."[12]

Although he seems to believe that this revolution will be particularly significant, Venel uses the word "revolution" later in the same article to refer to earlier upheavals in chemistry:

> In 1723 the new course of chemistry, according to the principles of Newton and Stahl, brought us Stahlianism and caused the same revolution in our chemistry that the reflections on attraction that Mr Maupertuis published in his discourse on the different shapes of stars brought about in physics, forcing us to accept Newtonianism.[13]

Thus, for Venel, even the introduction of new scientific ideas to France from abroad may be spoken of as a revolution. While the bar for describing a scien-

---

[11] Although there were other collaborators for each of the projects, these four men were the core, writing Guyton (1787), critiquing Kirwan (1788), and co-founding the journal *Annales de Chimie.* It is likely that Lavoisier's wife Marie-Anne Paulze-Lavoisier was involved in more than just the translation of Kirwan (1788) but she was not even credited in that work at the time, so it is difficult to estimate the true extent of her contributions (Kawashima 2008).

[12] Venel (1753, 409). All translations are my own.

[13] Venel (1753, 437). The works to which Venel refers are Sénac (1723) and Maupertuis (1732). The former contains next to nothing recognizably Newtonian, since Isaac Newton published none of his chemical knowledge beyond the occasional aside. For his unpublished work on chemistry, see Newman (2012). The latter is also discussed in the Preliminary Discourse to the *Encyclopédie,* since the victory of empiricism over Cartesian rationalist science was a major theme of the project.

tific innovation as revolutionary was quite low at this time, it is itself noteworthy that the state of chemistry was such that it was quite plausible to argue that the best path to progress was through revolution.

Nevertheless, Venel's prediction differs from the actual chemical revolution in significant ways: He maintains that it would not borrow from physics, nor from philosophy. He declares that the imposition of physics on chemistry will not lead to progress as they are concerned with different objects of study. Furthermore, Venel insists that philosophy cannot help chemistry in the way it has helped physics.

———————

Antoine-Laurent Lavoisier himself used the term "revolution" to describe the changes he was making in chemistry, having interpreted Venel's prediction as a call to arms—"The importance of the goal compelled me to return to this work, which seemed to me aimed to cause a revolution in physics and in chemistry. I believed that all that had been done before me must only be regarded as guidelines."[14] By the early 1790s Lavoisier considered the Chemical Revolution complete, since the *Académie Royale des Sciences* was on his side and all younger chemists were learning oxygen chemistry, but he admitted that many holdouts remained in Germany and Britain.[15]

But Lavoisier did not follow Venel's advice to the letter: he appropriated the idea of a revolution but he brought physics' methods to bear on the practice of chemistry and used cutting-edge empiricist philosophy to structure his system and help define chemical concepts. As reflected in his description of this revolution as being "in physics and in chemistry," he borrowed from physics a methodology centered on quantification, in the form of the balance sheet and the calorimeter.[16] It was these experimental innovations that enabled Lavoisier to make the discoveries and arguments that overthrew phlogiston chemistry. By subjecting familiar phenomena to quantitative experimental methods, phlogiston-based explanations could be shown to be self-contradictory and oxygen theory held up as the only viable theory. Moreover, Lavoisier used the superior precision of his instruments to great rhetorical effect: early in his career he used extremely precise weights of gases as an integral part of his argument on the composition of water; toward the end of his career, in the *Elements of Chemistry*, he gave weights in pounds to seven decimal places and provided tables for others to convert weights from grains to pounds up to nine decimal places.[17]

———————

[14] "Memorandum of February 20, 1773," in Guerlac (1961, 230). (N.B. this is twenty years after Venel's article.) For the dating of this memorandum, see Guerlac (1976, 1).

[15] Lavoisier (1997a, 1997b). For the holdouts, see Boantza and Gal (2010); Hufbauer (1982, ch. 7); Allchin (1992). There was also a number of British counter-revolutionaries (including Sir Humphry Davy and Count Rumford), whom Chang warns us not to treat in the same way as the holdouts (Chang 2010, 64–67).

[16] Lundgren (1990).

[17] Golinksi (1995) describes how William Nicholson (1753–1815) criticized Lavoisier's unrealistic degree of precision and failure to round the results of his calculations to the appropriate number of significant figures.

Even in his landmark "Reflections on Phlogiston," Lavoisier praises Stahl's empirical discoveries at length. He describes how Stahl's system was able to furnish explanations for an impressive range of phenomena and contrasts this with later modifications to phlogiston chemistry—particularly those of Pierre-Joseph Macquer (1718–1784) and Antoine Baumé (1728–1804)—which save certain phenomena at the expense of the coherence of the system:

> [Macquer and his fellow chemists have] established a great number of unique doctrines in which only the name phlogiston is conserved—each of them has attached a vague idea to this word, which no one has defined rigorously, and irreconcilable and contradictory properties have been united in the same entity without realising it.[18]

Yet reminding his audience of the original enunciation of phlogiston theory seems to have been intended primarily as a wedge to allow Lavoisier to expose disagreements among contemporary phlogiston chemists. Lavoisier's characterization assumes that any theory using the phlogiston concept must be a tightly crafted conceptual network and insinuates that its plausibility depends on the interconnectedness of its concepts.

3 Philosophical Revolutionaries

Similarly, the epistemology of Condillac empowered Lavoisier to break with the historicist tradition within chemistry and lay the metaphysical foundations of a new system that would last. Étienne Bonnot, abbé de Condillac (1715–1780) represented the French branch of the empiricist movement in philosophy. Like all empiricists, Condillac insisted that knowledge is built entirely from sense-experience.[19] He further claimed (in his mature writings, at least) that human capacities are also learned, as the ideas necessary for skills must first be acquired through experience of using separate senses in concert.[20] Condillac's later work dealt with the active role that language plays in the formation of knowledge and understanding.[21]

Partly inspired by Leibniz, he argued that the best way to maximize human reason is to make our use of language more algebraic. The decomposition of an idea into its simplest parts as well as the reverse—the synthesis of more complex things from simpler ones—was how Condillac believed all sciences should proceed.[22] In this way he differed from Venel, who thought that each science had a unique way of conceptualizing its object. For Condillac, there might be slight differences between the languages needed for each science, as their aims are different, but they will all be very similar, since they are all built on the same basic human sense-data. A well-made language should be sufficient to turn experimental obser-

---

[18] Best (2015a, D634).
[19] Condillac (1821c).
[20] Condillac (1821d).
[21] Condillac (1821a, Part II, ch. 2–4).
[22] Albury (1986, 205–207).

vations into reliable scientific discoveries—"taken to the highest degree of perfection, [methods of communication] will lead us as surely as algebra....To avoid error we need only know how to make use of the language we speak."[23] In short, producing scientific knowledge from sense-data is like balancing equations.

Much of Lavoisier's fascination with the quantification of chemical phenomena dates back to the beginning of his career[24] and thus may predate his reading of Condillac. Yet, his insistence on using the most precise instruments imaginable[25] accorded very well with the Condillacian outlook. One innovation attributable to the Condillacian approach is the introduction of the balance-sheet method to chemistry, for which Lavoisier is rightly lauded.[26] We cannot say that Lavoisier introduced the very idea of quantitative analysis to chemistry, as quantitative techniques had been used in certain domains of chemistry and, as far back as the early seventeenth century, Jan Baptista van Helmont had argued for a quantitative approach to science more generally.[27] What changed in the late eighteenth century was that Enlightenment sensibilities had primed the scientific community to value all forms of quantification.[28]

Although Lavoisier was inspired to produce a revolution in chemistry by Condillac and Venel, a number of his contemporaries had come to a similar conclusion independently. Guyton, in particular, was calling for nomenclature reform before Lavoisier expressed any interest.[29] Yet his first published attempt at a new system of nomenclature was based on phlogiston theory and he remained in that camp until around February 1787.[30] Along with Fourcroy and Berthollet, Lavoisier and Guyton formed a cadre to spearhead a major reform of chemical nomenclature.[31] Soon after, they struck a major blow at the British school of phlogiston chemistry, in the form of a French translation of Richard Kirwan's (1787) *Essay on Phlogiston*, interspersed with scathingly critical notes.[32] By the time they founded the journal *Annales de Chimie* in 1789 it was clear that these young turks had become the new regime.

———————

When Lavoisier himself spoke of a "revolution" in chemistry he meant the overthrow of phlogiston-theory—but this accomplishment was only part of a

[23] Condillac (1821b, 135).
[24] Palmer (1998).
[25] Daumas (1950; 1955, ch. 6).
[26] Bensaude-Vincent (1992, 2008); Roberts (1992).
[27] Newman and Principe (2005, 80–86).
[28] Not all philosophers saw this high degree of precision as desirable. Joseph Priestley worried that Lavoisier's use of his immense wealth to drive his research came at the cost of reproducibility (Golinski, 1999).
[29] Guyton (1782).
[30] Crosland (1962, 174).
[31] Guyton et al. (1787).
[32] Kirwan (1788). The translation itself and a translator's preface were written by Marie-Anne Paulze-Lavoisier; the mathematicians Pierre-Simon de Laplace and Gaspard Monge also contributed essays. The next English edition included those critiques translated into English by William Nicholson and Kirwan's rebuttals (Kirwan 1789).

larger revolution. It was because his allies in the fight against phlogiston were also reformers of chemical nomenclature that oxygen theory gained widespread acceptance. And only because of a gradual change in the ontological assumptions surrounding chemical principles (that began generations before) were these innovations locked in, allowing us to speak of *the* Chemical Revolution.

It could be argued that the Scholastic doctrine of the four Aristotelian elements—earth, fire, air, and water—was never a very fruitful approach to chemistry. Nevertheless, many practicing chemists did pay lip service to the idea that these elements were truly fundamental right up to the generation before the Chemical Revolution. Centuries earlier, actual practitioners of the chemical arts had already proposed alternative ensembles of simple "principles." From the early sixteenth century on, the Paracelsian school believed that all chemical changes were due to the combination of mercury, sulfur, and salt. Conversely many later French alchemists, particularly those influenced by Étienne de Clave, held that there were five principles—spirit, oil, salt, phlegm, and earth. Proponents of these systems usually admitted that Aristotle's four elements were the absolute building blocks but that their principles, built from the elements, were fundamental enough to explain all chemical phenomena.

Mineralogist Johann Joachim Becher's reinterpretation of the Paracelsian triad in terms of Aristotelian elements led him to divide earth into three types—*terra lapida, terra pinguis,* and *terra mercurialis.*[33] Thus, in attempting to reconcile principles and elements he was forced to deny earth its original place as one of the four unique and distinct elements. Similarly, natural philosophers in Britain started to undermine the status of air as an element by investigating not just its novel properties but also distinct types. Lavoisier put the nail in the coffin of the Scholastic tradition of Aristotelian elements with his experiments demonstrating that water could be decomposed into two gases, which could not be further decomposed,[34] thus finishing the work of British aerial philosophers and Helmontian chemists.[35]

Lavoisier's most famous discovery was that metals' gain in weight when calcined was caused by their absorption of a portion of the air, which led him to reject the existence of phlogiston altogether and develop a theory of combustion based on the absorption of oxygen from the air by the material being burned.[36] This seems particularly simple for the combustion and reduction of metals but the conceptual change that came with this inversion was really quite radical—under phlogiston-theory ores were regarded as simple substances and metals were compounds; after the revolution metals came to be seen as the simpler substances and ores were the compounds. Although

---

[33] I.e., stony earth (≈sophic salt); fatty earth (≈sophic sulphur); mercurial earth (≈sophic mercury) (Becher 1968).

[34] Meusnier and Lavoisier (1862). He used Cavendish's term *inflammable air* and the new name *vital air,* which he later renamed "hydrogen" and "oxygen."

[35] Newman and Principe (2002, 296–299).

[36] Best (2015a).

most metals did not change their names, the chemical concepts before and after were quite different and the chemistry of salts had to be completely rewritten.[37]

A pessimistic assessment of Guyton's *Method of Chemical Nomenclature* might argue that it did little more than bring the existing paper technology of affinity tables in line with oxygen chemistry. Indeed, that work was soon superseded by the ample tables of Lavoisier's *Elements of Chemistry*, which expanded the tableaux enormously. The change in nomenclature was a significant contribution because it came at the same time as an empirico-pragmatic conception of elements, making the Chemical Revolution a shift to an open system.

Unlike the triad and pentad systems of principles and the four Aristotelian elements, Lavoisier's system was open-ended. Whereas a principle-based chemical theory would provide a procrustean explanation of a compound's properties in terms of principles drawn from an exhaustive list of either three or five, only a couple of Lavoisier's substances—oxygen and caloric—acted as generic principles. The rest of the constituent ingredients of any compound were diverse substances chosen from an ever-expanding list. In this way, it was an enlightenment project par excellence, as the system was designed to grow in a way that could accommodate new phenomena.[38] Numerous organic compounds that were known and readily isolable from plant and animal sources were not well-enough characterized to be given their modern compositional names, as inorganic compounds were. Hence, radicals such as acetate, tartrate, and oxalate were treated (provisionally) as simple substances. (Similarly, other substances known to be decomposable were given names in the new system because they were known to undergo many reactions without decomposing.)

These principle-based systems used in alchemy and early-modern chemistry were based on a notion of "principle" very different from the modern concept of an element. Whether it be Aristotelian elements or Paracelsus' triad of principles, these chemical entities were not common substances but *hypostatical* principles.[39] In the early eighteenth century these principles became decreasingly relevant to the practice of chemistry since the coalface of research, particularly in the *Académie Royale des Sciences*, had turned increasingly toward

---

[37] Ursula Klein has emphasized the continuity of reference between conceptually different classes, arguing that this shift "may earn the designation 'revolution,' if other criteria for scientific revolutions are chosen than Kuhn's" (Klein and Lefèvre 2007, 185). I shall argue below that the Chemical Revolution was in fact more revolutionary than the Kuhnian model demands.

[38] Bensaude-Vincent (2002, §"1787: A 'Mirror of Nature' to Plan the Future").

[39] This term, first used by Robert Boyle in his criticism of Paracelsian chemistry, means that they are *ideal* substances, "real in the sense that they existed in matter, but not real in the sense that they could be handled and observed" (Boas 1958, 86). The five-principle system of Étienne de Clave is often interpreted in the same way, since his theory is very similar in structure to Paracelsus's, but Mi Gyung Kim has argued that these principles should be read as the products of analysis by distillation rather than transcendental (2001, 369–370).

the analysis of plant and animal materials, wherein the most important concepts concerned tangible chemical products, such as the distinction between fatty oils, essential oils, and empyreumatic oils.[40] A century before the Chemical Revolution chemistry had undergone a significant shift from distillation being accepted as the primary technique of analysis to dissolution being the preferred first step. Over a number of generations there was some debate as to whether the products of distillation had been present in the mixture or if they were artefacts of the fire.[41]

By the middle of the eighteenth century, the empirico-pragmatic definition of principles had become widely accepted. *Empirical* because the identity of these substances is established by laboratory experimentation without theoretical speculation (cf. the contemporary concept of empirical formulae of compounds); *pragmatic* because the conception defines elements for all practical purposes and prescribes agnosticism with respect to fundamental reality or future decomposition.[42] This was most clearly enunciated by Pierre-Joseph Macquer:

> But this analysis and this decomposition of bodies is limited: we can only push it to a certain point, beyond which all our efforts are useless. In whatever way we try, we are always stopped by substances that we find inalterable, that we cannot further decompose and that act as barriers beyond which we cannot go any further. It is to these substances that we must, I believe, give the name of principles or elements, at least that is what they truly are for us.[43]

Forty years later Lavoisier founded his system on an almost identical definition—"if we apply the term *elements* or *principles of bodies,* to express our idea of the last point that analysis is capable of reaching, then all the substances that we have not yet been able to decompose are for us elements."[44] These titans of chemistry were asking the same ontological question but formulated radically different theories. Macquer concluded that the most fundamental substances were the four Aristotelian elements, whereas Lavoisier—having decomposed water into hydrogen and oxygen—applied this definition to an open-ended list of simple substances.

---

[40] Holmes (1989).

[41] Holmes (1971); Boantza (2010).

[42] Cf. historian Mi Gyung Kim (2001) who follows her sources in calling this an "analytic ideal," in contradistinction to a "philosophical ideal," terms which seem to correspond to a posteriori and a priori knowledge, respectively. The subjects of Kim's study use the term "analytic" to mean a posteriori because they seek something like "the last point of chemical analysis." But the word "analytic" is not appropriate usage for philosophy of chemistry since it is not analytical. If one subscribed to a Kantian analytic–synthetic distinction, defining principles according to their properties would be analytic. The last point of analysis could never be the analytic definition of an *element*, since one remains agnostic about the possibility of further decomposing the substance by other means at some point in the future.

[43] Macquer (1749, 2).

[44] Lavoisier (1864, 7).

It is this new conception of chemical agents as nothing more than simple substances that marks Lavoisier's most profound break with earlier chemical theories because it insists that the essence of a substance is nothing more than the properties of a very pure sample. While Macquer continued to pay lip-service to chemistry's heritage of transcendental forms,[45] Lavoisier made it clear that he was only concerned with (what we would call) natural kinds. Just as Darwin changed the concept of *species* to incorporate the idea of a lineage, so Lavoisier made the concept of an *element* commensurate with tangible substances.[46]

4 Not Such a Kuhnian Revolution

The changes to chemistry brought about during the period of the Chemical Revolution are myriad, some sufficiently radical to have been called revolutionary, others mere reforms. Can a conjunction of reforms make a revolution? The overthrow of phlogiston chemistry and the instauration of oxygen theory is generally held to be either the most central feature of the Chemical Revolution or its totality. The conflict was won not with Lavoisier's experiments alone (these were completed in the early 1770s) but by constructing an oxygen theory that the rest of the community would find superior to phlogiston. It is debatable whether the most famous of Lavoisier's contributions—the overthrow of phlogiston—was by itself sufficient to constitute a revolution but, together with his other, more subtle contributions, they certainly did. Lavoisier's team did not simply persuade other chemists that oxygen was a better explanation of the phenomena, they made it so by providing a superior nomenclature (based on oxygen), which dovetailed with a metaphysics of simple substances.

While historians throughout the nineteenth and early twentieth centuries invariably depicted the reforms of chemistry at the end of the eighteenth century as a significant upheaval, it was in the mid-twentieth century that philosophically informed models of scientific revolutions changed the way we see scientific theory change in general and the Chemical Revolution in particular. Prima facie, the fact that this scientific revolution was so self-aware seems to fly in the face of Thomas Kuhn's claim of "the invisibility of revolutions".[47] We might concede to Kuhn that no one of these reforms alone *guaranteed* a change in worldview, thus this revolution was not visible as such until the end and some philosophers of chemistry might maintain that this is the most appropriate stance to take.

Yet we should also ask how Lavoisier could feel entitled to declare a revolution before it was achieved. This question just shows how Kuhn's concept, as

---

[45] Albeit reconceived as simple substances—"simple and homogeneous bodies" (Macquer 1749, 3).

[46] For some of the complications that follow from this, see Hendry (2012) and Best (2015b).

[47] Kuhn (1996, ch. 11).

an analyst's category,[48] is biased toward the old guard, emphasizing as he does the appearance of anomalies and making their resolution the main criterion for initial success of a new paradigm. Used as an actors' category by Lavoisier, "revolution" was a victors' category precisely because he did not use the word in its narrow sense as a success term—for him it was revolutionary all along.

————————

While an important theoretical shift, the introduction of oxygen theory should not be considered a revolution in and of itself, as the chemical *ancien régime* put up very little fight in France and the direst foes of Lavoisier's alliance— Joseph Priestley, Richard Kirwan, James Keir, and James Watt—were themselves moderates, quite willing to reform phlogiston theory.[49] These four British natural philosophers[50] agreed with Lavoisier about what was at stake for the science of chemistry but differed over how best to resolve the issues. They did not see Lavoisier's experiments as crucial tests of their theory—Priestley had produced those new gases well before Lavoisier, providing an explanation that many chemists considered satisfactory[51] and was unimpressed with Lavoisier's quantitative methodology. Nor did they consider his ideas incommensurable with their own—Keir tried to incorporate Lavoisier's empirical discoveries into phlogiston theory.[52] The experimental results that Lavoisier produced were far from sufficient to convince all his contemporaries that phlogiston did not exist.[53] Although the experimental results were anomalies for them, they considered these phenomena to be puzzles to be solved through refinement of phlogiston chemistry.

Unlike a crucial experiment[54] designed to resolve disagreement between competing theories, Kuhn's model of revolution involves competition between worldviews, which are incommensurable ways of seeing the situation, which disagree even about what is at stake.[55] Kuhn characterizes scientific revolutions as gestalt switches between these incommensurable paradigms; thus for him it was the paradigm shift from Stahlian to Lavoisian chemistry that made the refutation of phlogiston a revolutionary event.[56] While he eschews any claim that Lavoisier's chemistry was objectively true or even that it was predictively superior on all fronts, he does insist that it was a completely different worldview, relying upon concepts fundamentally incommensurable with those of Stahlian alchemy.

————————

[48] Collins (2008).
[49] Holmes (2000); Boantza (2008); Stewart (2012).
[50] For James Watt's claim to this title, see Miller (2009, §1). Another term, which Priestley, in particular, preferred is "aerial philosopher" (McEvoy 1978).
[51] Priestley (1774).
[52] Keir (1779).
[53] Kirwan (1789); Priestley (1803).
[54] Conant (1948) portrays Lavoisier's experiment on the reduction of mercuric oxide as an example of a crucial experiment.
[55] Kuhn (1996, ch. 10 "Revolutions as Changes of World View").
[56] At least for early and middle Kuhn (1996, 53–56, 69–72; 1977).

This account can only explain Lavoisier's success if Kuhn is correct in portraying phlogiston theory as a paradigm in crisis, struggling to overcome anomalies. But, in order to make this argument, Kuhn characterizes Stahlian alchemy as a closed conceptual network, under which claims about phlogiston and metal calces were true *sui generis,* thanks to the interdependence of its key concepts. While Stahl's own system can be read in this way, in fact phlogiston chemistry was a broad church and the general idea of an inflammable principle was widespread in French chemistry before Stahl became famous there.[57] As we have seen, Lavoisier was well aware and went to great lengths to show that the phlogiston theories that dominated French chemistry up to the middle of the eighteenth century were not even all that Stahlian.[58] Hence this revolution cannot have been from one well-defined model to another; it was a reaction against a set of diverse chemical theories that employed similar but distinct concepts.

Another way in which the Chemical Revolution might fail to fulfil the criteria of a Kuhnian revolution is with respect to change in classification. In Kuhn's later work he came to see radical change in taxonomic systems as an essential part of a scientific revolution.[59] This has led some historians, using naïve theories of denotation, to argue that the correlation of classes—such as calces with oxides—"constitutes commensurability rather than incommensurability: the kinds of the one system are directly translatable into the kinds of the other."[60] Indeed, an equally significant change to chemical taxonomy came half a generation after the Chemical Revolution, with Joseph Proust's law of definite proportions (1797–1806). The eventual acceptance of this principle meant that a clear line was drawn between chemical compounds and mere physical mixtures, with alloys finding themselves placed among the mixtures. The revolutionary cadre of the 1780s had no such principle in their system and Berthollet was a prominent opponent of Proust's law.

Yet post-Gottlob Fregean philosophers are far more interested in the sense than the extension of words. Attempts have been made to find semantic translation manuals but it must be acknowledged that this is not a straightforward process.[61] While the continuity that material history finds between different taxonomies leads some to downplay differences before and after the Chemical Revolution, the philosophy of chemistry must hold up this epistemic break as a profound rupture.

---

[57] Especially in the work of Wilhelm "Guillaume" Homberg and his protégé Étienne-François Geoffroy, who only became aware of Stahl's phlogiston after their famous experiments with burning lenses (Geoffroy 1708) and seems to have had no knowledge of Becher's system at that stage.
[58] Best (2015a).
[59] Sankey (1998); Buchwald and Smith (1997).
[60] Klein and Lefèvre (2007, 184–185).
[61] Kitcher (1978); Sankey (1991a, 1991b).

5 The Ultimate Revolution in Chemistry

Despite his belief that particular facts of chemical composition are provisional and tentative, Lavoisier built a system whereby chemical theories could endure, thus creating a worldview that was more flexible and more permanent than Kuhn's model allows. This revolution was more than a theoretical paradigm shift—technical and linguistic changes constituted an integral part of the Chemical Revolution.[62]

On top of the relatively simple theoretical shift from phlogiston to oxygen chemistry, Lavoisier his collaborators provided a new system of naming that was closely tied to an epistemic stance with regard to elements and substances. By producing a new system, they gave chemistry what Rudolf Carnap would call a new "linguistic framework."[63] While Lavoisier had already shown that oxygen theory better saved the (salient) phenomena than phlogiston, this linguistic framework offered practical superiority by being more than a set of rules—it was a whole lexicon. This linguistic framework was popular because actual names of "simple substances" and the compounds they could form were laid out ad nauseam in tables in Lavoisier's *Elements of Chemistry*. In line with the Enlightenment spirit, chemists recognized this system as objectively superior since it was an example of performative rationality—Lavoisier imposed rationality on the field by actualizing the system that Guyton and others had demanded.

The new linguistic framework was also more permanent because of its open-endedness. For earlier generations, upon the isolation of any new substance, a principle-based theory of chemistry would conceptualize it as a compound of the standard principles and explain that substance's properties in terms of their canonical properties. If new phenomena were discovered that seemed inexplicable in this way (e.g., electrical conductivity and resistivity), such a system could cope only through some reconceptualization of the principles, by attributing new properties to them. The advent of a new open-ended chemical lexicon minimized the need for such revisions and provided a fertile breeding ground for research into the discovery of new simple substances.

―――――

Using the hypostatical conception of chemical principles which existed before the Chemical Revolution meant that chemical agents were *defined* by their characteristic properties; their names functioned as descriptions in the way that most proper nouns do not.[64] Thus pre-revolutionary chemical theories could conceivably have been overthrown at any time by the discovery of a phenomenon that could not be explained by properties attributable to the standard principles, the discovery of a new substance that could not be decomposed or by the decomposition of a purported element.

―――――

[62] Wise (1993).
[63] Carnap (1956, 206–213). Cf. Kuhn (2000a, 76–77; 2000b); Da Silva (2013).
[64] Indeed, this approach is best interpreted as an example of Bertrand Russell's theory of definite descriptions (1905).

But with the revolutionary lexicon firmly in place such discoveries were no longer problematic. It is widely recognized that a change in the furniture of the world need not trigger a change in a name[65] and this is also the case when our beliefs about the facts change, so long as we treat scientific entities as natural kinds. Names do not necessarily reflect properties; in modern Western cultures at least, proper nouns are not meant to be descriptions of what they denote.[66] This approach is extended beyond personal names when we recognize that even natural kind terms function in this way: natural kinds get their names by baptism, in much the same way that people and places do.[67]

Saul Kripke uses the word "rigid" as a technical term to describe these kinds of designators, defining rigid designators as those terms that designate the same thing in every possible world.[68] This rigidity certainly accords with common usage of natural kind terms (as well as proper names), even when the name is ostensibly a description[69] and Kripke observes that the correct denotation of the name can lose all its (mere) connotations if "the property by which we identify it originally, that of producing such and such a sensation in us, is not a necessary property but a contingent one."[70]

Natural kind terms, like proper names, need not change simply because the natural kind has been reconceptualized.[71] Questions of necessary truths need not concern philosophers of chemistry in the way they do analytic philosophers of language. Nevertheless, we may treat actual scientific theories before and after a revolution in much the same way as they treat possible worlds. Rather than considering which terms refer to the same thing in every possible world, we should call some term an *invariant* designator if it denotes the same natural kind in every chemical theory.[72]

It is the intersection of this new definition of an element with the program of nomenclature reform that makes the system introduced by Lavoisier, Guyton, Fourcroy, and Berthollet one of invariant designation. What their *Method of*

---

[65] "If sand should choke up the mouth of the river [Dart], or an earthquake change its course, and remove it to a distance from the town [of Dartmouth], the name of the town would not necessarily be changed." (Mill 1974, VII.33 (bk I, ch. II, §5)).

[66] E.g., a woman may be named Bianca or Melanie no matter what her complexion; a baker may be named Smith or Taylor.

[67] Kripke (1980, Lecture III).

[68] Kripke (1980, 48). He suggests "nonrigid" or "accidental" for those designators that do not.

[69] E.g., men of all names possess an Adam's apple; a playmate's My Little Pony is never called "Her Little Pony."

[70] Kripke (1972). E.g., blueberries and blackberries are still referred to as such even when actually red in color because they are still green.

[71] Of course some people and places do sometimes change their names (e.g., Cassius Clay and Constantinople) but that change is not automatic. The corollary to my claim here is that noninvariant denotations (specifically definite descriptions) must *a fortiori* be amended as soon as new information is brought to bear.

[72] This term is analogous to Paul Feyerabend's term "invariance of meaning." Feyerabend was attacking the assumption of Ernest Nagel (and his Logical Positivist confreres) that the meanings of theoretical terms are sufficiently fixed to allow for straightforward reduction between scientific theories.

Chemical Nomenclature did was re-baptize all known substances. Most compounds were renamed; most elements retained their old names but were thoroughly reconceptualized by being placed in a table of simple substances. It was with these changes that they ushered in an era of commensurability whereby later generations of chemists could refine chemical theories without a shift in worldview. Thanks to that generation's commitment to a pragmatic definition of elements, the linguistic reforms of the Chemical Revolution took root and later theoretical changes did not necessitate further changing the names of substances. Using the names of elements to refer to ordinary substances (rather than some essence that they are supposed to embody) underwrites further progress in chemistry.

To put it in Gottlob Fregean terms, while the sense (or intension) of the names of various chemical substances continued to change as chemical theories evolved or were superseded, the names did not. For example, when Lavoisier moved ammonia from the simple substances column and reclassified it as a compound of nitrogen, he did not rename it "hydride of azote"; the name it was given in 1787 stuck. Thus the new names themselves trumped the systematicity of the *Method*. After the Chemical Revolution there was no longer any doubt that the reference (or extension) of these terms remained the same. A generation after the Chemical Revolution, Humphry Davy discovered that certain acids do not contain oxygen.[73] Within a principle-based system of chemistry a discovery of this magnitude would have caused a revolution of its own, which would have necessitated a new system of naming to eliminate the supposed acid principle, oxygen.[74]

I am not here claiming that Lavoisier's reforms instituted a theory of reference perfect enough to resist skeptical challenges such as the *qua* problem.[75] My more modest claim is that Lavoisier and his contemporaries in fact employed a notion of elements as natural kinds and that this encouraged many subsequent generations to embrace a realist attitude and simply assume continuity of reference. Although there is little consensus between philosophers of language and rival causal and hybrid theories of reference abound, the natural-kind–based theory of reference tacitly introduced as part of the Chemical Revolution was robust enough to convince the practitioners of the science that they were referring to the same substances across theoretical divides and obviated the need for the more profound kind of scientific revolution in which the metaphysical foundations are shaken. The fact that chemists could thereafter

---

[73] E.g., hydrochloric (Davy 1810).

[74] Lavoisier coined the French word *oxygène* from the Greek prefix ὀξυ-, meaning "acid" and the suffix -γενής, meaning "former." For details of oxygen's role in his theory of acidity, see Lavoisier (1783).

[75] Devitt and Sterelny (1999, §5.3). Robin Hendry (2005, 2010) does make an argument to that effect, which relies on Lavoisier using a Lockean rather than a Kripkean theory of reference. This seems a fair characterization of Lavoisier's approach early in his career when he preferred terms like "vital air" and "the matter of heat and of light" (Best 2016) but these became "oxygen" (even before his alliance with Guyton) and "caloric." This opacity of meaning surely facilitated the invariant interpretation.

continue to use the same terminology meant that they could assume that later theoretical innovations were refinements of chemistry's understanding of these substances, rather than a wholesale replacement of one a priori model with another. This change in theory of meaning was so long-lasting that contemporary chemists can now take it for granted that a reconceptualized substance refers to the same stuff that it did before, no matter how great the theoretical changes have been.[76]

Another reason the practical changes that the Chemical Revolution effected were also more permanent than any other scientific revolution was also partly because the widespread use of quantitative techniques (both experimental and bookkeeping) changed chemists' epistemological expectations. Although Lavoisier borrowed designs for much of his gas apparatus from other chemists, his own bespoke instruments gave quantitative results with a precision unlike any before seen.

Lavoisier's quantitative techniques were more than a new standard of "best practices." What makes the shift to quantification far more significant is that, once the epistemic superiority of these techniques was widely accepted, it introduced a ratchet effect to laboratory development. Any innovation in experimental technique became more easily scrutinized, thanks to the numerical nature of existing data. Thus, Lavoisier's use of the balance-sheet method was the nail in the coffin of a priori theoretical explanation and purely qualitative experimentation[77] because it introduced an expectation of precision that, in turn, contributed to a positivist idea of scientific progress that would dominate French thought for the next century.[78]

Conclusion

Unlike most revolutions in science, the Chemical Revolution had its own Isaiah in G.F. Venel, who predicted that a revolution in that science would elevate it to the same rank as physics. The *abbé* de Condillac predicted that all sciences would be reformed through the adoption of more rigorous and precise languages. A.L. Lavoisier had both of these in mind first, when he used quantitative experimental techniques to overthrow phlogiston chemistry and, second, when he and his allies created a new system of chemical nomenclature based on his oxygen theory. Thus these changes were revolutionary a priori.

This open system of chemistry, together with the explicitly empirical and pragmatic definition of an element meant new substances were thereafter

---

[76] Cf. Best (2015b).

[77] Even organic chemistry, which to this day has a strong emphasis on qualitative characterization of compounds, incorporated quantitative analysis at least as early as Liebig's invention of the *kaliapparat* in 1831 (Rocke 2001).

[78] Paul (1976); Petit (1995); McEvoy (1997).

named by baptism. This ensured commensurability across later theoretical changes and meant that the Chemical Revolution was the last true revolution that chemistry would ever need. Understood in this way, the Chemical Revolution is far more than a Kuhnian revolution.[79]

References

Albury, William Randal. (1986). "The order of ideas: Condillac's method of analysis as a political instrument in the French Revolution." In John A. Schuster and Richard R. Yeo (eds.), *The Politics and Rhetoric of Scientific Method* (Australasian Studies in History and Philosophy of Science, 4). Dordrecht: Reidel, 203–226.

Allchin, Douglas. (1992). "Phlogiston after oxygen," *Ambix*, 39, 110–116.

Anderson, Wilda C. (1984). *Between the Library and the Laboratory: The Language of Chemistry in Eighteenth-Century France*. Baltimore: The Johns Hopkins University Press.

Anderson, Wilda C. (1989). "Scientific nomenclature and revolutionary rhetoric," *Rhetorica*, 7, 45–53.

Becher, Johann Joachim. (1968). "Actorum laboratorii chymici monacensis, seu physicae subterraneae." In *A Source Book in Chemistry, 1400–1900* (Henry M. Leicester and Herbert S. Klickstein, eds.). Cambridge, MA: Harvard University Press, 55–58.

Bensaude-Vincent, Bernadette. (1992). "The balance: Between chemistry and politics," *The Eighteenth Century: Theory and Interpretation*, 33 (3, The Chemical Revolution: Context and Practices), 217–237.

Bensaude-Vincent, Bernadette. (2002). "Languages in Chemistry." In *The Modern Physical and Mathematical Sciences* (The Cambridge History of Science, 5) (Mary Jo Nye, ed.). Cambridge: Cambridge University Press, 174–190.

Bensaude-Vincent, Bernadette. (2008). "Lavoisier, disciple de Condillac," *Matière à penser: Essais d'histoire et de philosophie de la chimie*. Nanterre: Presses Universitaires de Paris Ouest, 127–150.

Berthelot, Marcellin. (1893). *La Chimie au moyen âge*, 3 vols. Paris: Imprimerie Nationale.

Best, Nicholas W. (2015a). "Lavoisier's 'Reflections on phlogiston' I: Against phlogiston theory," *Foundations of Chemistry*, 17, 137–151.

Best, Nicholas W. (2015b). "Meta-Incommensurability between theories of reference: Chemical evidence," *Perspectives on Science*, 23, 361–378.

Best, Nicholas W. (2016). "Lavoisier's 'Reflections on phlogiston' II: On the nature of heat," *Foundations of Chemistry*, 18, DOI: 10.1007/s10698-015-9236-x.

---

[79] I thank Noretta Koertge, Jordi Cat, Stephen Friesen, Scott Hyslop, and Patrick McNeela for their helpful discussions during the preparation of this chapter. I am also grateful to Michael Rings and Chad Gonnerman who provided valuable feedback on early drafts and two anonymous reviewers for their thoughtful comments on the penultimate draft. Parts of this essay are informed by research done at the Chemical Heritage Foundation in Philadelphia in the summer of 2010. I am grateful to this organization and to the Philadelphia Area Center for the History of Science (now the Consortium for History of Science, Technology and Medicine) for their assistance both scholarly and pecuniary.

Boantza, Victor D. (2008). "The phlogistic role of heat in the chemical revolution and the origins of Kirwan's 'Ingenious Modifications... Into the Theory of Phlogiston'," *Annals of Science*, 65, 309–338.

Boantza, Victor D. (2010). "Alkahest and Fire: Debating Matter, Chymistry, and Natural History at the Early Parisian Academy of Sciences." In *The Body as Object and Instrument of Knowledge: Embodied Empiricism in Early Modern Science* (Studies in History and Philosophy of Science, 25) (Charles T. Wolfe and Ofer Gal, eds.). Dordrecht: Springer, 75–92.

Boantza, Victor D. and Gal, Ofer. (2010). "The 'absolute existence' of phlogiston: The losing party's point of view," *The British Journal for the History of Science*, 44, 317–342.

Boas, Marie. (1958). *Robert Boyle and Seventeenth-Century Chemistry*. Cambridge: Cambridge University Press.

Buchwald, Jed Z. and Smith, George E. (1997). "Thomas S. Kuhn, 1922–1996," *Philosophy of Science*, 64, 361–376.

Butterfield, Herbert. (1931). *The Whig Interpretation of History*. London: G. Bell and Sons.

Butterfield, Herbert. (1957). *The Origins of Modern Science: 1300–1800* (New edn.) London: G. Bell and Sons.

Carnap, Rudolf. (1956). "Empiricism, Semantics, and Ontology." *Meaning and Necessity: A Study in Semantics and Modal Logic*. (Enlarged edn.) Chicago: University of Chicago Press, 205–221.

Chang, Hasok. (2009). "We have never been whiggish (about phlogiston)," *Centaurus*, 51, 239–264.

Chang, Hasok. (2010). "The hidden history of phlogiston: How philosophical failure can generate historiographical refinement," *Hyle*, 2010, 47–79.

Cohen, I. Bernard. (1985). *Revolution in Science*. Cambridge, MA: Harvard University Press.

Collins, Harry. (2008). "Actors' and Analysts' Categories in the Social Analysis of Science." In *Clashes of Knowledge: Orthodoxies and Heterodoxies in Science and Religion* (Peter Meusburger, Michael Welker, and Edgar Wunder, eds.). Dordrecht: Springer, 101–110.

Conant, James Bryant. (1948). "The Overthrow of Phlogiston Theory: The Chemical Revolution of 1775–1789." In *Harvard Case Histories in Experimental Science* (1) (James Bryant Conant, ed.). Cambridge, MA: Harvard University Press, 67–115.

Condillac, Étienne Bonnot de. (1821a). *De la logique* (Œuvres Complètes de Condillac, 15 Étude de l'histoire et Logique). Paris: lecointe et Durey.

Condillac, Étienne Bonnot de. (1821b). *De l'art de penser* (Œuvres Complètes de Condillac, 5 Art de Penser et Art d'Écrire). Paris: Lecointe et Durey.

Condillac, Étienne Bonnot de. (1821c). *Essai sur l'origine des connaissances humaines*, (Œuvres Complètes de Condillac, 1). Paris: Lecointe et Durey.

Condillac, Étienne Bonnot de. (1821d). *Traité des sensations* (Œuvres Complètes de Condillac, 3 Traités des Sensations et des Animaux). Paris: Lecointe et Durey.

Crosland, Maurice. (1962). *Historical Studies in the Language of Chemistry*. Cambridge, MA: Harvard University Press.

Crosland, Maurice. (2009). "Lavoisier's achievement: More than a chemical revolution," *Ambix*, 56, 93–114.

Daumas, Maurice. (1950). "Les Appareils d'expérimentation de Lavoisier," *Chymia*, 3, 45–62.

Daumas, Maurice. (1955). *Lavoisier: Théoricien et experimentateur*. Paris: Presses Universitaires de France.

Davy, Humphry. (1810). "Researches on the oxymuriatic acid, its nature and combinations; and on the elements of the muriatic acid," *Philosophical Transactions of the Royal Society*, 100, 231–257.

Debus, Allen G. (1998). "Chemists, physicians, and changing perspectives on the Scientific Revolution," *Isis*, 89, 66–81.

Devitt, Michael and Sterelny, Kim. (1999). *Language and Reality: An Introduction to the Philosophy of Language* (2d edn.) Oxford: Blackwell.

Geoffroy, Étienne-François. (1708). "Experiments upon metals, made with the burning-glass of the Duke of Orleans," *Philosophical Transactions of the Royal Society*, 26, 374–386.

Golinski, Jan. (1995). "'The Nicety of Experiment': Precision Measurement and Precision of Reasoning in Late Eighteenth-Century Chemistry." In *The Values of Precision* (M. Norton Wise, ed.). Princeton, NJ: Princeton University Press, 72–91.

Golinski, Jan. (1999). *Science as Public Culture: Chemistry and Enlightenment in Britain, 1760–1820*. Cambridge: Cambridge University Press.

Gough, J. B. (1988). "Lavoisier and the fulfillment of the Stahlian Revolution," *Osiris*, 4, 15–33.

Guerlac, Henry. (1961). *Lavoisier—The Crucial Year: The Background and Origin of His First Experiments on Combustion in 1772*. Ithaca, NY: Cornell University Press.

Guerlac, Henry. (1976). "The Chemical Revolution: A word from Monsieur Fourcroy," *Ambix*, 23, 1–4.

Guyton de Morveau, Louis-Bernard. (1782). "Sur les dénominations chymiques, la nécessité d'en perfectionner le système, et les règles pour y parvenir," *Observations sur la physique, sur l'histoire naturelle et sur les arts*, 19, 370–82.

Guyton de Morveau, Louis-Bernard, et al. (1787). *Méthode de nomenclature chimique*. Paris: Cuchet.

Hendry, Robin Findlay. (2005). "Lavoisier and Mendeleev on the elements," *Foundations of Chemistry*, 7, 31–48.

Hendry, Robin Findlay. (2010). "The Elements and Conceptual Change." In *The Semantics and Metaphysics of Natural Kinds* (Helen Beebee and Nigel Sabbarton-Leary, eds.). New York: Routledge, 137–158.

Hendry, Robin Findlay. (2012). "Chemical substances and the limits of pluralism," *Foundations of Chemistry*, 14, 55–68.

Holmes, Frederic Lawrence. (1971). "Analysis by fire and solvent extractions: The metamorphosis of a tradition," *Isis*, 62, 129–148.

Holmes, Frederic Lawrence. (1989). *Eighteenth-Century Chemistry as an Investigative Enterprise*. Berkeley: Office of History of Science and Technology, University of California.

Holmes, Frederic Lawrence. (2000). "The 'Revolution in Chemistry and Physics': Overthrow of a reigning paradigm or competition between contemporary research programs?," *Isis*, 91, 735–753.

Hufbauer, Karl. (1982). *The Formation of the German Chemical Community (1720–1795)*. Berkeley: University of California Press, 96–116.

Kawashima, Keiko. (2008). "Paulze-Lavoisier, Marie-Anne-Pierette." *Complete Dictionary of Scientific Biography* (24) (Noretta Koertge ed.). Detroit: Charles Scribner's Sons, 44–45.

Keir, James. (1779). *A Treatise on the Various Kinds of Permanently Elastic Fluids or Gasses* (2d edn.) London: T. Cadell.

Kim, Mi Gyung. (2001). "The analytic ideal of chemical elements: Robert Boyle and the French didactic tradition of chemistry," *Science in Context*, 14, 361–395.

Kim, Mi Gyung. (2003). *Affinity, That Elusive Dream: A Genealogy of the Chemical Revolution*. Cambridge, MA: MIT Press.

Kirwan, Richard. (1788). *Essai sur le phlogistique et sur la constitution des acides* (Louis-Bernard Guyton de Morveau, et al. eds.). Paris: Rue et Hôtel Serpente.

Kirwan, Richard. (1789). *An Essay on Phlogiston and the Constitution of Acids* (New edn.) London: J. Johnson.

Kitcher, Philip. (1978). "Theories, theorists and theoretical change," *Philosophical Review*, 87, 519–547.

Klein, Ursula and Lefèvre, Wolfgang. (2007). *Materials in Eighteenth-Century Science: A Historical Ontology*. Cambridge, MA: MIT Press.

Kripke, Saul A. (1972). "Naming and Necessity." In *Semantics of Natural Language* (Donald Davidson and Gilbert Harmon, eds.). Dordrecht: Reidel, 253–355.

Kripke, Saul A. (1980). *Naming and Necessity*. Cambridge, MA: Harvard University Press.

Kuhn, Thomas S. (1977). "The Function of Measurement in Modern Physical Science." *The Essential Tension: Selected Studies in Scientific Tradition and Change*. Chicago: University of Chicago Press, 178–224.

Kuhn, Thomas S. (1996). *The Structure of Scientific Revolutions* (3rd edn.) Chicago: University of Chicago Press.

Kuhn, Thomas S. (2000a). "Possible Worlds in History of Science." In *The Road Since Structure* (James Ferguson Conant and John Haugeland, eds.). Chicago: University of Chicago Press, 58–89.

Kuhn, Thomas S. (2000b). "Afterwords." In *The Road Since Structure*. (James Ferguson Conant and John Haugeland, eds.). Chicago: University of Chicago Press, 224–252.

Lavoisier, Antoine-Laurent. (1783). "General Considerations on the Nature of Acids, and on the Principles of Which They Are Composed." In *Essays on the Effects Produced by Various Processes on Atmospheric Air* (Thomas Henry, ed.). London: Johnson, 96–118.

Lavoisier, Antoine-Laurent. (1864). *Traité élémentaire de chimie*, ed. Jean-Baptiste Dumas (Œuvres de Lavoisier, I) Paris: Imprimerie Impériale.

Lavoisier, Antoine-Laurent. (1997a). "Lavoisier à Franklin, 2 février 1790." In *Œuvres de Lavoisier—Correspondance* (VI: 1789–1791) (Patrice Bret, ed.). Paris: Académie des Sciences, 109–111.

Lavoisier, Antoine-Laurent. (1997b). "Lavoisier à Chaptal, mars 1790." In *Œuvres de Lavoisier—Correspondance* (VI: 1789–1791) (Patrice Bret, ed.). Paris: Académie des Sciences, 119.

Lundgren, Anders. (1990). "The Changing Role of Numbers in 18th-Century Chemistry." In *The Quantifying Spirit in the Eighteenth Century* (Tore Frängsmyr, J.L. Heilbron, and Robin E. Rider, eds.). Berkeley: University of California Press, 245–266.

Macquer, Pierre-Joseph. (1749). *Élémens de chymie théorique* (1ère edn.) Paris: J.-T. Hérissant.

Maupertuis, Pierre-Louis Moreau de. (1732). *Discours sur les différentes figures des astres, d'où l'on tire des conjectures sur les étoiles qui paroissent changer de grandeur, et sur l'anneau de Saturne, avec une exposition abbrégée des systèmes de M. Descartes et de M. Newton.* Paris: Imprimerie Royale.

McEvoy, John G. (1978). "Joseph Priestley, 'Aerial Philosopher': Metaphysics and methodology in Priestley's chemical thought, from 1762 to 1781," *Ambix*, 25, 1–55, 93–116, 53–75.

McEvoy, John G. (1997). "Positivism, whiggism, and the Chemical Revolution: A study in the historiography of chemistry," *History of Science*, 35, 1–33.

McEvoy, John G. (2010). *The Historiography of the Chemical Revolution: Patterns of Interpretation in the History of Science.* London: Pickering & Chatto.

Meusnier, Jean-Baptiste and Lavoisier, Antoine-Laurent. (1862). "Mémoire où l'on prouve, par la décomposition de l'eau, que ce fluide n'est point une substance simple, et qu'il y a plusieurs moyens d'obtenir en grand l'air inflammable qui y entre comme principe constituant," *Œuvres de Lavoisier* (II—Mémoires de chimie et de physique) Paris: Imprimerie Impériale, 360–372.

Mill, John Stuart. (1974). *A System of Logic Ratiocinative and Inductive*, ed. J.M. Robson (8th edn., Collected Works of John Stuart Mill, 7–8) Toronto: University of Toronto Press.

Miller, David Philip. (2009). *James Watt, Chemist: Understanding the Origins of the Steam Age.* London: Pickering & Chatto.

Newman, William R. (2008). "The Chemical Revolution and its chymical antecedents," *Early Science and Medicine*, 13, 171–191.

Newman, William R. (2015). "The Chymistry of Isaac Newton," <http://chymistry.org>, accessed October 31, 2015.

Newman, William R. and Principe, Lawrence M. (2002). *Alchemy Tried in the Fire: Starkey, Boyle, and the Fate of Helmontian Chymistry.* Chicago: University of Chicago Press.

Newman, William R. and Principe, Lawrence M. (2005). "Alchemy and the Changing Significance of Analysis." In *Wrong for the Right Reasons* (Jed Z. Buchwald and Allan Franklin, eds.). Dordrecht: Springer, 73–89.

Palmer, Louise Yvonne. (1998). "The Early Scientific Work of Antoine Laurent Lavoisier: In the Field and in the Laboratory, 1763–1767" Doctoral dissertation, Yale University.

Paul, Harry W. (1976). "Scholarship and ideology: The chair of the general history of science at the Collège de France, 1892–1913," *Isis*, 67, 376–397.

Petit, Annie. (1995). "L'héritage du positivisme dans la création de la chaire d'histoire générale des sciences au Collège de France," *Revue d'histoire des sciences*, 48, 521–556.

Priestley, Joseph. (1774). *Experiments and Observations on Different Kinds of Air.* London: J. Johnson.

Priestley, Joseph. (1803). *The Doctrine of Phlogiston Established and that of the Composition of Water Refuted* (2nd edn.) Northumberland, PA: A. Kennedy.

Principe, Lawrence M. and Newman, William R. (2001). "Some Problems with the Historiography of Alchemy." In *Secrets of Nature: Astrology and Alchemy in Early Modern Europe* (William R. Newman and Anthony Grafton, eds.). Cambridge, MA: MIT Press, 385–431.

Pyle, Andrew. (2001). "The Rationality of the Chemical Revolution." In *After Popper, Kuhn, and Feyerabend: Recent Issues in Theories of Scientific Method* (Australasian Studies in History and Philosophy of Science, 15) (Robert Nola and Howard Sankey, eds.). Dordrecht: Kluwer, 99–124.

Roberts, Lissa. (1992). "Condillac, Lavoisier, and the instrumentalization of science," *The Eighteenth Century: Theory and Interpretation,* 33 (3, The Chemical Revolution: Context and Practices), 252–271.

Rocke, Alan. (2001). *Nationalizing Science.* Cambridge, MA: MIT Press.

Russell, Bertrand. (1905). "On denoting," *Mind,* 14, 479–493.

Russell, Colin. (1988). "'Rude and disgraceful beginnings': A view of history of chemistry from the nineteenth century," *The British Journal for the History of Science,* 21, 273–294.

Sankey, Howard. (1991a). "Translation failure between theories," *Studies in History and Philosophy of Science,* 22, 223–236.

Sankey, Howard. (1991b). "Incommensurability, translation, and understanding," *The Philosophical Quarterly,* 41, 414–426.

Sankey, Howard. (1998). "Taxonomic incommensurability," *International Studies in the Philosophy of Science,* 12, 7–16.

Sénac, Jean-Baptiste. (1723). *Nouveau cours de chymie suivant les principes de Newton et de Sthall [sic].* Paris: J. Vincent.

Silva, Gilson Olegario da. (2013). "Carnap and Kuhn on linguistic frameworks and scientific revolutions," *Manuscrito,* 36, 139–190.

Stewart, John. (2012). "The reality of phlogiston in Great Britain," *Hyle,* 18, 175–194.

Thagard, Paul. (1990). "The conceptual structure of the Chemical Revolution," *Philosophy of Science,* 57, 183–209.

Thomson, Thomas. (1830). *The History of Chemistry,* 2 vols. London: Colburn and Bentley.

Venel, Gabriel-François. (1753). "Chymie." In *Encyclopédie, ou dictionnaire raisonné des sciences, des arts et des métiers* (3) (Denis Diderot and Jean le Rond d'Alembert, eds.). Paris: Briasson, David, Le Breton & Durand, 408–421.

Wise, M. Norton. (1993). "Mediations: Enlightenment Balancing Acts, or the Technologies of Rationalism." In *World Changes: Thomas Kuhn and the Nature of Science* (Paul Horwich, ed.). Cambridge, MA: MIT Press, 207–256.

| Philosophical Issues in
(Sub)Disciplinary Contexts
The Case of Quantum Chemistry

KOSTAS GAVROGLU & ANA SIMÕES

IN A WAY, QUANTUM chemistry was "born" as a philosophical problem: It was, of course, chemistry, but owed its scientific status to physics; it was physics with the promise of explaining all of chemistry. Thankfully, following P. A. M. Dirac's verdict (1929), this state of affairs, at least some years after 1929, was for a future world, an almost utopian world. In the meantime, chemists, physicists, and mathematicians for about half a century defying Dirac's soothing call that all is well, but only on principle, brought about a new subdiscipline and all the methodological, epistemological, and philosophical problems that go along with the formation of any subdiscipline. In this chapter we put forward a proposal as to how we can write the history of an "in-between" discipline such as quantum chemistry, suggesting that this proposal can be extended to other "in-between" disciplines. Then, we address the role of theory in chemistry, and specifically in quantum chemistry, including the issues surrounding the ontological status of theoretical entities, and proceed to discuss the implications of the introduction of computers in quantum chemistry and the concomitant reconceptualization of experiment. Finally, we reappraise the question of reductionism from the perspective of the practitioners of quantum chemistry.

1 Revisiting the History of Quantum Chemistry

From the very beginning of the period when chemical problems were examined quantum mechanically, everyone involved in the subsequent developments tried to understand the *chemical character* of what was begotten in the encounter(s) of chemistry with quantum mechanics. Was quantum chemistry the subdiscipline for all those chemical problems formulated in the language of physics which could be dealt with by a straightforward application of quantum mechanics with, of course, the ensuing conceptual readjustments?

Was it the case that chemical problems could be dealt with only through an intricate process of appropriation of quantum mechanics by the chemists' culture? Furthermore, the development of quantum chemistry brought about new entities whose ontological status was continuously under negotiation: exchange energy, resonance, and orbitals were some of the more intriguing entities. Research papers, university lectures, textbooks, meetings, conferences, presidential addresses, inaugural lectures, and even correspondence among chemists and physicists became the forum for the discussion of these issues. By attempting to provide answers to these seemingly pedantic, and often implicitly posed, questions, various individuals or groups of individuals attempted to legitimize methodological outlooks and define the status of quantum chemistry. They attempted, that is, to achieve a consensus about the degree of relative autonomy of quantum chemistry with respect to both physics and chemistry and, hence, about the extent of its nonreducibility to physics.

Terminologically it appeared that there was a consensus that quantum chemistry had always been a "branch" of chemistry—despite the immense difficulties that many of the protagonists encountered in their attempts to convince chemists that talk about quantum chemistry was, in fact, talk about chemistry. Its history, however, shows that what appeared to be nominally so, was also the result of the failures of the different cultures predominantly expressed by physicists and applied mathematicians to appropriate quantum chemistry to their own cultures and practices.

Writing the history of a discipline is writing about the becoming of a culture specific to that discipline. Throughout the intricate processes of legitimation of (any) in-between discipline, the ideas, practices, institutions, and their interrelationships form the culture (with its associated subcultures) of those who identify themselves as constituting the community of practitioners of that particular discipline. It seems that a host of philosophical issues have a different relevance for and are expressed in different terms by different disciplinary cultures. Issues in philosophy of science have often been debated as if they have an exclusive reference to physics, and since physics was considered as the scientific field par excellence, the philosophical discussions were presented as having a transdisciplinary scope. A host of discussions concerning philosophical issues in the sciences rely on the assumption that there is a hierarchy among the sciences, physics being at the "top" of this hierarchy. And this assumption conditions the way philosophical problems are formulated. There have been attempts to formulate the philosophical problems in ways that are discipline-sensitive. We wonder whether it may be possible to reorient such a point of view. Perhaps philosophical issues should not even be considered as "belonging" to a discipline as such, but to its practitioners, since it is the discourse these practitioners form that accommodates these issues. By attempting to study the possibilities provided by such an approach, it is important to historicize the philosophical problems appearing in quantum chemistry. During the process of delineating a (new) in-between discipline, those who play any role in its becoming do not only devise and appropriate ideas, techniques and practices, but *contextualize philosophical problems* in order to make additional differentiations with respect to the "parent disciplines." The whole problem of reductionism is on a totally different footing when discussed within the context of (quantum) chemistry than it is the case when discussed within physics. The same holds true when (quantum) chemistry forces us to dramatically reassess the role of theory and its relations to experiment, a problem which bears radical differences when compared to the analogous question in physics. This is why it may be interesting to have more disciplinary histories and test the extent that they may be useful probes for revealing different contextualizations of issues in the philosophy of science (Gavroglu and Simões 2013).

The reference that quantum chemistry is an "in-between" subdiscipline does not imply that it is in a state of limbo, or in a continuous search for identity. In quantum chemistry—but in many other such subdisciplines as well—historically the original fluid state concerning its identity (is it chemistry or is it physics or, at times, is it applied mathematics?) gave way to a rather impressive stability, because there was progressively a consensus around the paradigm to be adopted by the relevant community, and, most importantly, because the issues related to its identity—methodological, conceptual, technical, institutional—through continuous reconceptualizations, and negotiations were at the core of everyday practices, and, in the end, became part of a culture whose strength derived from its ability to accommodate diversity.

It was shown elsewhere (Gavroglu and Simões 2012a) that the history of quantum chemistry can be narrated around six interrelated clusters of issues that manifest the particularities of its evolving (re)articulations with chemistry, physics, mathematics, and biology, as well as its institutional positioning. The first cluster involves issues related to the historical becoming of the epistemic aspects of quantum chemistry: the multiple contexts that prepared the ground for its appearance; the ever present dilemmas of the initial practitioners as to the "most" appropriate course between the rigorous mathematical treatment, its dead ends, and the semiempirical approaches with their many promises; the novel concepts introduced and the intricate processes of their legitimization. Though it may appear that there was a consensus that quantum chemistry had always been a "branch" of chemistry, this was not so during its history, and different (sub)cultures (physics, applied mathematics) attempted to appropriate it. The historical development of quantum chemistry has been the articulation of its relative autonomy both with respect to physics as well as with respect to chemistry, and we argued for the historicity of this relative autonomy. The second cluster of issues is related to disciplinary emergence: the naming of chairs, university politics, textbooks, meetings, and networking, as well as alliances the practitioners of the new discipline sought to build with the practitioners of other disciplines, were quite decisive in the formation of the character of quantum chemistry. The third cluster of issues is related to the contingent character of quantum chemistry. Quantum chemistry could have developed differently; the particular form it took was historically situated, at times being the result of not only technical but also cultural and philosophical considerations. What is important to understand is not what different forms quantum chemistry could or might have taken, but, rather, the different possibilities open for developments and the difficulties that at each particular historical juncture formed barriers that dissuaded practitioners from pursuing these possibilities. Throughout the first 50 years of its history, the criteria for assessing the "appropriateness" of each approach being developed gravitated among a rigorous commitment to quantum mechanics, a pledge toward the development of a theoretical framework where quasi-empirical outlooks played a rather decisive role in theory building, and a vow to develop approximate techniques for dealing with the equations. Such criteria were not, strictly speaking, solely of technical character, and the choices adopted by the various practitioners at different times were conditioned by methodological, philosophical, and ontological commitments and even by institutional considerations. The fourth cluster of issues is related to a rather unique development in the history of quantum chemistry: the rearticulation of the practices of the community after the early 1960s, which was brought about by one artifact—the electronic computer. Calculations, which had been impossible to perform, appeared at long last to be manageable. The fifth cluster of issues is related to philosophy of science. The issues that have been raised throughout the history of quantum chemistry played a prominent role in philosophical elaborations and discussions of reductionism, scientific realism, the role of theory, including its descriptive or predictive character, the role of pictorial representations and of mathematics, the role of semiempirical versus ab initio approaches, and the status of theoretical entities and of empirical observations. The sixth cluster is of a quasi-methodological and quasi-cultural character. The history of quantum chemistry displays instances that we approach in terms of "styles of reasoning."

There are many other (sub)disciplines (mostly "in-between" just like quantum chemistry). They include astrobiology, atmospheric chemistry, atmospheric physics, biogeochemistry, biogeography, biological chemistry, biomaterials, biophysics, chemical oceanography, chemical physics, colloid chemistry, electrochemistry, environment chemistry, geochemistry, geomagnetism, nuclear chemistry, photo chemistry, physical chemistry, physical geography, radiation chemistry, surface science, terrestrial ecology, and so on. Many of the clusters we discussed could be particularly relevant to the history of each one of these (sub)disciplines—some of which are still lacking even a simple chronology of their development. Thus, it might be the case that these six clusters of issues—the epistemic content of a (sub)discipline, the social processes involved in disciplinary emergence, the contingent character of its various developments, the dramatic changes brought about by the digital computer, the philosophical concerns of the protagonists, and the importance of styles of reasoning in assessing different approaches to the becoming of a particular (sub)discipline—may form a framework for weaving the narrative strands of the history of various in-between disciplines. Surely there is much more to their history and surely it will be the case that new philosophical problems will emerge through the history of these (sub)disciplines. Let us stress that this is neither a prescription of how to do disciplinary history nor an algorithm to be applied for every different (sub) discipline nor is the case that each one of these clusters is equally suggestive for understanding the history of every (sub)discipline.

These six clusters of issues—and most importantly their multifarious inter-relationships—comprise a way to articulate the constitutive characteristics of the culture of quantum chemistry. None of the issues related to each of the six clusters can be understood independently of the way each one of them has been expressed through, influenced by, adapted to, and juxtaposed with all the other issues, eventually redetermining them in the arduous process of the formation of a "standard" mode of practice in quantum chemistry. And by discussing the complex of the issues related to each one of these clusters and their relationships, we attempted to substantiate our claim that the history of the emergence and establishment of quantum chemistry could be told as the emergence and establishment of a new culture progressively adopted and propagated by the ever increasing practitioners of this "in-between" discipline—some of whom having started their careers as physicists, some as chemists and some as mathematicians.

2 The Role of Theory in Chemistry

It appeared that developments in quantum chemistry inaugurated discussions on a cardinal issue: the status of theory in chemistry. Many quantum chemists became actively involved in clarifying what chemists (should) mean by theory and in what respects specific theories differed from those of physics. For generations chemistry was identified as a laboratory science, and chemists were content with (empirical) rules. In ways that bear amazing similarities with the case of J. H. van't Hoff's chemical thermodynamics, the theoretical schemata of quantum chemistry were enthusiastically embraced by some and were barely tolerated by most—but, since they bore fruits those who ignored them could not do so for long. Much of the history, and to a large extent the philosophy, of chemistry shies away from discussing the role and character of theory in chemistry—as opposed, of course, to the case of physics. In contradistinction to the physicists, chemists have been happy with expressing allegiance to more than one theory or theoretical schemata—something close to anathema for physicists. Chemists were always open in making a rather liberal use of empirically determined parameters in constructing their theoretical schemata and, often, their schemata appeared to be "propped up" expressions of the rules they had already devised. For physicists the predictive strength of a theory was of paramount importance. Philosophers of science have attempted to understand the intricate balance between the descriptive, the explanatory, and the predictive power of (mainly) the physicists' theories. Meanwhile, the chemists were rather happy trying to explain to their colleagues how they would be using the theories they were devising or "borrowing," often realizing that these were theories that the physicists would snub, and most philosophers of science simply ignore.

It was G. N. Lewis, one of the most forceful advocates of chemical thermo-dynamics and someone whose musings over the mechanism of the covalent bond found an explanatory framework within quantum chemistry, who, hav-ing in mind the recent developments in quantum chemistry, contrasted the different features of theories in chemistry and physics in 1933. He presented structural organic chemistry as the paradigm of a chemical theory, as an ana-lytical theory in the sense it was grounded on a large body of experimental material from which the chemist attempted to deduce a body of simple laws which were consistent with the known phenomena. He called the paradigm of a physical theory a synthetic theory to stress that the mathematical physicist starts by postulating laws governing the mutual behavior of particles and then "attempts to synthesize an atom or a molecule" (Lewis 1933, 17). He main-tained that an inaccuracy in a single fundamental postulate may completely invalidate the synthesis, while the results of the analytical method can never be far wrong, resting as they do upon numerous experimental results.

But theories in chemistry needed a reappropriation of a number of concepts that had their origins in the physicists' *problématique*. Lewis' work in thermo-dynamics was indicative of the feasibility of such a process of reappropriation. His aim (and he was in tandem with van't Hoff) was to convince chemists of the deep significance of thermodynamics for the study of chemical systems, at a time when thermodynamic potentials "belonged" basically to the physicists, and were the physicists' prerogatives. The few chemists who had heard about them could hardly see how they could be applied to complex, real chemical systems.

Let us remember that the formulation of chemical thermodynamics did not automatically lead to its adoption by chemists. There ensued a stage of adapt-ing chemical thermodynamics to the exigencies of the chemical laboratory (Gavroglu and Simões 2012b). Chemical thermodynamics had to appeal to the chemists not only because it provided a theory for chemistry, but also because it formed a framework sufficiently flexible to include parameters which could be unambiguously determined in the laboratory. An aim shared by both van't Hoff and Lewis was the definition of entities which could be of practical use to experimentalists by avoiding a direct reference to entropy, a concept much more atuned to the physicists' needs. They both made efforts to propose *visu-alizable* entities, something which was not independent of the special relations of each with particular *laboratory practices*. For Lewis, thermodynamics could be assimilated in chemistry only if it became possible to work with concepts which could be unambiguously related to situations one meets in the labora-tory, rather than seeking the extension of such concepts, originally defined for ideal systems, to problems occurring in the laboratory. Thermodynamics could lose all its appeal to chemists if it remained a theory formulated in terms of entities which could not be unambiguously measured in the laboratory. For example, it was notoriously difficult to exactly determine partial pressures and concentrations—the entities in terms of which most of the equations of chem-ical thermodynamics were formulated.

Lewis proposed basing chemical thermodynamics on the notion of escaping tendency or fugacity, which he considered as being closer to the chemists' culture, both more fundamental than partial pressure and concentration and exactly measurable. He hoped that this new concept would become the expression for the tendency of a substance to go from one chemical phase to another. After discussing fugacity (whose experimental determination involved the difficult measurements of osmotic pressures), Lewis proposed that chemical thermodynamics be reformulated in terms of the *activity* of a substance. This measured the tendency of substances to induce change in chemical systems; it was defined as the fugacity divided by the product of the gas constant and the absolute temperature (Lewis 1901–1902, 1907).

In 1907 Lewis published a paper titled "Outlines of a new system of thermodynamics in chemistry" in which, among other things, he explicitly articulated his overall approach to chemical thermodynamics. He started by stating that there were two basic approaches in thermodynamics. The first uses entropy and thermodynamic potentials and had been employed by Willard Gibbs, Pierre Duhem, and Max Planck; the second approach, in which the cyclic process was applied to a series of problems, had been used by van't Hoff, Wilhelm Ostwald, Walter Nernst, and Svante Arrhenius. The first method was rigorous and exact and had been used mainly by physicists, whereas chemists preferred the second. According to Lewis, the main reason for the chemists' preference was the difference between the physicists' notion of equilibrium and that of the physical chemists. Although many aspects of the proposed theory may have been similar to the respective physical theory, Lewis' aim was to articulate not so much *the* theory of physical chemistry, but rather the theory *of* physical chemistry by emphasizing the significance of the unambiguously measured quantities for the chemist. Lewis' work repeatedly attempted to formulate thermodynamics on what he considered to be an axiomatic basis where the emphasis was on defining concepts and procedures which would appear *convenient* to the chemists. Lewis became one of the first, together with van't Hoff, to convince chemists of the importance of *theories* in chemistry, and that chemical thermodynamics provided such a possibility. He later did the same in the context of quantum chemistry. More significantly, Lewis tried to convince chemists of the usefulness, even the indispensability, of *mathematical* theories in chemistry, be it chemical thermodynamics or quantum chemistry.

In discussing the ways quantum chemists went about constructing their theories, it is necessary to discuss not only the problems which arise in their appropriation of physics, but also the resistances expressed in having *overtly* mathematized theories. It appears that since the last quarter of the nineteenth century, chemists were expressing their views about the elusive meaning of the term "overtly."

In one of his early papers, Linus Pauling acknowledged his debt to Lewis and showed how his theory came to explain Lewis' schema of the shared electron-pair bond. A comprehensive theory of the chemical bond based on the concept of resonance emerged from the "Nature of the Chemical Bond" series, which was completed by 1933 (Pauling 1931a, 1931b, 1932a, 1932b, 1939; Pauling and Sherman 1933a, 1933b; Pauling and Wheland 1933). Resonance—originally a physical concept—became crucial in the formulation of a chemical theory. In fact, Pauling believed that the task of the chemist should be "to attempt to make every new discovery into a general chemical theory."[1] The concept of resonance played a fundamental role in the discovery of the hybridization of bond orbitals, the one-electron and the three-electron bond, and the discussion of the partial ionic character of covalent bonds in heteropolar molecules. Furthermore, the idea of resonance among several hypothetical bond structures explained in "an almost magical way" the many puzzles that had plagued organic chemistry.[2] Resonance established the link between Pauling's new valence theory and the classical structural theory of the organic chemist which Pauling classified as "the greatest of all theoretical constructs."

> The theory as developed between 1852 and 1916 retains its validity. It has been sharpened, rendered more powerful, by the modern understanding of the electronic structure of atoms, molecules and crystals; but its *character* has not been greatly changed by the addition of bond orbitals, the theory of resonance, partial ionic character of bonds in relation to electronegativity, and so on. It remains a *chemical* theory, based on the tens of thousands of chemical facts, the observed properties of substances, their structure, their reactions. It has been developed almost entirely by induction (with, in recent years, some help from the ideas of quantum mechanics developed by the physicists). It is not going to be overthrown.
>
> (PAULING 1970, 998, emphasis ours)

It was as succinct a statement about the historical role of the newly emerging valence theory as there could be. Pauling was not willing to break ranks with the chemists. He argued that his was not a new theory, but a way of modernizing the very framework of chemists, which he viewed as being determined by structural theory. It was but part of a well-entrenched theoretical tradition of chemistry. Structural theory was a solid chemical theory and developments in the form of resonance theory did not alter its character—despite "some help from the ideas of quantum mechanics developed by the physicists." Pauling spoke as a chemist to fellow chemists. His was a striving for ideological hegemony among the chemists. And he was perfectly suited for this role by virtue of not having been tricked by the siren song of the physicists' quantum mechanics. *His* use of quantum mechanics did not shadow the chemists' tradition as expressed by structural theory: It further augmented it.

Well into the 1970s, the period when it became clear that computers were bringing dramatic changes to quantum chemistry, E. Bright Wilson, the co-author of *Introduction to Quantum Mechanics with Applications to Chemistry*

---

[1] Ava Helen and Linus Pauling Papers, Special Collections, Oregon State University, Box 242, Popular Scientific Lectures 1925–1955, "Recent Work on the Structure of Molecules," Talk given to the Southern Section of the American Chemical Society, 1936.
[2] Ava Helen and Linus Pauling Papers, Special Collections, Oregon State University, Box 242, Popular and Scientific Lectures 1925–1955, "Resonance and Organic Chemistry," 1941.

(1935) with Pauling, wrote a paper examining the impact of quantum mechanics on chemistry. He posed the following questions: Is quantum mechanics correct? Is ordinary quantum mechanics good enough for chemistry? Why should we believe that quantum mechanics is in principle accurate, even for the lighter atoms? Can quantum-mechanical calculations replace experiments? Has quantum mechanics been important for chemistry? Can many-particle wavefunctions be replaced by simpler quantities? Based on the ways in which computers were being used in quantum chemistry, and worried about the lack of new ideas during the last twenty years, Wilson speculated on the possibility that the "computer age will lead to the partial substitution of computing for thinking." But he hoped for "new and better schemes," and he still believed that qualitative considerations would continue to dominate the applications of quantum chemistry. This was, after all, because of the special methodology of chemistry:

> Chemistry has a method of making progress which is uniquely its own and which is not understood or appreciated by non-chemists. Our concepts are often ill-defined, our rules and principles full of exceptions, and our reasoning frequently perilously near being circular. Nevertheless, combining every theoretical argument available, however shaky, with experiments of many kinds, chemists have built up one of the great intellectual domains of mankind and have acquired great power over nature, for good or ill.
>
> (WILSON 1976, 47)

Wilson was encapsulating the development of quantum chemistry in an amazingly succinct, yet shocking, way. There was no attempt to polish the narrative or to turn the protagonists into heroes. Nor was there any attempt to be humble. And the message was clear: The history may have been messy, but the result was unique. From the very beginning, among the chemists there was an ambivalent attitude toward any new proposal of "*how* to do quantum chemistry" or rather, "*what* to do with quantum mechanics when doing quantum chemistry." To many physicists, the chemists' pragmatism appeared flippant. To some chemists or chemically oriented physicists, the physicists' mania to do everything from first principles appeared unnecessarily cumbersome and tortuous. Disagreement over technical issues, more often than not, had its origins in differences of methodological and ontological commitments. Different cultural affinities brought about further murkiness, yet produced more and more new results. And throughout these developments many chemists were attempting to convince chemists that quantum chemistry was a different ball game altogether: One needed to be convinced that chemistry will have different theoretical schemata, and that this state of affairs would be the constitutive aspect of the subdiscipline.

3 Experiments through Computers

Quantum chemistry brought not only a reassessment of the role of theory but also of the role of experiment. Postwar developments in computers led

quantum chemists to rethink the status of experimental practices and to reconceptualize the notion of experiment. Of course, as is always the case and despite the often-assumed autonomy of experiments, theory and experiment did
have various ties between them, even in this new framework.

In the symposium *Aspects de la Chimie Quantique Contemporaine* held
during 1970 in Menton, France, Roald Hoffman, then at Cornell University
and future Nobel Prize winner (1981), offered an analysis of the "meager
achievements" of quantum chemistry in the field of the chemical reactivity of
molecules in their excited states, and outlined the ways to circumvent it. He
sharply distinguished between two types of theoretical chemistry. He called
"interpretative theoretical chemistry" to the promising search for the "theoretical framework used to relate the experimental measurement of some physical
observable to a microscopic parameter of a molecule." Opposing this type of
theoretical chemistry were the "electronic structure calculators," deemed to be
not very successful, and prone to many extremes. He expressed a worry about
a trend whereby chemists are encouraged not to do laboratory experiments,
but their substitutes through computer calculations, something, according to
Hoffman, to be surely avoided.

> If we consider a calculation on a molecule as a numerical experiment and focus
> on the observables that are measured (predicted) by such a numerical experiment
> on a small molecule of the size of butadiene, then I would bet that the experimen-
> talist will be able to predict (on the basis of his experience, reasoning by analogy)
> more correctly the outcome of his theoretical colleague's numerical experiment
> than the theoretician could predict his experimental friend's laboratory observation.
>
> (HOFFMAN 1971, 134)

Hoffman had no doubts that in the methodological approach of "interpretative
theoretical chemistry" laid the future success of quantum chemistry. When
properly applied it produced results of far more lasting value, "the hard facts of
true molecular parameters," than the ephemeral approximate calculations.

A few years later, in another conference forum, H. C. Longuet-Higgins
voiced similar concerns. He talked about three kinds of chemistry: experimental, theoretical and computational. He asserted that even though most
chemists tend to think of molecular computations as belonging to theoretical
chemistry, it could be argued that such computations were really experiments.
Conventional experiments are carried out on real atoms and molecules, "computational experiments are performed on more or less 'modest' and unreli-
able models of the real thing." So the chemist who does computations is
obliged to have a convincing explanation for why the numbers come out as
they do. If not, there may be doubt as to whether they "may not be artefacts of
his basic approximations." This, he considered, was the substance of most
objections to heavy computations of molecular properties by ab initio methods. If such methods have been well attested for a given class of problems,
then it is not unreasonable to "attach weight to the computational solution of
a further problem in that particular class. Unfortunately, the most interesting

problems are usually those with some element of novelty" (Longuet-Higgins 1977, 348).

Experimental practices were redefined not so much within the more traditional framework involving instrumentation and laboratories, but almost exclusively within the framework of mathematics. Soon afterward, the idea of a mathematical laboratory materialized and was explored successfully by quantum chemistry groups. Quantum chemists were not only apt users of the new instruments but played a role themselves both in developing hardware and software and in producing special codes for the numerical calculations of molecular quantities. They had previously enrolled expert human computers for their calculations; now they became themselves computer wizards. Computers and these laboratories emerged simultaneously and reshaped the culture of quantum chemistry.

From the early days of the war the Mathematical Laboratory took shape at the University of Cambridge. In the fall of 1950, J. C. Slater established the Solid-State and Molecular Theory Group at M.I.T., which was initially housed in the premises of the new Research Laboratory of Electronics. R. S. Mulliken's Laboratory of Molecular Structure and Spectra was created in 1952. C. A. Coulson founded the Mathematical Institute within the School of Mathematics also in 1952, with special premises for people to meet and discuss; a decade later he was a member of a committee that started the first University Computing Laboratory (Altmann and Bowen 1974, 88–89). In 1958, the laboratory of P. -O. Löwdin's Quantum Theory Group was inaugurated. As exemplified by all these cases, many opted to associate their new groups to sites they deliberately chose to call *laboratories*. They were not, of course, experimental laboratories. Much like them, however, they were churning out numbers; had a hierarchical structure; were populated by scientists with different expertise; had people visiting; had technicians; and they could accommodate distinctive practices and characteristic cultures. The new laboratories became the sites where successive generations of computers were adapted to the needs of quantum chemistry. Built, tested, used and superseded by more powerful ones, they were often supported by contracts with military agencies eager to profit from them in the upcoming era of Big Science.[3]

4 Old Problems (Re)Considered

Throughout the history of quantum chemistry it appears that in almost all the cases, the reasons for proposing new concepts or engaging in discussions about the validity of the various approaches were:

---

[3] In fact, human computers, often females or students, gave way to computing machines, which became increasingly fast and potent. From the ENIAC, the electronic numerical integrator and computer, built in 1945; to the EDVAC, the electronic discrete variable arithmetic computer, built in 1952; and the UNIVAC, the universal automatic computer, generations of computers replaced former ones at an amazing pace.

1. To circumvent the impossibility to do analytical calculations.
2. To create a discourse with which chemists would have an affinity.
3. To make compatible two languages, the language of classical structure theory and that of quantum mechanics.

Perhaps it may be argued that the involvement in such discussions of almost all those who did pioneering work in quantum chemistry—either in their published papers, in their correspondence, or in their public appearances—had to do with *legitimizing the epistemological status of various concepts in order to be able to articulate the characteristic discourse of quantum chemistry*. Of course, the process of legitimization is not only related to the clarification of the content of the proposed concepts and the correctness of certain approaches. The process itself is a rigorously "social" process, involving rhetorical strategies, professional alliances, institutional affirmations, presence in key journals and conferences, and so on. Interestingly many of the philosophical repercussions appear to have been the unintended implications of such strategies. The relations between the epistemological status of the proposed concepts in the discourse being formed and the philosophical aspects of these concepts is no trivial matter and many times the validity of the former cannot be assessed without recourse to the latter—even if such a recourse has been done by many quantum chemists in a philosophically naïve manner. It was the successes of quantum mechanics in chemistry that induced some chemists and some philosophers to bring to the fore a number of philosophical issues about chemistry, or to discuss problems other philosophers of science had been discussing, but, now, within the context of chemistry. Reductionism turned out to be one of the pivotal issues.[4]

Concentrating on the period after the advent of quantum mechanics, chemists dealt with the conceptual difficulties of their discipline in a manner that was rather different from the ways physicists did. Physicists are used to working with abstract entities, and the questioning (by them or by philosophers) of the ontological status of these entities, such as space, time, mass, and fields, to name just a few, is not something which comes as a surprise to them. It has become part of their culture. Even if a number of physicists were willing to acknowledge the existence of these problems, such an acknowledgement

---

[4] In discussing the philosophical issues of quantum chemistry, it would not be right to put all the emphasis on the question of reductionism. The question of realism has also been intensely discussed by quantum chemists, since coming to a consensus concerning the ontological status of a number of entities (resonance and bonds being two of the more important ones) was dictated by practical necessities and not by philosophical sensitivities. Let us remember that for many decades chemists considered chemistry an exclusively laboratory science. When there were the first attempts at the end of the nineteenth century to adopt a "theory" for chemistry, whether that was the chemical thermodynamics as developed by van't Hoff or the various mathematical models for the structure of the atoms and the various forces between and within atoms, or to subscribe to models explaining the differences between the physical and chemical atom, the chemists put up a strong and obstinate fight to stick to their exclusively laboratory science.

did not interfere with their everyday practices; they continued happily with their calculations and experiments. Some physicists were willing to hear the views of the philosophers of science who studied these problems. Few were willing to intervene in these discussions. Not so with the chemists. Throughout the history of quantum chemistry, chemists never shied away when they came in contact with a philosophical problem. Interestingly, what we consider as a philosophical problem was a "real" problem to chemists with repercussions for their practice, which they had to deal with. They discussed it as chemists, and though a realization of such problems did not block their going on with their calculations and experiments, there was a widespread feeling that these are issues that they—as practicing chemists—have to come to grips with. In our book we have many such examples, including the nature of the various sorts of chemical bonds and molecular orbitals, of resonance, of the various kinds of theoretical schemata proposed, including the role of empirical parameters and approximations, as well as the role of visualizability or the lack of it, among others (Gavroglu and Simões, 2012a). Perhaps, this may be one—one of many—parameters which may explain the late rise of philosophy of chemistry: chemists collectively did not delegate the discussion of these kinds of problems to others as the physicists did. And perhaps their active involvement in the philosophical issues of their discipline may also account for the great success of philosophy of chemistry.

Perhaps Dirac's claim should be considered as nothing more than a physicist's projection of what physics can do for chemistry, yet the question remains as to how the chemists' practice had come to terms with reductionism. Interestingly, in a survey of all chemical papers citing Dirac's statement in the next fifty years, it was concluded that most took it as a historical injunction, not a philosophical prediction about the reducibility of chemistry to physics. That is, they took it as a failed *historical prediction* about the future of chemistry, revealing Dirac's inability to forecast the importance of relativistic effects in chemistry and the coming promise of exact computations (Simões 2002).

The interesting problem is not whether theories are reducible, but whether the ontology of chemistry is reducible to that of physics. On a trivial level there is much in favor of reductionism: Both physics and chemistry deal with atoms and electrons. They comprise the ontological stratum of all the phenomena involved in these disciplines. Again, in a trivial manner, there is a serious difficulty with emergence: It is almost impossible to "build" the phenomena related to both disciplines starting from the building blocks. Hence, such an asymmetry brings in serious complications in the discussion of the philosophical problem. R. Bishop (2005) insists upon a different point. Much of what is associated with reductionism is the claim that physics is the only science offering the possibility of a complete description of the physical world. If that is so, then reductionism will eventually dominate. But is this epistemic claim about physics historically tenable? Might it be the case that reductionism is a historically (not even epistemologically) contingent claim? If despite these objections, one insists on introducing the problem of reductionism, is it not the case that the ultimate statement of reductionism is that all chemistry is explainable in terms of spin—a purely quantum mechanical notion? It may just be the case that reductionism cannot be satisfactorily discussed independent of the character of theory in chemistry.

Let us now raise a different but correlated question: whether reductionism may be a misplaced category if one wants to discuss a number of philosophical issues in chemistry. Perhaps reductionism is a physicist's analytical tool and not a chemist's. Might it be the case that the whole notion of reductionism expresses a trend that is dear to the physicists' own culture rather than that of the chemists? Though physicists took for granted the reduction of chemistry to physics and did little about it, the chemists did not have the luxury of waiting for history to fulfill such an agenda. The benign neglect by chemists of their doomsday so clearly planned by the physicists is, certainly, worth taking note. For reductionism may have been a program, but it was nearly impossible to realize it because, as became evident right at the beginning, one could not deal analytically with any of the other elements except hydrogen and helium, and those only in grossly approximate terms. Let us note that what we want to articulate are not the philosophical considerations of reductionism—the discussion of which in the case of quantum chemistry, and, thus, quantum mechanics, has been greatly enriched by contributions of Hans Primas (1980, 1983), Jeff Ramsey (1997), Eric Scerri (2007), and J. van Brakel (2000), among others. Rather, we articulate the ways such considerations marked the culture of quantum chemists, the way the awareness of such a problem by the community of (quantum) chemists—usually in naïve philosophical terms—has sipped through their practices. Though a number of them had expressed their "worries," reductionism *in any of its variants* was certainly not a paralyzing factor for their everyday practices.

Perhaps one of the intriguing aspects of reductionism is that much of the discussion depends on the theoretical framework with respect to which such a discussion is realized. But how has this problem appeared, for example, in the context of another theory in chemistry, that of chemical thermodynamics? What can one say about reductionism in this case? Is the claim of reductionism valid in this case? What does the unquestionable validity of thermodynamics, and its applications in chemistry, tell us about the relationship between physics and chemistry? Much like the quantum mechanical case we have a "similar" ontology in physics and chemistry when viewed through chemical thermodynamics. But how can we formulate the problem of reductionism within the framework of chemical thermodynamics by taking into consideration its descriptive and explanatory strengths and weaknesses at the time when chemical thermodynamics was being projected as *the* theory for chemistry, much before the all-embracing role of quantum mechanics came to the fore? It appears that the chemists at the time were not so much disturbed by the fact that a theory of physics was being "translated" to cater to their needs, but by the generalized use of mathematics in chemistry—and this appears to be independent of whether one subscribed to Ostwald's energetics or to British atomism. As we have already noted, chemical thermodynamics, though sharing many common features with thermodynamics, differed from it in various respects. The differences were not only because chemical thermodynamics included new and, at times, arbitrary parameters, but because *chemical theories were formulated by chemists with fundamentally different cultural outlooks in comparison to those of the physicists.*

Compared to physicists, these chemists expressed a different culture when it came to formulate a theory and to impose their demands on such a theory—such as the constitutive and regulatory role of empirical data in theory building. Can one, then, pose the question whether reductionism may be dependent on the subculture of chemists and physicists? Is it not the case that the question has up to now been formulated almost exclusively in terms of the physicists' culture? Different scientific communities impose different explanatory demands on their theories, and this constitutes an important cultural characteristic of each community. It may, perhaps, be the case that the formulation of many of the philosophical issues in the different scientific (sub)disciplines is dependent on the specific cultural characteristics of the particular communities which comprise the practitioners of each of these (sub)disciplines. And it may thus be the case that the discussion of these issues may be greatly enriched through such a perspective when the history of the (sub)disciplines does not only come to the fore in order to clarify theoretical or technical issues, but also in order to articulate practices and cultural strands.

Let us now formulate a more general question: What was *at stake* every time such philosophical/theoretical issues were raised during the history of quantum chemistry? When quantum chemists were trying to articulate their views about, say, "reductionism," or when they were disagreeing on how a "proper" theory is to be built, what was the character of the ensuing discussions? In what respects did such discussions bring about changes to entrenched mentalities and start articulating the strands of the emerging subcultures of the emerging (sub)disciplines? Was there, in other words, a philosophical agenda on the part of some of the quantum chemists, or might it have been the case that these issues and discussions were not necessarily aiming at the clarification of their philosophical implications? One cannot claim, of course, that quantum chemists were unaware of the philosophical character of many of the concepts they were introducing or of the philosophical implications of the discussions they were having. But at the same time, one cannot claim that it was philosophy that they had in mind when they were proposing the concepts or when they were pursuing their discussions.[5]

Conclusions

Discussing the role of theory in chemistry and the subsequent efforts to clarify issues that pertain to the theoretical framework of chemistry provides sufficient

---

[5] This situation slowly changed after the 1970s when some participants began discussions about issues involving philosophical aspects and later even began contributing to philosophy of chemistry journals.

material in order to bring to surface the *theoretical particularity* of chemistry—the character of its theories, and their differences from what are considered as theories in physics. If anything is clear, it is that chemical theories are not "incomplete" physical theories, they are not "pretheoretical" schemata that will reach maturity when the physicists decide to deal with them "properly" or when chemists (or anyone else) will decide to deal with them in the physicists' way. Theories in chemistry—and, more particularly, theories in quantum chemistry—have had an autonomy of their own, they were continuously adapted to the chemists' culture, re-forming it in the process. These theories may have been the result of intricate processes of appropriation and reappropriation of physical theories, but, at the end of the day, they themselves "became" chemical theories.

The specific role of mathematics in physics creates a number of philosophical problems that can be unambiguously formulated. There are no intrinsic limitations as to how deep physics can probe. Whether it studies the planets, billiard balls, atoms, nuclei, electrons, quarks, or superstrings, it is "still" physics. Both chemistry and biology are particularly sensitive to such changes of scale and this intrinsic characteristic reflects itself in the character(istics) of their theories. Coulson (1960, 174) made an intriguing point: "Chemistry itself operates at a particular level of depth. At that depth certain concepts have significance and—if the word may be allowed—reality. To go deeper than this is to be led to physics and elaborate calculation. To go less deep is to be in a field akin to biology." Coulson did his utmost to convince chemists—and, perhaps, physicists—that in quantum chemistry, the *role* of theory was not something static, and it had a lot to do, among other things, with the demands of the community, of its decisions concerning the "appropriate depth" at which quantum chemistry will operate. It appears that the history of (quantum) chemistry has been, also, a history of the attempts of chemists to establish the autonomy of its theories with respect to the "analogous" physical theories. In a way, chemists had been obliged to follow such a path. Otherwise chemists would be continuously living in an identity crisis, and would never be sure whether chemistry should be doing the describing and physics the explaining. Quantum chemists have passionately debated these issues, and the myth of the reflective physicist and the more pragmatic chemist is, if anything, historically untenable.

After the Heitler-London paper (1927) on the homopolar bond of hydrogen, the reductionist program for chemistry—as envisioned by Dirac in 1929—, appeared to be a viable program. The paper implied something else as well: that the chemical bond could be perceived as being explained almost exclusively in terms of the electron spin. But spin, after all, was under the jurisdiction of the physicists. And though physicists felt that the new quantum mechanics had also taken care of chemistry, the chemists themselves did not appear to have been under any panic that their identity was being transformed and they were being turned into physicists. Nor did they feel that their very existence was being threatened, since it appeared that what they had been doing could now be done much better by the physicists. The attempts to overcome cultural resistances within the chemical community on how to appropriate quantum mechanics, and the different views on how to form the appropriate discourse became hotly debated issues among the chemists (Gavroglu and Simões 2012a). And almost none of the discussions were taking place under the specter of reductionism, but they formed part of rhetoric for legitimizing the autonomy of quantum chemistry.

In this respect, the metaphor of the artificial fiber proposed by Coulson (1953, 37) is quite suggestive. To underline "how much the validity of the scientist's account depends on the degree of interlocking between its elements," Coulson called attention to the fact that "the strength of an artificial fiber depends on the degree of cross-linking between the different chains of individual atoms." In the same manner, one might argue that the explanatory success of quantum chemistry throughout successive developmental stages rested on the degree of interlocking among constitutive elements—chemical concepts, mathematical notions, numerical methods, pictorial representations, experimental measurements, virtual experiments—to such an extent that it was not the relative contribution of each component that mattered, but the way in which the whole was reinforced by the *cross-linking* and *cross-fertilization* of all elements.

The success of quantum chemistry did not rest on its becoming a subdiscipline of physics. Its success depended not only on epistemological but also on social aspects of this *cross-fertilization*. It involved the establishment and permanent negotiation of alliances among members of a progressively more international community of practitioners, intense networking, and adjustments and readjustments within the community, both at the individual, institutional, and educational levels.

The "globality/homogeneity" of a particular (sub)discipline when it has reached a mature stage may, perhaps, be understood better in terms of the cultures of (sub)disciplines. If we regard the beginnings of "in-between" disciplines as processes where (technical) success has to be accompanied by an intricate strategy of legitimization, then the different outlooks, be they different methodological priorities, or different philosophical viewpoints or different ontological commitments, coalesce into articulating the different cultures of how to actually practice quantum chemistry. Of course, there were issues related to personal antagonisms, to priorities, to genuine disagreements, to different interpretations—but did not all these define each separate culture for doing quantum chemistry? But the consensus was achieved by eclecticism, by members of the "second generation" for whom neither philosophical sophistication nor historical consciousness was a necessary condition for continuing to practice what they inherited: as the first generation moved away from center stage, there were progressively fewer disputes concerning the character of quantum chemistry, not because someone scored a clear victory, but because at the beginning there were diverging trends which through eclecticism eventually gave way to a process of confluence. What started as a patchwork evolved into a seamless whole. A theoretical framework which accommodated a number of epistemologically and, at times, socially induced, partitions, started progressively to look smoother since the reasons for continuing to sustain the partitions—be they technical, institutional, or personal—began to wane. And they began to wane as there were more and more technical successes, as the discipline started to be institutionally visible and assertive, as professional successes helped deflate the egos of the protagonists. In other words, legitimization, theoretical homogeneity, and cultural homogeneity were interdependent processes.

The simultaneous and legitimate coexistence of many theories, the reconceptualization of experiments because of the role of computers, the ever-present issue of reductionism, and the ontological status of theoretical entities have comprised some of the philosophical problems that practicing quantum chemists have been systematically discussing. From our point of view, the point of view of historians, what may be of additional interest is not only the way these problems were treated in a philosophically informed manner, but the way these problems were perceived as philosophical by the practicing quantum chemists who did not delegate their "solutions" to philosophers but convinced each other that their clarification was part of the way of doing quantum chemistry.

It is, perhaps, fitting to close this chapter with a question full of all kinds of philosophical implications and which was posed by a modern chemist: Are all chemists now potential quantum chemists?[6]

Acknowledgments

We thank Grant Fisher and Eric Scerri for inviting us to take part in this volume, and an anonymous referee for many useful comments.

References

Altmann, S.L., and E.J. Bowen. (1974). "Charles Alfred Coulson 1910–1974," *Biographical Memoirs of the Fellows of the Royal Society*, 20, 75–134.
Bishop, R. (2005). "Patching physics and chemistry together," *Philosophy of Science*, 72, 710–722.
Coulson, C.A. (1953). *Christianity in an Age of Science*. London: Oxford University Press.
Coulson, C.A. (1960). "Present state of molecular structure calculations," Conference on molecular quantum mechanics, University of Colorado at Boulder, June 21–27, *Reviews of Modern Physics*, 32, 170–177.
Dirac, P.A.M. (1929). "Quantum mechanics of many-electron systems," *Proceedings of the Royal Society* A, 123, 714–733.

---

[6] This is a question posed by an anonymous referee of our chapter, whose comments reflected his/her deep knowledge of modern day chemistry.

Gavroglu, K. and A. Simões. (2012a). *Neither Physics nor Chemistry. A History of Quantum Chemistry.* Cambridge, MA: MIT Press.

Gavroglu, K. and A. Simões. (2012b). "From physical chemistry to quantum chemistry," *Hyle,* 18, 45–68.

Gavroglu, K. and A. Simões. (2013). "Historical and Philosophical Perspectives on Quantum Chemistry," Book Symposium on K. Gavroglu and A. Simões, *Neither Physics nor Chemistry. A History of Quantum Chemistry. Metascience,* 22, 523–544.

Heitler, W., and F. London. (1927). "Wechselwirkung neutraler Atome und homöopolare Bindung nach der Quantenmechanik," *Zeitschrift für Physik,* 44, 455–472.

Hoffman, R. (1971). "Chemical Reactivity of Molecules in their Excited States." In *Colloque International sur les Aspects de la Chimie Quantique Contemporaine,* 8–13 July 1970, Menton,France, org. R. Daudel, and A. Pullman. Paris: Éditions du Centre Nationale de la Recherche Scientifique, 133–153.

Lewis, G.N. (1901–1902). "The law of physico-chemical change," *Proceedings of the American Academy of Arts and Sciences,* 37, 49–69.

Lewis, G.N. (1907). "Outline of a new system of thermodynamic chemistry," *Proceedings of the American Academy of Arts and Sciences,* 43, 259–293.

Lewis, G.N. (1933). "The chemical bond," *Journal of Chemical Physics,* 1, 17–28.

Longuet-Higgins, H.C. (1977). "Closing remarks," *Faraday Society Discussions,* 62, 347–348.

Pauling, L. (1931a). "The nature of the chemical bond. Application of results obtained from the quantum mechanics and from a theory of paramagnetic susceptibility to the structure of molecules," *Journal of the American Chemical Society,* 53, 1367–1400.

Pauling, L. (1931b). "The nature of the chemical bond. II. The one-electron bond and the three-electron bond," *Journal of the American Chemical Society,* 53, 3225–3237.

Pauling, L. (1932a). "The nature of the chemical bond. III. The transition from one extreme bond type to another," *Journal of the American Chemical Society,* 54, 988–1003.

Pauling, L. (1932b). "The nature of the chemical bond. IV. The energy of single bonds and the relative electronegativity of atoms," *Journal of the American Chemical Society,* 54, 3570–3582.

Pauling, L. (1939). *The Nature of the Chemical Bond and the Structure of Molecules and Crystals: An Introduction to Modern Structural Chemistry.* Ithaca, NY: Cornell University Press.

Pauling, L. (1970). "Fifty years of progress in structural chemistry and molecular biology," *Daedalus,* 99, 988–1014.

Pauling, L., and J. Sherman. (1933a). "The nature of the chemical bond. VI. The calculation from thermochemical data of the energies of resonance of molecules among several electronic structures," *Journal of Chemical Physics,* 1, 606–617.

Pauling, L., and J. Sherman. (1933b). "The nature of the chemical bond. VII. The calculation of resonance energy in conjugated systems," *Journal of Chemical Physics,* 1, 679–686.

Pauling, L., and G. Wheland. (1933). "The nature of the chemical bond. V. The quantum mechanical calculation of the resonance energy of benzene and naphthalene and the hydrocarbon free radicals," *Journal of Chemical Physics,* 1, 362–374.

Primas, H. (1980). "Foundations of Theoretical Chemistry." In *Quantum Dynamics of Molecules, the New Experimental Challenge to Theorists* (R. G. Woolley, ed.). New York: Plenum Press, 39–113.

Primas, H. (1983). *Chemistry, Quantum Mechanics and Reduction.* Berlin: Springer.

Ramsey, J.L. (1997). "Molecular shape, reduction, explanation and approximate concepts," *Synthese,* 111, 231–251.

Scerri, E.R. (2007). "The ambiguity of reduction," *Hyle,* 13, 67–81.

Simões, A. (2002). "Dirac's claim and the chemists," *Physics in Perspective,* 4, 253–266.

van Brakel, J. (2000). *Philosophy of Chemistry. Between the Manifest and the Scientific Image.* Leuven: Leuven University Press.

Wilson, E.B. (1976). "Fifty years of quantum chemistry," *Pure and Applied Chemistry,* 47, 41–47.

| Reduction and Explanation

CHAPTER 4 | # Reality without Reification

Philosophy of Chemistry's Contribution to Philosophy of Mind

MARINA PAOLA BANCHETTI-ROBINO &
JEAN-PIERRE NOËL LLORED

CHEMIST AND PHILOSOPHER OF chemistry Joseph E. Earley has recently argued that, in order to resolve some of its most seemingly intractable problems, philosophy of mind should take into consideration the work currently being done in philosophy of chemistry.[1] This is because there exist obvious parallels between questions that inform philosophy of chemistry and the so-called hard problem of consciousness in philosophy of mind. As David Chalmers describes it, the hard problem of consciousness is that of explaining the relationship between physical phenomena, such as brain states, and experience (i.e., phenomenal consciousness, mental states, or events with phenomenal qualities or "qualia").[2] The "hard problem" is related to the problem of the reduction of mental states to brain states and of the emergence of mental phenomena from physical phenomena. Similar issues are encountered in philosophy of chemistry, such as the reduction of higher-level chemical phenomena to lower-level physical states and the emergence of the higher-level phenomena from the lower-level states. An important and related concern that arises in both philosophical subfields, particularly when dealing with emergence, is the question of "downward causation," that is, the question of whether the higher levels, such as chemical properties or mental states, exert downward causal influence over the lower levels, such as fundamental physical states or brain states. Given the parallels between these two fields, Earley argues that there are three different ways in which philosophy of chemistry can be of assistance to philosophy of mind. The first is by "developing an extended mereology applicable to chemical

---

[1] Earley, Joseph E., "How philosophy of mind needs philosophy of chemistry," *Hyle*, 14, 1 (2008), 1.

[2] Chalmers, David, "Facing up to the problem of consciousness," *Journal of Consciousness Studies*, 2, 3 (1995), 200–219.

combinations."[3] The suggestion is that, if successful, such an extended mereology may also be applicable to the whole-parts relationships between complex systems such as the brain (and its associated mental phenomena) and individual brain states. A second way is by "testing whether 'singularities' prevent reduction of chemistry to microphysics."[4] If chemical "singularities" indeed prevent such reduction, one might extrapolate that mental "singularities" might also prevent the reduction of mental states to electrochemical interactions in the brain. Finally, a third way is by "demonstrating 'downward causation' in complex networks of chemical reactions"[5] and by extending this demonstration from complex chemical networks to neurological systems and their associated mental states.

In agreement with Earley, we propose that the "hard problem" of consciousness should be approached in a manner similar to that used to address parallel problems in philosophy of chemistry. Our contribution will thus first scrutinize the ways chemists and quantum chemists think about and use (1) different levels of organization, and (2) chemical relations and *relata*. We will then investigate the problem of "downward causation" as it relates to the question of emergence, though these issues are inextricable from the mereological questions. As Michel Bitbol has argued and as this chapter will concur, a solution to the "hard problem" of consciousness will require that we transcend traditional emergentism and its substantialist conception of mind. In this respect, we will show that, as paradoxical at it may seem, the science of the transformations of "substances," namely chemistry, enables us to go beyond susbstantialism. The nature of mental phenomena must, instead, be examined in terms of the relationality of wholes (the brain as a complex system and its associated mental states) and parts (specific brain states). As Rom Harré has pointed out in the context of philosophy of chemistry, this will require the development of a mereology that explains how parts and wholes may co-define each other. In this manner, downward causation, which clearly exists but cannot be explained by traditional emergentism (as we will show), must itself be rendered into a relational concept. We propose that an extended mereology for philosophy of mind must, therefore, be one that entangles the whole, its parts, and the environment in much the same way as the nonstandard mereology for quantum chemistry that is proposed by Harré and Llored.[6] This is not "a classical transitive mereology. It is not merely a holistic description within which the whole is necessary to define the parts or the kinds of parts involved. [Nor is it] merely a reductionist analysis that only needs the parts to define the whole."[7] Rather, as previously stated, the parts and the wholes co-define each other in

---

[3] Earley, Joseph E., "How philosophy of mind needs philosophy of chemistry," 1.
[4] Ibid.
[5] Ibid.
[6] Harré, Rom and Jean-Pierre Llored, "Mereologies as the grammars of chemical discourses," *Foundations of Chemistry*, 13 (2011), 63–76.
[7] Llored, Jean-Pierre, "Emergence and quantum chemistry," *Foundations of Chemistry*, 14 (2012), 265.

the context of chemistry. Yet, in order for such a mereology to be successfully developed, our understanding of parts and wholes, as independent concepts, must itself be altered. And this is where Bitbol's proposal for a non-substantialist account of levels of reality can be of immeasurable use. Before developing these points, however, we will begin by investigating the manner in which quantum chemists involve different levels of organization in their daily calculations.

1 First Enquiry: The Co-Definition of the Levels of Organization

Robert S. Mulliken, who actively participated in the advent of quantum chemistry, considers each molecule to be a self-sufficient unit. He asserts:

> Attempts to regard a molecule as consisting of specific atomic or ionic units held together by discrete numbers of bonding electrons or electron-pairs are considered as more or less meaningless, except as an approximation in special cases, or as a method of calculation [...]. A molecule is here regarded as a set of nuclei, around each of which is grouped an electron configuration closely similar to that of a free atom in an external field, except that the outer parts of the electron configurations surrounding each nucleus usually belong, in part, jointly to two or more *nuclei*.[8]

Rejecting the interpretation of the concept of valence as an intrinsic property of the atom, Mulliken opposes the use of "energy state" deduced from molecular spectra on the basis of an *electronic configuration*, that is, of a distribution of the molecular electrons in different orbits. Each orbit is delocalized over all the *nuclei* and can contribute, depending on each specific case, a stabilizing or destabilizing energy contribution to the total energy of the molecule. For Mulliken, *the atom does not exist in a molecule.* Thus, there is a key semantic shift from the concept of molecular *orbit* to that of molecular *orbital*. Molecular orbitals are wave functions that contain one electron and that can be delocalized either over all the nuclei or simply over a set of particular nuclei. The complete electronic wave function ψ is restricted to one of several types that depend on the symmetry of the nuclear skeleton.

Let us consider the simple case of a molecule containing two nuclei. Our starting point is thus the molecular wave function ψ which, depending on the case, can be usefully written as a linear combination of two atomic orbitals φ_1 and φ_2:

$$\psi = c_1\varphi_1 + c_2\varphi_2 \tag{1}$$

This equation seems to imply that the whole, which belongs to the "molecular level," is reduced to electrons and nuclei, that is to say, to a more "fundamental

<hr>

[8] Mulliken, R. S., "Electronic structures of polyatomic molecules and valence, I," *Physical Review*, 40 (1932), 55.

level." This conclusion, however, is at odds with Mulliken's holistic standpoint and does not, in fact, correspond to the manner in which we should understand his work. Let us, therefore, discuss how equation (1) should be understood in the context of Mulliken's holistic perspective. To do this, let us refer to Dirac's notation, where "H" is the molecular Hamiltonian:[9]

$$<\psi_i|H|\psi_i> \ = \ \int_{space} \psi_i^* H\psi_j \, d\tau = H_{ii} \text{ is the Coulomb integral}$$

$$<\psi_i|H|\psi_i> = \ \int_{space} \psi_i^* H\psi_j \, d\tau = H_{ij} \text{ is the exchange integral}$$

$$<\psi_i|H|\psi_i> \ = \ \int_{space} \psi_i^* \psi_j \, d\tau = S_{ij} \text{ is the overlap integral}$$

φ_1 and φ_2 can, at least partially, overlap in the space region that corresponds to the intersection of each atomic space. The word "orbital" takes all of its meaning from Max Born's probabilistic interpretation according to which the square of a molecular orbital corresponds to the probability density of finding a particular electron within the molecular space. Thus,

$$<\psi|H|\psi> = c_1^2 <\varphi_1|\ \varphi_1> + 2\ c_1 c_2 S_{12} + c_2^2 <\varphi_2|\varphi_2> \qquad (2)$$

Mulliken's interpretation of the weighting coefficients c_i^2, where i is equal to either 1 or 2, implies that these respectively represent the part of the electron density that belongs to nucleus 1 or to nucleus 2 only. On the other hand, the term "$2\ c_1 c_2 S_{12}$," namely "the overlap population," expresses the part of the electronic density that refers to the two atomic functions *at the same time* and, thus, to the *whole molecule*. This third term provides crucial information about the "strength" of the chemical bond. For instance, in the case of the molecule LiH, Mulliken proves that (1) a first fraction, equal to 0.111 of the whole electronic charge, "belongs" to the hydrogen atom; (2) another fraction, equal to 0.641, belongs to the lithium; and (3) the last (but not the least) fraction, equal to 0.248, belongs to both the hydrogen and the lithium.[10]

We can generalize this outcome to a molecule that contains more than two nuclei. However, in such a case, the calculation of the overlap population of each related pair of nuclei requires us to take into account the contribution of *all* the electrons belonging to *all* the molecular orbitals that are occupied, knowing that there are as many molecular orbitals as there are atomic orbitals in the linear combination. As a consequence, the overlap population between any set of two related nuclei depends upon the whole molecule. In other words, when one wants to study a local bond between particular *atoms* in the molecule, one has to integrate the characteristic of belonging to the whole molecule into the calculation. *The whole and the parts are thus co-defined.* Chemists need

---
[9] $d\tau$ is a volumic element and ψ_i^* the conjugate of the complex function ψ_i.
[10] Mulliken, R. S., "Electronic population analysis on LCAO-MO molecular wave functions. I," *The Journal of Chemical Physics*, 23, 10 (1955), 1833–1840.

both of these, at the same time. It seems, therefore, that there is no room for eliminativism and reductionism in such a story.

At this point, we would like to go a step further and consider what Mulliken calls "atomic population" and the manner in which the weighting coefficients of the linear calculations are determined. Mulliken defines the net charge of a particular atom within the molecule as the algebraic sum of its nuclear charge q_n and its electronic charge q_e. The latter is calculated by sharing, into two equal pieces, the charge of the electronic density around the particular bonded atoms that are under study. In the case of a "diatomic" molecule, Mulliken ascribes the fraction $c_1^2 + \frac{1}{2}(2c_1c_2S_{12})$ to the atom A corresponding to theatomic orbital φ_I, in order to calculate q_e. This atomic charge depends on the whole molecule, via the overlap population S_{12}. In the case of a more complicated molecule, one must use the expression

$$q_e(A) = \sum_{i,j} n_i c_{iA} c_{ij} S_{Aj} \qquad (3)$$

where (1) n_i refers to the number of electrons belonging to the molecular orbital ψ_i, (2) c_{iA} is the weighting coefficient of the atomic orbital that represents A and c_{ij} is that of the atom j *within the same molecular orbital* ψ_i, and (3) S_{Aj} is the overlap integral between both the atoms A and j in ψ_i.

Once again, equation (3) takes into account all of the molecular orbitals and all of the atoms j of the molecule. In brief, when one aims to determine each local atomic charge, one has to refer to the whole molecule. On the other hand, when one investigates molecular reactivity, one has to use atoms endowed with local charges, in order to identify reactive sites within each molecule of the chemical reaction. In this respect, reactivity at the "molecular level" and atomic charges at the "atomic level" are co-dependent. Their meanings co-emerge within a particular practice of calculation. As Isabelle Stengers claims, "as soon as the question of emergence is at stake, the whole and its parts must thus co-define [each other], and mutually negotiate what an explanation of the one from the other means."[11]

What happens when we examine the weighting coefficients c_1 and c_2 in our previous, simpler case? Let us consider a system (atom or molecule) in its fundamental state, and let us call E_o the system's corresponding eigenvalue for the energy. The exact solution of the Schrödinger equation cannot be calculated, and chemists thus have to employ a large panel of approximations. The average energy <E>, calculated from an approximate wave function ψ, is always superior or equal to E_o:

$$\langle E \rangle = \frac{\int_{space} \psi^* H \psi \, d\tau}{\int_{space} \psi^* \psi \, d\tau} \geq E_0 \qquad (4)$$

---

[11] Stengers, Isabelle, "La vie et l'artifice: Visages de l'émergence," in *Cosmopolitique II* (Paris: Éditions La découverte, 2003), 2007, as cited in and translated by Jean-Pierre Llored, "Emergence and Quantum Chemistry," 265.

The Variation Method is currently applied in order to reach the best solution possible. The key point is to start from a family of wave functions within which chemists believe that they will find the best approximation, according to their previous calculations and to their chemical expertise. They use a wave function that depends on at least one parameter, and they determine the value of the parameter(s) (in our present case, the coefficients c_1 and c_2), which leads to the lowest value of the average energy. The average energy is defined as follows:

$$<E(c_1, c_2)> = <\psi(c_1, c_2)\,|\,H|\,\psi(c_1, c_2)>|<\psi(c_1, c_2)\,|\,\psi(c_1, c_2)>$$

The integration of the linear combination into the previous expression leads to

$$<E(c_1, c_2)> = <c_1\varphi_1 + c_2\varphi_2\,|\,H\,|\,c_1\varphi_1 + c_2\varphi_2>|<c_1\varphi_1 + c_2\varphi_2\,|\,c_1\varphi_1 + c_2\varphi_2>$$

The normalization of the two atomic orbitals implies that

$$<\varphi_1\,|\,\varphi_1> = <\varphi_2\,|\,\varphi_2> = 1$$

which in turn implies that

$$<E(c_1, c_2)> = \frac{c_1^2 H_{11} + 2c_1 c_2 H_{12} + c_2^2 H_{22}}{c_1^2 + 2c_1 c_2 S_{12} + c_2^2} = \frac{N(c_1, c_2)}{D(c_1, c_2)} \tag{5}$$

We propose to call $N(c_1, c_2)$ and $D(c_1, c_2)$, the numerator and the denominator of equation (5), respectively. The minimization of the energy using the coefficients c_1 and c_2 implies that the partial derivatives of the average energy become equal to zero:

$$\frac{\partial \langle E \rangle}{\partial c_i} = \frac{1}{D}\frac{\partial N}{\partial c_i} - \frac{N}{D^2}\frac{\partial D}{\partial c_i} = \frac{1}{D}\left[\frac{\partial N}{\partial c_i} - \frac{N}{D}\frac{\partial D}{\partial c_i}\right] = \frac{1}{D}\left[\frac{\partial N}{\partial c_i} - \langle E \rangle \frac{\partial D}{\partial c_i}\right] = 0$$

As a consequence, this minimization leads to:

$$\langle E \rangle = \frac{\dfrac{\partial N}{\partial c_i}}{\dfrac{\partial D}{\partial c_i}} \tag{6}$$

We simply wish to clarify this term, in order to grasp what is at stake in the calculation of those coefficients. The numerator $N(c_1, c_2)$ is written in the following way:

$$N(c_1, c_2) = c_1^2 H_{11} + 2c_1 c_2 H_{12} + c_2^2 H_{22}$$

The partial derivative calculation, using a particular coefficient while fixing the other, implies that:

$$\frac{\partial N(c_1, c_2)}{\partial c_1} = 2c_1 H_{11} + 2c_2 H_{12} \quad \text{and} \quad \frac{\partial N(c_1, c_2)}{\partial c_2} = 2c_2 H_{22} + 2c_1 H_{12}$$

In brief, each partial derivative depends upon (1) the coefficients c_1 and c_2, (2) one integral (H_{11} or H_{22}), which refers to a unique atomic orbital (φ_1 or φ_2, respectively), and (3) one integral H_{12}, which deals with the energetic coupling between the two atoms inside the molecule. This situation is tantamount to saying that the two atoms intervene, particularly by means of their coupling, in the determination of each coefficient. We should bear in mind that the coupling exists once the molecule is created. In this respect, each coefficient depends upon the whole molecule and not solely upon its correlative atoms! We can draw the same conclusion from the study of the denominator.[12] In short, the minimization of the average energy leads to a *ratio* of two quantities, each depending on atoms but also on their interactions. In this respect, the weighting coefficients of the linear combination, from which one seems free to conclude that the "molecular level" is reduced to a more "fundamental level," requires the whole molecule to be determined![13]

It is thus possible to conclude, from a *technical* standpoint, that there is no reduction but only co-definitions and interrelations between different levels in this calculation. The overlap and the atomic populations, as well as the determination of the weighting coefficients, imply a mutual dependence of the whole molecule and its parts. Mulliken needs both, at the same time, in order to explain and to predict chemical transformation. If one tries to eliminate the whole by means of a linear combination of its parts, the whole reappears when determining the weighting coefficients. If one tries to eliminate the parts, they "reenter through the window," so to speak, because they enable chemists to render intelligible the notion of "reactive site." This reasoning is not purely holistic, as Mulliken seems to understand it. But it is also not purely reductionist, because of the codependence of the whole and the parts. It simply attempts *to negotiate* the meaning of the whole through its parts and vice versa. This interesting point may be of importance for philosophers of mind as well, because it allows for a different understanding of the relation between distinct "levels" of organization. Nevertheless, this is not where our story ends. In fact, if we now attempt to go beyond the aspect just discussed, another question inescapably arises: How can one justify the use of the Variation Principle and the minimization of energy that it implies?

The Variation Principle was elaborated *prior* to the development of quantum chemistry and it is employed, along with new variational methods, in most approaches used by quantum chemists. These include the method of molecular orbitals, Pauling's "Valence Bond theory," the "Density Functional Theory"

---

[12] $D(c_1, c_2) = c_1^2 + 2\,c_1 c_2 S_{12} + c_2^2$ for which the minimization implies that $\dfrac{\partial D(c_1, c_2)}{\partial c_1} = 2c_1 + 2c_2 S_{12}$ and $\dfrac{\partial D(c_1, c_2)}{\partial c_2} = 2c_2 + 2c_1 S_{12}$. The term $\dfrac{\partial D(c_1, c_2)}{\partial c_1}$ also depends on c_1, c_2, and the overlapintegral of the two atoms. This latter term does not concern two "isolated" atoms but their overlap within the volume of the whole.

[13] It is then possible to find the energy values by solving the system of equations $\dfrac{\partial N}{\partial c_i} = \langle E \rangle \dfrac{\partial D}{\partial c_i}$ where i can be equal to 1 or 2: $2c_1 H_{11} + 2c_2 H_{12} = \langle E \rangle [2c_1 + 2c_2 S_{12}]$ which leads to $[H_{11} - \langle E \rangle]c_1 + [H_{12} - S_{12}\langle E \rangle]c_2 = 0$ and $2c_2 H_{22} + 2c_1 H_{12} = \langle E \rangle [2c_2 + 2c_1 S_{12}]$ which implies that $[H_{12} - S_{12}\langle E \rangle]c_1 + [H_{22} - \langle E \rangle]c_2 = 0$. The two energy values then enable chemists to determine c_1 and c_2 knowing that $c_1^2 + c_2^2 = 1$.

(DFT), and the "Quantum Theory of Atoms in Molecules" (QTAIM), to name but a few of these approaches.[14] The Variation Principle is thus *extrinsic* to quantum chemistry. Moreover, the Hamiltonian operator used corresponds, more often than not, to that of an "isolated" molecule, even if one can include the effects of the solvent or those of an external electromagnetic field into the equation. In the case discussed here, Mulliken does not integrate such information about the environment into the Hamiltonian. He only includes nuclei and electrons without reducing, as we have just pointed out, the molecule to its parts. The *rational closure* of this method and many others rests on a use of the concept of energy that belongs to the sphere of other scientific practices, particularly to thermodynamics and thermochemistry. As Vemulapalli reminds us,

> We are led to conclude that it doesn't matter what the states of the parts are, but it does matter the surroundings soak up the excess energy of the molecule, increasing entropy, and make the molecule settle down into the lowest energy state. It is that part of the universe coupled to the system, and the varieties of interactions between the system (molecules) and the surroundings that determine the structure of the molecule. Holism thus appears as the root of the apparent reduction of properties of a molecule to its parts through coupling states. We are able to follow a reductionist program in calculating molecular properties, but what we are able to do is a gift of holism.[15]

One needs to refer to the second principle of thermodynamics in order to explain why the molecular system continuously eliminates its excess energy by interactions with its environment. An energy transformation into local entropy legitimates the use of the variation method. The molecular whole, its parts, and the surroundings are thus all required, at the very same time, in order to render all these methods intelligible. This typical calculation clearly illustrates what is involved in explaining a structure or a mechanism and in predicting a transformation. Quantum chemists use a lot of interrelated tools, within a large and sophisticated network that brings together mathematical functions and devices, empirical outcomes, computer engineering, quantum and classical physics, and thermochemistry, as well as theoretical and practical chemical expertise. Chemists have contrived specific methods within which the whole and its parts are constitutively co-defined.[16] When philosophers of mind such as Jaegwon Kim argue for reduction of a whole system to its parts within a functional analysis, they should bear in mind the whole set of "translations"

[14] Leach, Andrew A., *Molecular Modelling. Principles and Applications*, second edition (New York: Pearson/Prentice Hall, 2001); K. B. Lipkowitz and D. B. Boyd, *Reviews in Computational Chemistry*, 10 (New York: Wiley-VCH Publishers, 2003).

[15] Vemulapalli, G.K. "Property reduction in chemistry. Some lessons," in *Chemical Explanation. Characteristics, Development, Autonomy*, Joseph E. Earley (ed.), *Annals of the New York Academy of Sciences*, 988, 1 (2003), 97.

[16] Llored, Jean-Pierre, "Emergence and Quantum Chemistry," 265.

and "articulations," to use Michel Callon's phrase,[17] involved in the techno-scientific network to which they refer. There is no "pure" and "linear" reduction from top to bottom. This is because no scientific discipline, whether it is physics, chemistry, biology or any other, can extract any relevant meaning or achieve any efficacy independently of the other disciplines.

The point that we have defended so far can be extended to quantum chemical methods that deny any overlap between atoms, as it is typically the case in Bader's "Atoms in Molecules" theory.[18] This approach is based on a topological description of the electronic density of the molecule. As Chérif Matta states, "[i]t is the topology of the electron density that determines the boundaries of an atom which in turn determine its shape which in turn determines its properties inside a molecule."[19] Matta and Bader thus define a reductionist program, according to which the molecule becomes the sum of "topological atoms." "The time has arrived for a sea change in our attempts to predict and classify the observations of chemistry, time to replace the use of simplified and arbitrary models with the full predictive power of physics, as applied to an atom in a molecule."[20] Nevertheless, Paul Popelier, another great expert in this field, qualifies the previous statement by claiming that

[w]e recollect that the gradient vector fields *naturally* partitions the molecules into atoms, i.e. the gradient of ρ carves the atoms by the term *molecular atoms* as opposed to free or isolated atoms. Thus, every molecule falls apart into non-overlapping molecular atoms.... Every type of nucleus appears inside thousands of possible molecular atoms. In fact, there are millions of carbon (molecular) atoms because each atom is cut of a particular chemical molecular environment of which there are as many as there are molecules. In a manner of speaking, every molecular atom is endowed with properties it inherits from the molecule of which it is a part. In other words, the atom reflects the features of its particular chemical environment.... There are literally many millions of molecular atoms because there are millions of molecules which all give rise to a set of constituent atoms. Nevertheless the sometimes bewildering shapes of atoms have been criticized as being contrary to chemical intuition. This should not be disconcerting, rather it could be interpreted as an expression of the richness of chemistry. Indeed, the amazing variety of atoms is a result of quantum systems cutting themselves into fragments, each leaving behind on the fragments detailed finger-

[17] Callon, Michel, "Four Models for the Dynamics of Science," in *Handbook of Science and Technology Studies*, S. Jasanoff, G. Markle, J. C. Peterson, and T. Pinch (eds.) (London: Sage, 1995), 29–64.

[18] Bader, Richard, F. W. *Atoms in Molecules: A Quantum Theory* (Oxford: Oxford University Press, 1990).

[19] Matta, Chérif. F. *Applications of the Quantum Theory of Atoms in Molecules to Chemical and Biochemical Problems*, Ph.D. Thesis (Hamilton: McMaster University, 2002), 28.

[20] Bader, Richard F. W. & Chérif F. Matta, "Atoms in molecules as non-overlapping, bounded, space-filling open quantum systems," *Foundations of Chemistry* (November 2012).

prints of the total molecule. Is it possible, then, to find exactly the same atom more than once coming from different molecules?[21]

Chemists reply negatively to this last question, due to the dependence on the environment within the very definition of any part. Once again, the molecular level and the atomic level are co-defined. Furthermore, Bader and Matta also use variational methods for minimizing energy which, in turn, implicitly strengthen the codependence of the whole, its parts, and the environment. We claim that this codependence puts Kim's analysis of downward causation into question. To defend this claim, we will now discuss the manner in which chemists think about chemical relations and *relata*.

2 *Second Enquiry:* Relata, *Relations, and Affordances*

The British emergentist philosopher George Henry Lewes once asserted that

> In this pinch of table salt there is no appearance of the soft metal sodium, or the pungent gas chlorine, which the mental eye of the chemist sees there, and which all men of science would declare to be really there, supporting their assertion by dragging out both metal and gas, and presenting them to Sense. I, on the contrary, maintain that neither metal nor gas *is* there; and my assertion is supported by the fact that so long as the salt remains *salt* no trace of *gas* or *metal* can be perceived. To prove his assertion that these elements are really present, underlying the appearances, the chemist has to completely alter the whole group of relations, and for that group substitute a different group, *then*, indeed, metal and gas will appear.[22]

Relations allow chemists to define chemical entities and properties, while operations allow them to obtain pure chemical bodies. Those bodies then enter into new relations and result in new compounds that, once purified, allow chemists to widen and deepen their classification by analogy. The process is open-ended and depends on the modes of access, either instrumental or cognitive. In this context of scientific practices, *relata* do not exist prior to relations, and relations are not achievable without purifying operations and purified chemical bodies. Relata and relations are thus *constitutively co-defined*. They *depend on one another*, within an *ordered and evolving network*. Relata are not independent from the relations but, rather, depend on selective operations or on chemical transformations.[23] Chemical purity is not an intrinsic property

---

[21] Popelier, Paul. *Atoms in Molecules. An Introduction* (London: Prentice Hall, 2000), 35–49. Popelier prefers to use the expression "molecular atoms" instead of that of "topological atoms" in order to insist on their context-dependence. The italics are used by Popelier. We express our own insistence by underlining two groups of words.

[22] Lewes, George Henry, *Problems of Life and Mind. First Series: The Foundations of a Creed*, Vol. 2 (Boston: Osgood R. and Company, 1875), 51. The italics are in the original.

[23] Llored Jean-Pierre and Michel Bitbol, "From Chemical Practices to a Relational Philosophy of Chemistry," in *The Philosophy of Chemistry: Practices, Methodologies, and Concepts,*" Llored Jean-Pierre (ed.) (Newcastle upon Tyne: Cambridge Scholars Publishing, 2013), 385–415.

of matter but is the result of transformations from composites. It is also a *condition of possibility* that enables chemists to synthesize new chemical bodies and to extend relational taxonomies.[24] *Therefore, and perhaps surprisingly, the relational work of chemists does not require a substantialist framework of thought but merely a pragmatic use of networks of interdependencies.*[25] There is a strong indication that this work can be extended from the philosophy of chemistry to allow for the development of a non-ontological philosophy of mind, and we will return to this point later in this essay.

Returning to the notions of *relata* and relations in chemistry, we shall illustrate the points already made with the following example. Let us take the examples of NMR and column chromatography with regard to the following chemical transformation:

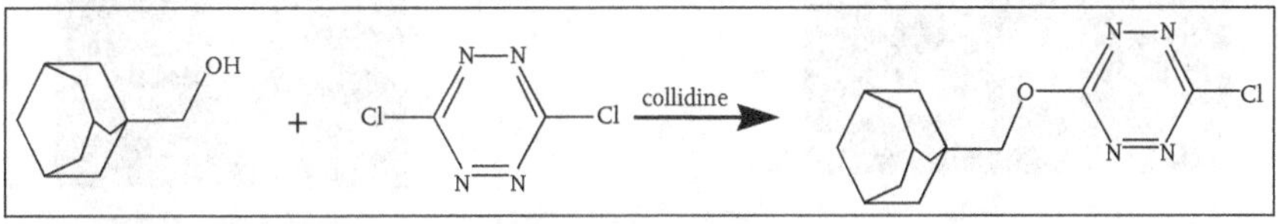

FIGURE 4.1 Chemical reaction studied in this chapter.

The first step, after the chemical synthesis, is the purification of the compound that separates from initial reagents, solvents, and secondary products. Chemists accomplish this purification in two different phases: a solid phase, called the stationary phase (in this example, the silica in the column), and a mobile phase (in this example, a liquid that is cooled and then percolated through the solid). The mobile phase is often a mixture of solvents with a particular global polarity. This mobile phase dissolves our initial mixture and allows it to go through the silica column by means of gravity. The different chemical bodies can interact with the silica differently, depending on their own polarity. They are thus separated, insofar as they don't have the same velocity. The column and the mobile phase thus *afford* the different parts of the whole mixture. Therefore, parts isolated from wholes are "affordances," to use Rom Harré's term, where "affordances" are understood to be "displays produced by specific manipulations of actual locally constructed apparatus"[26] (figure 4.2).

The separation involved here is not that of a mere aggregate that results from an addition of parts. It is the result of an interaction with a column and a liquid, and it depends upon chemical polarities. When chemists perform purification by means of column chromatography, they develop a parts/wholes

---

[24] Schummer, J. (1998). "The chemical of chemistry I: A conceptual approach," *Hyle*, 4, 2, 29–162.
[25] Llored, Jean-Pierre, "Emergence and Quantum Chemistry," *op.cit.*, 269.
[26] Harré, Rom, "Hinges and Affordances. New Tools in the Philosophy of Chemistry," in *The Philosophy of Chemistry: Practices, Methodologies, and Concepts*, Jean-Pierre Llored (ed.) (Newcastle upon Tyne: Cambridge Scholars Publishing, 2013), 580–595.

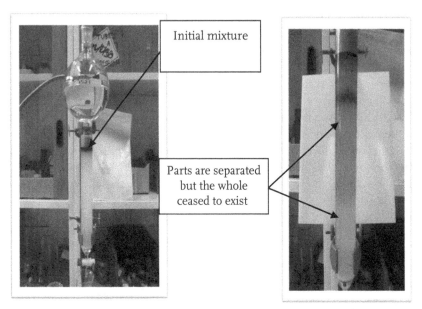

Initial mixture

Parts are separated
but the whole
ceased to exist

FIGURE 4.2 Experiment carried out by Benoît Marcillaud, student at the École Normale Supérieure de Cachan (France). Reproduction authorized by Professor Pierre Audebert, who supervised this work within the PPSM laboratory.

discourse. That is to say, they develop a contextualized mereology that takes into account the constitutive relation with the instrument. The resultant parts/whole dependence does not arise "from nowhere" but is, instead, the result of a chemical operation. *Parts/wholes relations are not intrinsically defined but, rather, are defined through chemical operations (i.e., operationally defined).* Once the purification occurs, chemists can analyze their product by means of NMR (nuclear magnetic resonance). In line with Joseph Earley, we consider NMR to be a good example of how to think about the manner in which an ingredient is transformed when it is included into a molecular whole.[27] We would like now to develop this point further by using Harré's concept of affordances.

With specific reference to the chemical transformation discussed above, the "spectrum" that results from NMR technology displays the radio frequency response of the different atoms ^{1}H inside the molecule. Different parts emerge from the spectrum. Each signal represents a set of atoms ^{1}H that are chemically equivalent, according to the ^{1}H NMR magnetic scale. This method *affords* different sets of atoms of hydrogen. Every signal depends on the locality, that is, on the environment of each particular nucleus of proton. The set of different signals enables chemists to identify a chemical body and, thus, to assess its degree of purity (figure 4.3).

However, chemists can also afford a complementary description of equivalent parts, that is to say, of equivalent *sets of nuclei of carbon* using NMR [13]C. *The*

---

[27] Earley, Joseph E., "How philosophy of mind needs philosophy of chemistry," *op. cit.*, 10.

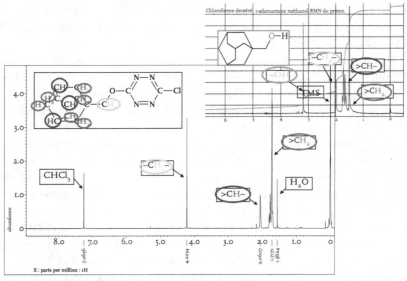

FIGURE 4.3 Experiment carried out by Benoît Marcillaud, student at the École Normale Supérieure de Cachan (France). Reproduction authorized by Professor Pierre Audebert, who supervised this work within the PPSM laboratory.

parts are now different from those obtained by NMR ^{1}H. A whole/parts structure can thus be afforded by the conjoint use of a magnetic field and a wave. The chemical shift on the axes of abscises provides information about the local environment relative to a reference, the tetramethylsilane (TMS). The parts of a structure can be afforded relative to a mode of access, an environment, and a chemical standard. *Parts are, thus, constituted by the interaction between the whole and the apparatus. They are not intrinsic. Changing the mode of access, that is, changing the type of NMR, amounts to changing the kinds of parts that originate from the same whole.*[28]

One can take this discussion a step further, however. For example, when chemists synthesize a solid sample of $CaCO_3$, they need to saturate a solution and precipitate the solid. They can achieve this precipitation by various means, the most common of which are solvent evaporation and chemical reactions that produce insoluble species. Starting from a different number of ingredients, particles will grow to attain different final sizes and morphologies. Thus, the final materials may appear completely different, depending on whether a reactive material is added all at once or gradually. By adding a small amount of fine material to be precipitated (i.e., seeds), one can better control the apparently chaotic nucleation step. For example, adding calcite seeds allows for the precipitation of pure calcite. On the other hand, without seeds, one obtains a mixture of calcite and vaterite with a larger particle size distribution and

---

[28] Llored, Jean-Pierre & Rom Harré, "Developing Mereology of Chemistry," in *Mereology and the Sciences*, Claudio Calosi, and Pierlugi Graziani (eds.) (Dordrecht: Springer, 2014).

various morphologies. The whole ($CaCO_3$) depends on the process used and on the time employed. Neither the device nor the history of the chemical reaction can be eliminated from the final results.[29]

We know what the initial "parts" of the mixture (say, the reagents) are. But, depending on the process used, the "whole" that is obtained may vary. Chemists also know how to separate initial parts from the new whole by means of various operations. In our example above, the structure of the $CaCO_3$ obtained will be different, depending on the environment, the device, and the time employed for the precipitation. Though the compounds that are obtained from the two different approaches share the same formula ($CaCO_3$), they are different wholes. Their composition is the same, but the environment, the procedure, and the time involved in the production of the compound have influenced the relatedness of their parts. Thus, once again, the whole, its parts, and their environment are intertwined within a process. The mode of operation cannot be eliminated from the final product, because the mode determines the whole and its correlative parts. The structure of the crystals may differ if the chemical device changes, and they can even differ within the same particular chemical device, depending on the size of the crystals, which itself depends on the environment. *Therefore, chemists have to hold the composition, the structure, the parts, the whole, the environment, and the device together within the same coherent explanation.*

There is no basic level to be found here but only codependent levels. It is important to realize that this situation is totally novel in the history of chemistry, and it remains a challenge for contemporary chemists. According to us, however, philosophers of mind should also take seriously this interrelationship between relata, relations, and affordances, as they attempt to understand the relationship between brain states and consciousness. To draw out the implications of the above discussion for philosophy of mind, we will now examine the issue of downward causation and how chemical mereology can help resolve this seemingly intractable problem.

3 Third Enquiry: Chemical Mereology and Downward Causation

As mentioned at the beginning of this chapter, Michel Bitbol has proposed a novel and innovative approach to the problems of emergence and downward causation that involves reconceptualizing consciousness and mind in nonsubstantialist terms. Bitbol's proposal is informed by recent developments in quantum physics. However, we will show that developments in quantum chemistry add even further support to the solution proposed by Bitbol. Before we discuss Bitbol's proposal, we must first examine the current state of philosophy

---

[29] Aimable A., Brayner R., Llored, J.P., Sarrade S., Rozé M., "Chemistry and Interfaces" in *Philosophy of Chemistry: Practices, Methodologies and Concepts*, J.-P Llored (ed.) (Newcastle upon Tyne: Cambridge Scholars Publishing, 2013), 172–201.

of mind and the manner in which emergence and downward causation have been traditionally understood. Put simply, the relationship between supposedly emergent mental states and submergent brain states has been so problematically construed by traditional emergentism that this position has inadvertently become the reductionist's best friend. For example, Jaegwon Kim concludes that, given the weakness of the traditional emergentist position, there is really no credible middle ground between reductive materialism and Cartesian-type substance dualism. To the extent that the latter is clearly unacceptable, the former must be the only viable alternative. In his discussion, Kim correctly explains that, according to emergentism, "complex systems aggregated out of…material particles begin to exhibit genuinely novel properties that are irreducible to, and neither predictable nor explainable in terms of, the properties of their constituents. It is evident that emergentism is a form of what is now standardly called 'non-reductive materialism', a doctrine that aspires to position itself as a compromise between physicalist reductionism and all-out dualisms."[30] As Kim describes them,[31] the central doctrines of emergentism are

1. emergence of complex higher-level entities;
2. emergence of higher-level properties;
3. unpredictability of emergent properties;
4. unexplainability/irreducibility of emergent properties; and
5. causal efficacy of the emergent entities/properties upon the submergent entities/properties.

According to Kim then, to accept emergentism thusly construed is to accept both the inability to explain how material complex systems can exhibit novel "higher-level" properties and the inability to predict what novel properties will be exhibited by such systems while, at the same time, claiming that such novel properties exhibit causal powers that affect the submergent base. Kim spends a great deal of time and effort arguing that emergentism cannot be an acceptable position, and he takes issue most strongly with the concept of downward causation, a central tenet of emergentism but one that he finds to be incoherent. By way of a long and complex argument, Kim concludes that, paradoxically, a higher-level property M can only have causal efficacy if it can be reduced to its lower-level base. However, since emergent properties, if they exist, are by definition not reducible to their lower-level bases, they are causally and hence explanatorily inert.[32]

Ultimately, Kim's synchronic conception of causality requires that there be causal closure and that such closure can occur only at the most fundamental or "lowest" ontological level. He states that "the mental…will not collapse into

---

[30] Kim, Jaegwon, "Making sense of emergence," *Philosophical Studies*, 95 (1999), 4.
[31] Ibid., 20–22.
[32] Kim, Jaegwon, *Physicalism or Something Near Enough* (Princeton, NJ: Princeton University Press, 2005), 65.

the biological...for the simple reason that the biological is not causally closed. The same is true of macro-level physics and chemistry. It is only when we reach the fundamental level of microphysics that we are likely to get a causally closed domain."[33] For Kim, as Earley points out, only "elementary-particle-level explanations account for mental functioning...[and only] the kind of reductive physicalism that he and others have developed 'is a plausible terminus for the mind-body debate'."[34] In other words, as mentioned above, the causal and explanatory "closure" that Kim requires must occur at the most ontologically fundamental level. No real "closure" is available at any other level than that of the most fundamental "particles," not even at the cellular, molecular, or atomic levels. This concept of causal closure as only existing at the lowest microphysical level is in keeping with the extreme reductionist program described and defended by Putnam and Oppenheim, who claim that "whenever it *can* be shown that things of a given level existed before things of the next higher level came into existence–some degree of indirect support is provided to the particular special case of [micro-reduction as an instrument for the unity of science]."[35] We claim, however, that chemists have reached entirely different conclusions than those of Putnam and Oppenheim and, in order to address those conclusions, we must further scrutinize Kim's argument and its flaws.

Kim supposes that, at a particular time t, a whole W displays an emergent property P and that T is completely decomposed into a basic set of parts $a_1,..., a_n$, each having or being a property P_i. He considers the relation R $(a_1... a_n)$ that holds those parts together and then analyzes the "synchronic reflexive downward causation" according to which for any part a_i, the fact that W displays M at t *causes* the fact that a_i displays P_i at t.[36] He recalls that those same parts *must cause* W having P at the same time. Kim then refers to "the causal-power actuality principle," in order to show that the synchronic reflexive downward causation collapses. According this "principle," at time t, the whole W can exercise the causal powers it has in virtue of having property P, if and only if W already possess P at t. The logical problem is that this argument cannot escape the fact that a_i does not already have P_i at t, which in turn implies that the assumed emergence base of W having P at t has vanished and, eventually, that there is no emergence at all.

Kim is able to reach these conclusions via an articulation between a particular metaphysical principle and a certain kind of mereology, in which the whole is decomposable into parts. His argument, even if Kim does not express this explicitly, needs to satisfy the ceteris paribus clause, in order to gain its legitimacy. Kim's reasoning does not take into account the surroundings in

---

[33] Ibid.

[34] Earley, Joseph E., "How philosophy of mind needs philosophy of chemistry," *op. cit.*, 6.

[35] Putnam, Hilary and Paul Oppenheim, "Unity of science as a working hypothesis," in *Concepts, Theories, and Mind-Body Problem*, Herbert Feigl, Michel Scriven, and Grover Maxwell (eds.) (Minneapolis: University of Minnesota Press, 1958), 23.

[36] Kim, Jaegwon, "Making Sense of Emergence," *op. cit.*, especially §VII.

which the whole and the parts exist and, thus, cuts them off from the rest of the world. Kim never envisions that both the whole and its parts might *constitutively* depend on the mode of access (apparatus, solvent, chemical device, and so on). Thus, there is no other solution for him than to claim that the parts are intrinsic to the whole, rather than being afforded by a mode of access as we have been claiming throughout this chapter. Kim uses an analytical reasoning style that consists of cutting the whole into *virtual* parts. Following this line of reasoning, he envisions the relation R between the parts and concludes that, in virtue of "the causal-power actuality principle," if a_j does not already have P_j at t, it is impossible for W to possess it, ceteris paribus as we would like to add.

But what does occur during the decomposition from W to parts a_i? If any a_i verifies R (R being a holistic characteristic), it already belongs to the whole and, consequently, displays P_i. Otherwise, the whole could not exist. In this case, there is no problem with synchronic downward causation. However, we would like to ask: "What counts more in Kim's reasoning: Is it the relation R or the set of properties P_i?" The answer is far from being clear, insofar as the relation R $(a_1 \ldots a_n)$ is supposed to be functionally nonreducible. Nevertheless, there is a second possibility, that is, that the parts have implicitly changed during the recomposition, in order to recuperate the whole. The parts have moved from the virtual state of being isolated from one another to the state of being interdependent within the whole W, in virtue of the relation R. In this case, Kim implicitly assumes that *the identity of the parts remains the same*, be they isolated or included into a whole. The trouble is that the ceteris paribus requirement collapses because the parts have been, at least partially, transformed.[37]

The kind of mereology that Kim uses does not fit the requirement of chemical activities according to which relata and relations are codefined. Thus, although Kim wishes to extend his argument regarding the reducibility of mind to a claim about the reducibility of material reality in general, his conclusion contradicts what practicing chemists have established empirically regarding the relationship between wholes, parts, and environment. The defense of an ontology of relata with intrinsic properties is incompatible with chemical practice, and this is why reifying relata and relations prevents Kim from envisioning a non-ontological form of downward causation. In the context of chemistry, what Kim asserts amounts to saying that the atom of carbon in the molecule CH_4 is strictly identical with the atom of carbon in the molecule HCOOH. But, this is precisely what Earley[38] and Popelier warn us against when they discuss the presence of salt in the sea and of "molecular atoms" in molecules, respectively. Those chemical parts are altered according to the environments both inside and outside the whole structure.

---

[37] Llored, Jean-Pierre, "Chimie, chimie quantique et concept d'émergence: étude d'une mise en relation," PhD dissertation. History, Philosophy and Sociology of Sciences. Ecole Polytechnique X, 2013. French. Available online at https://halshs.archives-ouvertes.fr/pastel-00922954/document
[38] Earley, J, "Why there is no salt in the sea," *Foundations of Chemistry*, 7 (2005), 85–102.

What can be said about the *allegedly unproblematic* version of diachronic reflexive downward causation? Let us suppose that W having P at t causes a_i to have Q_i at $t + \Delta t$. Kim asserts that there is no problem of circularity in this case, because a_i having Q_i at $t + \Delta t$ is not an ingredient of W at t. We should, nevertheless, develop this point further. Following Kim's line of reasoning, M exists at $t + \Delta t$ if and only if a_i has P_i at t and if a_i enters into the relation R $(a_1 \ldots a_n)$, which gives rise to W having P at t, ceteris paribus. One property of a_i has, nevertheless, changed due to W having P at t. This change does not eliminate R. However, Kim says nothing about the possible role of Q_i with regard to the emergence of P and the co-evolution of the other parts. If one parameter Q_i has changed from t to $t + \Delta t$, other "properties" Q_j of other parts a_j may have changed as well. Thus, Kim fails to meet the ceteris paribus requirement. The fact that he reifies the whole and its parts by abstracting them from their environment jeopardizes his argument, particularly if we think about chemical transformation through time.[39] Kim's reasoning relies on a specific type of metaphysics that considers matter to be passive and only moved by external forces. However, if we consider, as any chemist does, that matter and *matters* are active and able to interact with one another, Kim's metaphysics turns out to be of no avail, and we have to assign a new meaning to the ceteris paribus clause within this novel conceptual framework.[40]

The traditional conceptual framework embraced by Kim and many other reductionist philosophers is entirely upset in the context of chemical discourse. The chemical conceptual framework considers three levels: (1) the level "L-2," which contains the nuclei and the electrons; (2) the atomic level "L-1"; and (3) the molecular level. For chemists, M, M*, P, and P* are co-defined and depends upon the mode of access and the environment (see figure 4.4). For example, as we have already seen, we can infer different types of hydrogen parts within a molecule, depending on the kind of NMR being used. We can thus add that, in chemistry, we are no longer concerned with issues regarding own-being but with interactive processes of which we partake through our instrumental and social activities.

In contrast with the point of view embraced by chemists, Kim's underlying assumption is that things at the lower level are not altered by their participation in a composite at the higher level, and that no new properties emerge when things at the higher level come into existence.[41] However, as Harré and Earley point out, this is precisely what comes into question in the context of an extended non-classical mereology in which wholes and parts are codependent and mutually constitutive. As we have shown, this is typically the case within

---

[39] Llored, Jean-Pierre, "Chimie, chimie quantique et concept d'émergence: étude d'une mise en relation. PhD dissertation," *op. cit.*

[40] Llored, Jean-Pierre, "Investigating the meaning of the *ceteris paribus* in chemistry," in *Philosophy of Chemistry: Synthesis of a New Discipline*, 2nd edition, vol. 306, Eric Scerri and Lee McIntyre (eds.) (Dordrecht: Springer, 2015), 219–233.

[41] Llored, Jean-Pierre, "Chimie, chimie quantique et concept d'émergence: étude d'une mise en relation. PhD dissertation," *op. cit.*

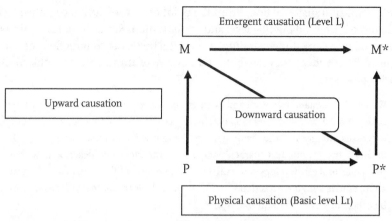

FIGURE 4.4 Relationships between levels of organization.

chemistry. Earley mentions several other problems with Kim's arguments, among which are Kim's approach to causality. Such a synchronic, rather than diachronic, approach "is inconsistent with the circumstance that human mental functioning necessarily involves networks of processes that have many different characteristic time-parameters. Chemists who work with quantitative models of open, non-linear, dynamic systems should be able to provide a rigorous account of how 'downward causation' operates in such systems an adequately respond to 'Kim's Challenge'."[42]

As Earley suggests, in order to respond to Kim and his reductionist allies, we must examine how we think about causation. But, to do this, we must also examine how we think about the notions of lower and higher levels of reality. We cannot emphasize enough how much we agree with Earley. One of the fundamental problems with the traditional approach to the question of levels of reality is that the problem is posed in ontological and substantialist terms. Despite the post-Kantian development of process philosophy, ontology has remained handicapped by its predilection for substance metaphysics, whether dualist or materialist, and this handicap has made it impossible to resolve fundamental problems, particularly in philosophy of mind. In traditional substantialist metaphysics, for example, levels of physical reality are thought of as ontologically determined, with a fixed lower level from which causal chains of events begin. This is where philosophy of chemistry's profound understanding of processes can contribute some interesting insights, particularly with regard to the co-dependence between relations, relata, and levels of organization. Many philosophers of chemistry are, in fact, practicing chemists who have come to conceive of both macro-level and micro-level realities more as dynamic processes than as fixed substances. Many of these philosophers are also keenly aware that it is fully consistent, as we also have shown, to develop a non-substantialist conceptual framework in order to think about active matter

---

[42] Earley, Joseph E., "How philosophy of mind needs philosophy of chemistry," *op. cit.*, 7–8.

and chemical transformation. From the point of view of such a cognitive context, there is no fundamental level and no reification. Rather, there is an always open and provisory understanding of chemical bodies as conjunctions of affordances. These ideas may serve the philosophy of mind very well indeed. As Earley points out,

> We now understand that each and every entity is involved in many temporal processes at once: vibrations, rotations, translations, chemical interactions, metabolic processes, integration and disintegration of patterns of neuronal activity.... Each of these interactions has characteristic time-parameters that describe how that particular process develops sequentially.... The notion of the state of a system 'at an instant' is a high abstraction—and it is not at all clear that that abstraction can be coherently applied to human minds.[43]

As has been emphasized in the above discussion, it is precisely the ontological and substantive construal of mind, body, and causality that present the greatest challenge to philosophy of mind. Thus, and as promised earlier, we will turn to Michel Bitbol's alternative proposal for a conception of emergence and of downward causation that is non-substantialist and non-ontological and that is inspired by phenomenology and by recent developments in quantum physics.

4 Toward Non-Ontological Conceptions of Emergence and Downward Causation

To the extent that the seemingly intractable problems in philosophy of mind are the result of the traditional construal of emergence and downward causation, Bitbol's alternative proposal promises to resolve these problems, in part, by avoiding them altogether. As Bitbol points out, if we think of emergence and downward causation in the traditional substantive and ontological sense, then the reductionist arguments are most convincing. First of all, when we apply the discourse of emergent properties to consciousness, we run into an obvious problem. Properties are features of the world that are "public," that can be accessed intersubjectively, and "that can be described in a third-person mode."[44] Consciousness, however, is not a property in this sense of property.[45] Informed by developments in phenomenology, Bitbol explains that consciousness is, instead, "a situated, perspectival, first-person mode of access. Any discourse about consciousness is thus bound to be first and second-personal in disguise (behind the curtain of a formally referential language)."[46] According to Bitbol, the "hard problem" of consciousness is "hard" and it is a "problem" precisely because this approach to the question of consciousness has ignored its "radical

---

[43] Ibid., 14.
[44] Bitbol, Michel, "Ontology, matter and emergence," 293.
[45] Ibid.
[46] Ibid., 293–294.

situatedness." Chemistry can help to reinforce Bitbol's point by insisting on the situatedness both of chemicals as affordances and of its own methods, which articulate heterogeneous practices, as was shown in our earlier discussion of molecular orbitals.

As already mentioned, Bitbol's alternative proposal is influenced by two different lines of research, phenomenology and quantum physics. We will first address the phenomenological perspective and its impact on Bitbol's construal of consciousness. Prior to his untimely passing in 2001, neuroscientist and phenomenologist Francisco Varela had already advocated a different approach to the question of consciousness, an approach of naturalization without reduction that he called "neurophenomenology." As Varela and his colleagues explained, the problem of the "naturalization of the mental" is "the problem of understanding how the essential properties of objects classified in a certain ontological division (the mental one) can be made to belong to objects classified in a separate and different one (matter, to put it briefly)."[47] Bitbol expands upon the significance of this neurophenomenological approach:

> [Varela] advocated an original (dis)solution to the 'hard problem' of consciousness which involved a consistently *methodological* approach rather than one more theoretical view. The basis of this approach was the remark according to which a third person, objective, description arises as an invariant focus for a community of embodied, situated, subjects endowed with conscious experience in the first place....Overlooking the effective primacy of situatedness, which is a common trend in our culture, leads to downplaying the status of consciousness. If one accepts that conscious experience is but a parochial path...towards an intrinsically objective reality of which we partake, then it is likely to be either completely dismissed (strong eliminativism), or reduced to a field of description which is easy to objectify (physicalist reductionism), or treated as an objective entity in its own right (substance or property dualism). Conversely, overrating the fact that third-person accounts are produced by...sentient beings...usually means indulging in skepticism, relativism, or subjective idealism.[48]

Following Varela's proposal, Bitbol argues that one cannot develop a credible account of emergence and "downward causation" unless one first reconceptualizes consciousness and causality in non-metaphysical and non-ontological terms. As previously discussed, a crucial reason for the intractability of the "hard problem" is that it has historically been formulated in terms of extreme, substantialist ontologies that originate, in one form or another, from pre-Socratic cosmologies in which both the mental and the material were reified. As Bitbol explains, emergentism in philosophy of mind was developed as the

---

[47] Roy, Jean-Michel, Jean Petitot, Bernard Pachoud, and Francisco J. Varela, "Beyond the Gap: An Introduction to Naturalizing Phenomenology," in *Naturalizing Phenomenology: Issues in Contemporary Phenomenology and Cognitive Science* (Stanford: Stanford University Press, 1999), 46.
[48] Bitbol, Michel, "Science as if situation mattered," *Phenomenology and the Cognitive Sciences*, 1 (2002), 181–182.

compromise between the two extreme ontologies that have dominated Western philosophy.[49] "The first of these ontologies is monist and materialist: it says that there exists nothing else than material elements and their properties. The second ontology is dualist: It says that there are two substances or two realms of being, mind and matter, or life and inanimate matter. Emergentism aims at finding an ontological 'middle course',"[50] all the while tacitly accepting the reification of these two ontological realms. As already explained, the most interesting construal of emergentism argues that higher-level emergent properties exhibit causal efficacy, not only on same-level and higher-level properties, but also on the lower-level entities and properties from which they have emerged. Bitbol argues, however, that it is precisely in this, its most interesting, form that traditional emergentism is also most theoretically weak.

> No convincing proof of there being genuinely, ontologically, emergent *properties* can be given. Even less so when one believes that the basic constituents of the world are little things (say the elementary particles) endowed with intrinsic properties.... Suppose that the world is indeed made exclusively of particles endowed with intrinsic properties. Then, it is clear by definition that, at the level of *being*, there is nothing more in a high-level feature than the system of underlying low-level properties. Nothing precludes that the high-level behavior manifests new and unexpected features with respect to the low-level; but under our very strong ontological assumptions, this is only a superficial *appearance* which does not support the further claim that the large-scale features are *really* new, *really* autonomous, let alone that they have intrinsic existence [or causal efficacy].... The appearance of new and autonomous features can be easily explained by ontological reductionists by assuming that our experimental or perceptive analysis is coarse-grained.[51]

Thus, the weakness of traditional emergentism is its inability to naturalize the mental without reducing it, and this is due precisely to its unquestioned commitment to the reification of both material and mental properties. Bitbol argues that, to begin solving the problem of consciousness, we must move away from the reifications, both of consciousness and of matter, that have grounded traditional dualisms, reductive materialisms, and traditional emergentism. Paraphrasing Ludwig Wittgenstein, Bitbol claims that "consciousness as experience *is not something*...it is neither an object, nor a property, nor even a phenomenon."[52] And this is precisely why one of the major errors committed by traditional emergentism is that of reifying the so-called mental properties, as if these were genuine, publicly available properties that are describable in the third person, as are other properties of the world. Precisely because consciousness is not fully describable in this manner, traditional emergentism cannot sustain the

---

[49] Bitbol, Michel, "Ontology, matter and emergence," *op. cit.*, 294.
[50] Ibid.
[51] Ibid., 294–295.
[52] Bitbol, Michel, "Is consciousness primary?," *NeuroQuantology*, 6, 1 (2008), 54.

truth of its fundamental assumption and falls victim to attacks by reductionist critics. In this sense, reductionists such as Kim are correct in claiming that there are no emergent "mental properties," if by this we mean properties that can be described objectively and in the third person, which is indeed what the traditional emergentist means by "mental properties."

Bitbol, however, points out another weakness of traditional emergentism, which is that it renders consciousness ontologically secondary to material reality. This is a result of insisting on reifying the mental, while at the same time avoiding ontological dualism by turning the mental into a by-product of material interactions. Thus, it is precisely in trying to save the "somethingness" of consciousness that traditional emergentism reduces consciousness (no pun intended) to "nothing." On the other hand, the claim that consciousness should not be reified, that it is not a "thing" (res), does not imply that it is not real. In fact, consciousness is not only real, it is both methodologically and existentially primary, so that even attempts to deny consciousness reinforce its primacy, as René Descartes so clearly points out in his *Meditationes de prima philosophia*. Consciousness, therefore, "*is not nothing!* For us, now, while we are reading/writing these lines, conscious experience might be *everything*. It is not something that we *have*, but it identifies with what we are in the first place. It is not something that can be known or described by us in the third person as if we were separated from it; but it is what we dwell in and what we live through *in the first person*."[53] As mentioned above, this is something that Descartes understood above all else, and this truly qualifies him as the real "father of phenomenology," although he is guiltiest of all by eventually misconstruing the nature of consciousness. For, after having understood the primacy of consciousness, he commits the fatal error of absolutely reifying and objectifying mind.

In fact, we wish to stress that we need both the first- and the third-person understanding of consciousness. These are co-dependent in the same way that relata and relations are co-dependent within chemistry and within the grammar of chemical discourses. The third-person description is simply a heuristic device that helps us conceptualize consciousness *as if* it were a real thing of the world. Such a device is a "purification," to use Bruno Latour's phrase.[54] The first person description is also a useful device, because it reintroduces human beings to the immanence of their lives. Such a device is a "mediation," to use Latour's vocabulary once again. We must articulate these two devices without reifying them, if we want a better grasp of our experiences.

In addition to his proposal of reconceptualizing consciousness in a non-ontological manner, Bitbol also proposes a reconceptualization of causality that will allow us to embrace downward causation without the obvious problems raised by Kim and other reductionists. Bitbol's reconceptualization of causality in diachronic and interventionist terms is itself entangled with his

---

[53] Ibid.
[54] Latour, Bruno, *We Have Never Been Modern. Essay on Symmetrical Anthropology* (Cambridge, MA: Harvard University Press, 1993).

proposal to reconceptualize our ontological commitments in a terms. For, as Bitbol argues, not only do we err when we reify consciousness but we also err when we reify the material. It is in reconceptualizing consciousness, materiality, and causality that Bitbol offers novel and non-traditional accounts of emergence and "downward causation" that do not fall prey to the criticisms of reductive materialism. In his alternative view, both "emergent" and "submergent" properties are considered as *relational properties*, and this ties into the proposal for an extended and non-classical mereology that accounts for the relationality, co-dependence, and mutual constitution of parts and wholes in chemistry. Bitbol's proposal also questions what it means for a property to be considered "fundamental" or to be considered "emergent." In fact, he argues that what is considered "fundamental" and what is considered "emergent" is itself a dynamic, methodological feature of the domain of study and of the domain of concepts, rather than a static, ontological feature of the "domain of *being*" or of the world itself.[55]

Bitbol's proposal here is inspired and influenced by recent developments in quantum physics. However, our previous discussion and that of other philosophers[56] help to strengthen and widen Bitbol's work within the framework of chemistry and quantum chemistry. In doing so, we support his proposal for an alternative to traditional emergentism, particularly when he states that

> If the whole process is *groundless throughout*, then there may be emergence without emergent properties. Not asymmetric emergence of high-level properties out of basic properties, but *symmetrical co-emergence* of microscopic low-level features and high level behavior. Not emergence of large scale absolute properties out of small scale absolute properties, but *co-relative* emergence of phenomena. Those phenomena, in turn, are to be construed as relative to a certain experimental context, with no possibility of separating them from this context. The notion of emergence thus gains credibility at the very same time it loses ontological content. [...] In the same way, there may be inter-relatedness without efficient causality; and there may be upward *and* downward causation without *any* causal power. In a non-substantialist framework of thought, the issue of downward causation is not one of inherent powers, but one of relations and actions.[57]

The first element of Bitbol's proposal for an alternative conception of causality deals with the notion of causal and ontological "closure," which is one of the key elements of Kim's attack on the notion of "downward causation" as being paradoxical. Kim's requirement for such closure, however, is meaningful only if one assumes that there is, in fact, a lowest level of elementary "particles" that are uniquely causally efficacious and from which all causality arises. As Bitbol

---

[55] Bitbol, Michel, "Ontology, matter and emergence," *op. cit.*, 302.
[56] Llored Jean-Pierre and Michel Bitbol, "From Chemical Practices to a Relational Philosophy of Chemistry"; Llored, Jean-Pierre, "Emergence and Quantum Chemistry," *op. cit.*
[57] Bitbol, Michel, "Ontology, matter and emergence," *op. cit.*, 304–305.

demonstrates, however, developments in quantum physics have shed doubt on the existence of such a foundational and fixed lowest ontological level. We wish to add that chemistry and quantum chemistry have also enabled philosophers to draw similar conclusions. All of these studies converge toward the development of a non-substantialist and non-ontological approach. Bitbol explains:

> The physical process may have no substantial roof of emergent properties, but it has no substantial ground of elementary particles either, according to the most straightforward reading of Quantum Mechanics. Actually, this claim of groundlessness is presently creeping into many other branches of physics, besides ordinary Quantum Mechanics. In modern cosmology, it has recently given rise to the idea that the observable features of the "elementary" particles are just as much determined by global features of the universe than the other way around. In Quantum Field Theory, the standard model of particle physics is often construed as a mere "effective field theory" of an underlying model. But this underlying model is likely to be no more "basic" than the standard model itself. Some authors noticed that it could itself be construed as the effective field theory of an even deeper model, etc. with no ultimate "basic" layer.[58]

Tian Yu Cao and Sylvan Schwebber have explained, at length, the conceptual and philosophical implications of renormalization procedure in Quantum Field Theory to show that construing the model as an effective field theory creates a dialectical tension regarding the atomicity of the "particles" that are being studied at any given time, because the "particles" that are described as "basic" and atomistic at one energy level can no longer be described as such at a different energy level. The atomicity of these "particles," then, is a feature of the energy level at which we are describing them, not a feature of the world itself, thereby putting into question the whole possibility of causal "closure" that is required by Kim, Putnam, Oppenheim, and other reductionists. We will proceed to quote Cao and Schwebber at some length in order to capture the strength of their point. It is important to note that, despite the seeming concessions to atomism at the beginning of this citation, Cao and Schwebber are not referring to the exact point model. Any atomicity is based on a specific energy level and is, therefore, merely transitional.

> By removing any reference to inaccessible, very high energy domains and the related structural assumptions, the renormalization procedure reinforces the atomistic commitments of QFT [Quantum Field Theory]. This is because the particles or the fields appearing in the renormalized theories do act as the basic building blocks of the world. But note that atomicity here no longer refers to the exact point model. To the extent that tone removes the reference to the inaccessible very high energy domain – which arises conceptually in the exact point model by virtue of the uncertainty principle—renormalization blurs any pointlike character...the quasi-point model seems to be merely a mathematical device

---

[58] Ibid., 302.

needed and useful in a transition period...[but] in physics (as presently formulated and practiced) the structural analysis of objects at any level is always based on (seemingly) structureless objects—genuine or quasi-point-like—at the next level....Thus the adoption of the quasi-point model seems to be unavoidable in QFT....The dialectics of atomism in this context can be summarized as follows: on the one hand, the structureless character of particles as we know them at any level is not absolute but contingent and context-dependent, justified only by relatively low energy experimental probing....On the other hand, with the unraveling of the structure of particles at one level, there emerge at the same time—as a precondition for the unraveling—(seemingly) structureless objects at the next level. And thus the original pattern of 'structured objects being expressed in terms of (seemingly) structureless objects' remains, and will remain as long as QFT remains the mode of representation.[59]

We can draw a quick parallel between what Cao and Schwebber are saying here and Mulliken's own ideas. In fact, Mulliken is at odds with the notion of an aggregation of atoms that refers to the general properties of masses and movements, that is, to any type of mechanics. The energy approach to molecular orbitals advocated by Mulliken does not consider atomicity as an intrinsic property of atoms. This idea recalls, to some extent, Pierre Duhem's reconceptualization of Aristotle's concept of *dunamis* in terms of thermodynamic potential.[60]

From the above discussion, we can conclude that, in the absence of a fixed ontological foundation, the ontological levels are co-defined and co-emergent so that one cannot speak of a "bottom" level or "ground" level of properties or entities that remains fixed at that level. Thus, if reality is without ground, then the notions of "lower level" and "higher level" properties are dynamic and relational notions in which to be at the "higher" level or the "lower" level is a function of the perspective adopted by the researchers when studying the properties in question. Unlike traditional emergence, then, the dependence between levels is symmetrical, rather than asymmetrical, and there is no "closure," to speak of, at the "fundamental" level since there is no such level per se.

What our best contemporary physics reveals is that there are no elementary 'particles', elemental levels, or some such particulars; everything is composed of quantum fields, of various scales and complexity....Quantum field theory shifts the basic ontology of the universe from micro-particles to quantum-fields-in-process. What have seemed to be 'particles' are now conceptualized as particle-

[59] Cao, Tian Yu and Sylvan S. Schweber, "The conceptual foundations and the philosophical aspects of renormalization theory," *Synthese*, 97 (1993), 70–71.
[60] Duhem, Pierre, *Le mixte et la combinaison chimique. Essai sur l'évolution d'une idée* (Paris: Fayard, 1985); Duhem, Pierre, *Mixture and Chemical Combination, and Related Essays*, translated and edited, with an Introduction by Paul Needham (Dordrecht: Kluwer Academic Publishers, 2002).

like processes and interactions resulting from the quantization of field processes and interactions.[61]

Conclusion

In line with Earley, Harré, and Bitbol, this chapter highlights how it is possible to widen the space of reflection in the philosophy of mind by taking chemistry and quantum chemistry as resources for thinking about fundamental problems in philosophy of mind. The constitutive role of the mode of access, the co-definition of relata and relations, the co-dependence of levels of organization, as understood from within chemical practices, help us address the problems of emergence, downward causation, and the relationship between brain states and consciousness. These ideas need to be further investigated, particularly from the perspective of the philosophy of chemistry, since this discipline is as concerned with the notions of levels of beings, relata, and relations between wholes and parts as is the philosophy of mind.

The development of a mereology adapted to chemistry can be of service, in this respect, for important future studies, while the shift from an Aristotelian notion of efficient cause to an "interventionist" concept of causality can be of great use in the philosophy of mind. As Michel Bitbol correctly asserts,

> Causal closure of a low level of organization (say the level of micro-physics) is perfectly compatible with the thesis that there are also efficient causes at an upper level of organization. Causal closure at one level shows only the efficacy and reasonable exhaustiveness of the procedures of intervention and access that define this level, in yielding a coherent picture *relative to these procedures*. Nothing then prevents one from obtaining another coherent picture, possibly with causal closure, at another level and with other procedures of intervention and access. These two internally coherent pictures and causal closures can *both* be valid, each at its own level, and relative to its own set of procedures of intervention and access. Far from being mutually exclusive (as an ontological conception of causation would require), they can be made mutually consistent.[62]

We must overcome our ontology's predilection for fixed and static substance metaphysics and develop a philosophy of mind that is in line with recent developments in cognitive sciences, quantum physics, chemistry, and quantum chemistry, and that could enable us to envisage consciousness also as a series of dynamic processes that are co-constitutive and relational. As Bitbol suggests, relata and relations can help us to achieve a performative description of the world, of the sciences, and of consciousness. This is not a proposal for a

---

[61] Campbell, Richard J. and Mark H. Bickhard, "Physicalism, Emergence and Downward Causation," *Axiomathes*, 21, 1 (2011), 45.

[62] Bitbol, M. (2010b). *Downward Causation without Foundations, Synthese* 180 (2010), 14. The italics are in the original. For further development, refer to Bitbol, M. *De l'intérieur du monde. Pour une philosophie et une science des relations* (Paris: Flammarion, 2010).

new ontology that will replace the traditional substantialist metaphysics. Rather, the co-dependence of relata and relations is not conceived as a feature of the world but as a condition for the possibility of our experience. Some of these processes can be described in terms of publicly accessible properties that are amenable to third-person descriptions, while the process that we call consciousness is unique in that it is subjective and amenable to first-person description. Rather than attempting to describe this process in exclusively third person and objective terms, we must realize that time has come to integrate the first-person subjective approach to consciousness in order to better investigate our lives from "the flux of our experience."

CHAPTER 5 | Who's Afraid of Supervenient Laws?

LEE MCINTYRE

OVER THE LAST FEW decades there has been much debate in the philosophy of science over the attractiveness–and potential costs–of supervenience.[1] As philosophers well know, supervenience burst onto the scene as the "Davidson debate" in the philosophy of mind began to raise some provocative questions over whether it was desirable to think of mental events as in some way irreducible to physical events, while still being firmly rooted in material dependence. After some initial misunderstanding over the question of whether supervenience was committing us to a sort of ontological break between the mental and the physical, it was finally settled that the autonomy one was after need not be metaphysical; an epistemological break would do just fine. After this the merits of supervenience could be clearly considered, for it allowed one to have it "both ways" in the dispute over mental states: mental explanations could be epistemically autonomous from physical ones (and thus probably not reducible to them), even while one preserved the notion of the ontological dependence of the mental on the physical (thus avoiding any embarrassing entanglements in supernatural or other spiritually based accounts of causal influence).

Davidson himself, of course, never really bought into the non-reductive materialist craze that he started, preferring to champion his own idiosyncratic view of anomalous monism, which allowed the mental to continue to exist as irreducible, even while he gave it no causal or explanatory work to do.[2] Since then Jaegwon Kim—the person who has done most to shed light on Davidson's view and demonstrate how the concept of supervenience could recast it as a more legitimate contender among the many proposals on the merits of non-reductive materialism—has appeared to repudiate his own earlier views about explanation and now wholeheartedly endorses a type of physicalist-based account that

[1] I thank Eric Scerri and an anonymous referee for this volume for their very helpful comments on this chapter.
[2] See Davidson's essays "Mental Events" and "Psychology and Philosophy" both reprinted in his *Essays on Actions and Events* (Oxford: Clarendon Press, 1980).

is even more conservative than Davidson's.[3] In his recent work, Kim has argued not only for the elimination of any mentally based causal descriptions (or laws) of human behavior, but also seems to call into question the very idea that in pursuing scientific explanation we need to pay much attention to secondary-level descriptions. For Kim, physical explanation seems explanation enough.

This loss, I feel, is an important one both for the philosophy of mind and the philosophy of social science, but it also has reverberations across the rest of philosophy, all the way to the philosophy of chemistry. For isn't it ironic that at the very moment when so many prominent philosophers of mind seem to be abandoning the merits of anti-reductionism, emergent explanation, and supervenience, the philosophy of chemistry seems to be discovering them?

The problem, of course, is that champions of reduction have always felt that they could turn to chemistry as a reliable ally in their quest to show that there is no legitimacy to secondary causal explanations, when materialist dependency is not in doubt. Given time, they argue, all such autonomous explanations would (and should) be reduced to the primary physical level. But much recent work in the philosophy of chemistry has questioned precisely this assumption. In his path-breaking book *The Same and Not the Same* (New York: Columbia University Press, 1995), Roald Hoffmann (the 1981 Nobel Prize winner in chemistry) has noted numerous examples wherein key chemical concepts are irreducible to the physical level and retain their explanatory (and causal) power only if we are prepared to describe the phenomena that they hope to explain at the chemical level of description.[4] Sadly for those materialist philosophers of mind who had hoped to find support in the traditional concept of a non-emergent, reductive relationship between chemistry and physics, the supervenience of some chemical properties has now been well established.

But what then of chemical laws? Perhaps, some have felt, we've got our hands full enough merely establishing the explanatory autonomy of chemical *concepts* to bother with the tricky question of whether there are autonomous chemical *laws*. But a better answer, I feel, is that through the concept of supervenience one can begin to understand how it might be possible that if chemical *properties* can be supervenient, why not the laws that instantiate the regularities that hold between those properties at the chemical level of description?

In this essay I aim to defend the idea that there are such supervenient laws of chemistry, conceived of as non-reductive causally efficacious relationships between the concepts that we use as explanatory entities at the (secondary) chemical level of description. Such laws are epistemologically emergent—even if they remain materially dependent upon physical relationships—and so indicate a role for supervenience in supporting autonomous chemical explanation.

---

[3] See Kim's "Mechanism, Purpose, and Explanatory Exclusion," in J. Tomberlin (ed.), *Philosophical Perspectives*, 3 (Atascadero, CA: Ridgeview, 1989), 77–108. For a critique of his views see Lee McIntyre, "Supervenience and Explanatory Exclusion," *Critica*, 34, 100 (2002), 87–101.
[4] See in particular his chapter "Fighting Reductionism," 18–21.

1 Supervenience

The concept of supervenience is one of the most misunderstood ideas in philosophy. The notion is thought to have originated in ethics, through the work of G.E. Moore, but it was Jaegwon Kim who brought it to prominence. Kim notes that, although Moore did not himself use the term "supervenience," he nonetheless recognized its metaphysical force when he wrote that,

> If a given thing possesses any kind of intrinsic value in a certain degree, then not only must that same thing possess it, under all circumstances, in the same degree, but also anything exactly like it, must, under all circumstances, possess it in exactly the same degree.[5]

Here one sees in nascent form the fundamental idea behind supervenience, which is one of consistency in property dependence.

If the debate had stopped with Moore, however, it would not be obvious how useful this concept might be to scientific explanation. Within Kim's work we find the clearest and most contemporary account of the supervenience relationship, which has had wide influence across the literature in the philosophy of mind and the philosophy of science. Kim has characterized supervenience as,

> belonging in that class of relations, including causation, that have philosophical importance because they represent ways in which objects, properties, facts, events, and the like enter into dependency relationships with one another, creating a system of interconnections that give structure to the world and our experience of it.[6]

Here the ontological character of supervenience is manifest, as is the budding idea of asymmetry in dependency relationships across different levels of organization. Phenomena at a higher level of organization, that is, would seem to have their properties in virtue of their material dependence upon relationships at a more basic level of organization, while still preserving the notion that there are two distinct levels.

An example here seems called for. Suppose that I am grading the final exams for my introductory logic class. I finish assigning point values and note that five of the exams are in the low 90s, five are in the low 80s, five are in the low 70s, and so on. Based on my desire to spread the grades out more or less evenly across the class, I decide on a curve by which 90 and above is an A, 80 and above is a B, 70 and above is a C, and so on. Later, I run across three exams that I forgot to grade. I do so and notice that the point values are 72, 89, and 89. After giving the 72 a C, I am in a quandary about what to do. I do not want to give more than six A's in the class, but since I have already decided on five A's, I am presented with the problem of what to do with the two exams that are

---

[5] Quoted in Jaegwon Kim, "Concepts of Supervenience," *Philosophy and Phenomenological Research*, 45, 2 (December 1984), 154.
[6] Ibid.

89 points. Clearly, I must treat like cases alike; if I am going to give one of them an A, I must do so for the other. Finally, I decide to stick to my original plan of giving all exams in the 80s a B. In the back of my mind, however, I know that had there been only one exam with an 89, and the other had been lower, I probably would have raised the 89 to an A. In either instance, one might say that the grade I gave the third exam supervened on the fact that it had 89 points, for while that was not enough in and of itself to *determine* the grade, the dependency relationship *was* strong enough that I had to treat like cases alike.

The strength of this relationship, however, has caused some to raise the stubborn question of whether supervenience entails reducibility or—in even stronger terms—whether "supervenience is just disguised reduction."[7] In his early works on supervenience, it seems clear that Jaegwon Kim thought that the answer to this question was "no." He writes,

> Explanation is an epistemological affair....[T]he thesis that a given domain supervenes on another is a metaphysical thesis about an objectively existent dependency relation between the two domains; it says nothing about whether or how details of the dependency relation will become known so as to enable us to formulate explanations [or] reductions.[8]

Kim elaborates on the issue when he argues in the same article that: "[Supervenience] acknowledges the primacy of the physical without committing us to the stronger claims of physical reductionism."[9]

Such a non-reductive account of supervenience makes clear its value for scientific explanation. Supervenience is a matter of ontology. It is a metaphysical relationship between two different levels of phenomena that is based on their asymmetric material dependence. Explanation, however, is an altogether different matter, where one may find it plausible to uphold the notion that there are different levels of explanation based on different levels of description of one and the same reality. Even though causes are rooted in ontology, they emerge as explanatory entities only at some level of description, while they

---

[7] Paul Teller, "Is Supervenience Just Disguised Reductionism?" *The Southern Journal of Philosophy*, Vol. 23 (1985), 93–99. See also Harold Kincaid, "Supervenience Doesn't Entail Reducibility," *The Southern Journal of Philosophy*, 25 (1987), 343–356.

[8] Kim, "Concepts of Supervenience," 175.

[9] Ibid., 156. In his later work on explanatory exclusion, Kim appears to take back this allowance for the epistemological autonomy of scientific explanation by raising worries about overdetermination. But this is to confuse ontological issues with epistemological ones. Kim has never revised his views on supervenience, however, which raises the question of how he might reconcile his earlier statements against reductive explanation with his later view that there is only one level of explanation for any given phenomenon (and that non-reductive materialism is incoherent). See here Kim's earlier cited "Mechanism, Purpose, and Explanatory Exclusion," but also his "Explanatory Realism, Causal Realism, and Explanatory Exclusion," *Midwest Studies in Philosophy*, 12, 225–239 and also his "The Myth of Non-Reductive Materialism," *Proceedings and Addresses of the American Philosophical Association*, 63, 3 (November 1989), 31–47. For more on my claim that Kim's earlier views on supervenience cannot be reconciled with his later views on explanatory exclusion, see my "Supervenience and Explanatory Exclusion."

may remain opaque at others. Reduction (understood as an epistemological notion) need not follow ontology this closely. In Kim's own words, "supervenience...says nothing about how successful we shall be in identifying causes and framing causal explanations; it is also silent on how successful we shall be in discovering causal laws."[10]

This interpretation of the supervenience relationship has given hope to many philosophers who have wished to find a respectable way to move beyond reductionism in their accounts of scientific explanation, by embracing the notion of non-reductive materialism.[11] If coherent, this view might afford a true conceptual autonomy in all of the "secondary" sciences like chemistry, where we may appreciate how explanatory concepts like "smell" and "transparency" could be epistemologically emergent from the microphysical relationships that underlie them, without violating material dependence or introducing other suspect notions of causal influence. Although it may break the great chain of continuity that is presumed to exist from the special sciences all the way to quantum mechanics, we are yet able to maintain the deep ontological connection between chemistry and physics. But we have no need of "ontological emergence;" supervenience is all that is needed to uphold the notion of material dependence coupled with epistemological autonomy in scientific explanation.[12]

The debate over supervenience will continue to rage within the philosophy of mind and the philosophy of science. Even so, many have already made their peace with the idea of the supervenience of concepts (or properties) within the secondary sciences. But what then of that most rigorous understanding of the relationship between properties and concepts in scientific explanation: laws?

2 Laws

The literature on the criteria for and existence of laws of nature has been fouled by sometimes unrealistically high standards for what constitutes a scientific law.[13] Various minimal lists have included such factors as that a law would have to be a well-confirmed general regularity, firmly embedded in a theory, confirmable by its instances, and not a priori true.[14] Others have gone beyond this to claim that a law would also have to be universal in scope,

---

[10] "Concepts of Supervenience," 175.
[11] Though Kim himself rejects this.
[12] For a contrasting argument in favor of the ontological interpretation of emergence, see O. Lombardi and M. Labarca, "The Ontological Autonomy of the Chemical World," *Foundations of Chemistry*, Vol. 7, 2 (2005), 125–148. See also R. Le Poidevin, "Missing Elements and Missing Premises: A Combinatorial Argument for the Ontological Reduction of Chemistry," *British Journal for the Philosophy of Science*, 56 (2005), 117–134.
[13] For a thorough discussion of this issue see the accounts by Armstrong (1983), Dretske (2004), and Tooley (2004) cited in the bibliography. For a more contemporary account see Bird (2002).
[14] One particularly intriguing account is that by Mario Bunge, "Kinds and Criteria of Scientific Laws," *Philosophy of Science*, 28, 3 (July 1961), 260–281, who examines scores of different types of laws and then reduces them to a manageable set of criteria.

exceptionless, immutable over time and space, yield accurate predictions, and support its counterfactuals. Naturally, such an expansive list may cause one to wonder how many of the current laws of physics would be able to survive such a rigorous test, leading some to question whether there actually exists a tidy set of necessary and sufficient conditions such that all and only the laws of nature could pass them.[15]

In an earlier work, I explored the notion of whether it was possible to defend a robust account of nomological explanation in the social sciences.[16] Rather than rehash that argument now for the "secondary science" of chemistry, let us instead focus on the core idea of what it means for an explanation to be "law-like," confident that if we cannot meet this then the entire notion of secondary laws must be called into question.

At base, what is a law but a robust and regular empirical relationship between two properties? Of course, it would be preferable to think that this relationship was based on some sort of causal relationship in nature, thus vindicating the idea that the connection between the two properties in question was real and not based on some sort of naive correlation. It is notoriously true, however, that no matter how good our scientific evidence may be in favor of any correlation, we can never be certain that the relationship we are studying is truly causal. As a line of skeptical philosophers from Plato down through Hume have so aptly taught us, we can never know reality in and of itself but only through sensory evidence, which admits of no extra-empirical interrogation of its actual correlation with any particular state of nature. Likewise, any nomological relationship must be empirical. If it is based on a logical or other "necessary" connection then it is not "natural" enough to be a law of nature and runs the risk of being true merely in virtue of the meaning of our concepts.

This is to say that, even if it is a truism, we would do well to remember that when we are doing science there is always a degree of uncertainty. This is the price we pay for trying to enlarge our understanding of the world beyond the truths of logic and mathematics. But it is time, then, to examine the implications of such inductive problems for what it means for something to be a "law of nature" and admit that *even in the face of the most robust scientific evidence*, we can never be sure that *any* law is based on a true ontological connection, but instead must satisfy ourselves with knowing reality only at a remove that is mediated through our theories and descriptions.

Yet, if taken seriously, this idea should shake our understanding of laws to the very foundation of science. A law is more than just a causal connection in nature. This is because there is no way to know reality in and of itself. A law is, instead, a medium of explanation for what we hope to be an underlying natural

---

[15] Achinstein (1971, 1). Peter Achinstein, Law and Explanation: An Essay in the Philosophy of Science (Oxford: The Clarendon Press, 1971), 1.
[16] Lee McIntyre, *Laws and Explanation in the Social Sciences: Defending a Science of Human Behavior* (Boulder, CO: Westview Press, 1996).

regularity, yet this character of a law will not emerge unless we have unlocked it with the proper description. Even if reality is fixed, when we are trying to explain it our account must be mediated by a descriptive vocabulary that represents our theoretical point of view. When dealing with a causal law, there must of course be a contribution from nature, but what is sometimes overlooked is that there must also be a contribution from us as well. As Werner Heisenberg observed, "what we observe is not nature itself, but nature exposed to our method of questioning."[17]

As explanatory entities, laws are based on our descriptions of nature. They cannot just reflect a correlation between pre-existing natural kinds, for how—outside our theoretical and vocabulary-based empirical investigations—can we expect to know these natural kinds in the first place? This brings us to the well-known problem of "underdetermination," whereby any given set of empirical results is in principle compatible with an infinite number of possible theories. Ontology may rule out infinitely many theories that do not fit the evidence, but there are always an infinite number that are nonetheless ruled in. How many smooth curves can be drawn between a fixed set of data points? We may prefer the simplest ones, but no matter how many additional points we plot, the potential number of smooth curves remains infinite.

Many philosophers of science have been quick to dismiss this as a mere "logical" problem, and have refused to grapple with the implications of underdetermination for the question of scientific explanation.[18] Yet, when we do take underdetermination seriously, one finds that it has implications for our understanding of laws, theories, and perhaps even the concept of causality itself. How many good descriptions can fit a given reality? And how many theories or laws might be based on these? Until we can confidently solve the "logical" problem of underdetermination, the fact that we must honestly answer "an infinite number" should trouble those who have been so quick to embrace reductionism as an answer to our explanatory desiderata.

Based on this account, one should now be prepared to admit that in some sense *all* scientific laws are epistemically emergent. Such a commitment need not violate materialist ontology. Recall that supervenience affords the possibility of explanatory emergence even while embracing ontological dependence. But the point is that since laws are explanatory entities—and so are based not just on nature but also on our descriptions of nature (which may be infinite)—we must recognize that laws are not merely a matter of ontology but are instead the product of a dialogue between nature and our descriptions. Laws are not, in any coherent sense, a "blueprint" of reality any more than there is one ideal map that can be drawn for any given piece of terrain; it depends on our purpose

[17] "Physics and Philosophy: The Revolution in Modern Science." Lectures delivered at University of St. Andrews, Scotland, Winter 1955–1956.
[18] See my "Taking Underdetermination Seriously," *SATS: Nordic Journal of Philosophy*, 4, 1 (2003), 59–72.

and the inherent limits in our powers of representation.[19] There may be many such dialogues and many possible laws at different levels of description.

This means that laws at the primary level, no less than at the secondary level, must depend on descriptions. What many have thought of as the "gossamer" nature of laws at the secondary level is therefore no less true at the primary level. They depend on descriptions. A law is more than just a causal relationship in nature, but is instead an account of those relationships as described in a vocabulary that makes the nomological connection obvious. But different patterns may emerge at different levels of description. Therefore one reality may support a multiplicity of (nomological) explanations that are based on alternative accounts of the regularities that are supported by a single ontology. But, since scientific explanation is relative to our level of description, it is perfectly appropriate now to contend that we ought to search for many different regularities, based on different concepts and vocabularies. Why do this? Precisely because some descriptions will reveal laws and others will not. Any given ontology may admit of laws at several different levels of description, depending on the nature of our diverse explanatory interests.

An analogy here might make the point clearer. Consider the pattern of constellations that we view in the night sky. As we all know, there are different constellations claimed by different cultures, but each is based on the same "ontology" of stars in the heavens. To narrow the point even further, let's consider the three particular stars that constitute "Orion's Belt," based on the traditional Western constellation. It does not require a trained eye to notice that these three stars are perfectly aligned, almost as if they were data points begging for a straight line, which our imagination fills in as Orion's "belt." Yet, in actuality, these three stars are separated from one another by millions of miles and *only from the very narrow point of view here on Earth* do they appear to be a straight line. The point is that from almost any other perspective in the heavens we would see these stars differently, in a different pattern, or perhaps in no pattern at all. The fact that our eye sees them as straight and fits them into the line of "Orion's belt" is based not just on the fixed point of the stars in the heavens, but also on the perspective from which we are looking at them here on Earth. But, for all we know, there may be many other, equally smooth patterns formed by these three stars, from the perspective of other points of view. Perhaps creatures on another planet think that these stars fit into a pattern too, but see them as a triangle. The stars in Orion's Belt make up a pattern based on some points of view, and they do not from many others, even though all are based on one and the same set of stars. Patterns emerge, that is, based not merely on ontology but also on our point of view which, in scientific explanation, is read off as theories and descriptions.

The connection here with supervenient laws should be obvious. As explanatory (epistemological) entities, all laws are supervenient, dependent not just

---

[19] For an illustration of this, see my discussion in the following section about the periodic law and the periodic table.

on the ontology that supports them, but on the descriptions and theories that we bring to the study of that ontology.

3 An Example from Chemistry

If the claim that I have been making about the prevalence of supervenient laws is correct, then it should be easy to find examples. What is harder, however, is to find one that can serve as an exemplar of the virtues of the supervenience interpretation of nomologicality, which are redescription and anti-reductionism.

One such example is the periodic law of chemistry, which states that—when arranged according to atomic number—*after certain regular but varying intervals the chemical elements show an approximate repetition in their properties.* There is, of course, some debate among chemists over whether this deserves to be called a "law," given the varying length of chemical periodicity and the approximate nature of the repetition. This debate has historically taken a back seat, however, to the formidable task of trying to find an adequate way to represent these regularities (inexact though they may be) in visual form. This has led to the intricate and fascinating history of the invention of the periodic table, which has recently been retold and placed in its proper philosophical perspective by Eric Scerri.[20] It is well to remember, however, that although there is a deep relationship between the periodic law and the periodic table—which has led to a succession of theoretical breakthroughs in our understanding of what is behind chemical periodicity—the latter serves as a representation of the former and it is the periodic law that underlies the whole system. According to Scerri, there are more than 700 different versions of the periodic table, but only one periodic law![21] Indeed, the periodic law may usefully be thought of as one of the foundations of modern chemistry, comparable to evolution by natural selection in biology.[22] This, then, is no mere example of *a* law of chemistry, but arguably *the* law of chemistry, which serves as the organizing principle of the entire discipline.

First, it is important to establish the periodic law's claim to nomologicality: it is a well-confirmed general regularity, confirmable by its instances, embedded in a theory, and is not a priori true. The periodic law easily meets these minimal criteria. It may not, however, meet some of the inflated criteria (such as universality) that are sometimes said to be necessary for nomologicality. But that is all right. Neither do some of the most famous laws in biology (such as Dollo's Law of morphological irreversibility).[23]

But does variance of the chemical period and the inexactness of the repetition of the properties of the elements undermine its law-like status? Even those

---

[20] Eric Scerri, *The Periodic Table: Its Story and Its Significance* (New York: Oxford University Press, 2007).
[21] Ibid., 20–21.
[22] Ibid., 25.
[23] See my "Gould on Laws in Biological Science," *Biology and Philosophy*, 12, 3 (1997), 357–367.

who defend it as a law concede that it is approximate. Worse, as we learn through studying the history of the discovery of the periodic table, the periodic law did not readily fall out as a deductive consequence of some preexisting theory. Although it famously led to several accurate predictions (that arguably made the reputation of its discoverers) some intuition was needed in limning the nature of these predictions and some reliance on empirical data was used to determine them.

But the periodic law is redeemed by meeting one of the most difficult criteria of all: it is exceptionless. The anchor for the periodic law is that the elements must be arranged according to their atomic number. In previous iterations, this anchor has dragged from equivalent weight to atomic weight, to atomic number, based on the evolution of chemical understanding of the role played by valence and Bohr's eventual discovery of the atomic theory of matter. But that does not detract from the fact that the periodic law has always been embedded in a theory that stood behind this regularity. This was no "Bode's Law" of the elements, where the only guiding principle was slavish numeracy and the only method trial and error. Even if its governing theory occasionally had to change—as the ontology behind the elements was redescribed in different theoretical terms—it is important to realize that chemists always believed that there was a rationale behind the periodicity of the elements. The periodic law is a tool for explaining and predicting chemical phenomena, but it is certainly about more than just "saving the phenomena."

Is the periodic law also supervenient? Here one must examine two important factors that govern the supervenience relationship: the importance of redescription and the irreducibility of chemical properties. In his account of the early history of the periodic table, Eric Scerri makes clear that some degree of redescription was at work, as scientists cycled through various hypotheses about how to arrange the elements so that periodicity would make sense. There were many false starts and blind alleys, ranging from analogies with geometric figures and the periodicity of the planets, to hypotheses about the "law of octaves" in music. Indeed, such confusion led one crank sarcastically to suggest that the elements should perhaps be arranged in alphabetical order![24] But all of this fluidity in descriptive terms eventually led to the concept of equivalent weight, which morphed into the concept of atomic weight, after which the regularity between the elements began to take shape as a law: as one increased in atomic weight the properties of the elements underwent periodic (though inexact) repetition.

This finally resulted in the classic version of the periodic table, invented by Mendeleev and refined by Julius Lothar Meyer, who cemented the reputation of their system not only by accommodating all of the known elements, but also by predicting three new ones. Of course, there were some remaining anomalies— such as the backward pairing of tellurium and iodine—which would await later correction. But even here, the theoretical basis for the periodic law was preserved

---

[24] Scerri, 2007, 78.

as Mendeleev insisted that the atomic weights for these two elements must be incorrect, since the periodic law was exceptionless: The elements had to be arranged according to their atomic weight. Even though this theory was later revised, it was clear that the periodic law was firmly embedded in theory. All of that redescription had finally paid off in a law of nature.

Later, when Niels Bohr discovered the atomic theory of matter, the periodic law was placed on even more solid theoretical footing, as redescription once again led to a deeper understanding of the causal forces behind chemical periodicity. Now it was realized that the ordering of the elements should be made according to their atomic number, not atomic weight, based on the number of electrons in each element. This theory succeeded in explaining for the first time why some of the chemical properties of the elements actually occurred. But, as Scerri points out, it would be false to assume from this that the new theory explained everything about chemical periodicity or that the entire periodic system could now be reduced to a branch of physics.[25]

It is important to be clear here about the nature of the supervenience relationship between chemistry and physics. Recall that supervenience does not presuppose some ontological break between the primary and secondary level. Although supervenience is anti-reductionist in its epistemological commitments, this is based on explanatory considerations and does not require one to believe that chemical regularities—such as periodicity—emerge via some mysterious route of chemical causation. Chemical periodicity is firmly rooted in physical relationships. But this does not necessarily mean that the periodicity of chemical properties can be completely explained even by our best underlying physical theory, quantum mechanics.

In part, this is due to considerations drawn from the nature of scientific explanation. As Scerri writes, "every science can decide for itself the level at which it should operate and that the deepest foundations are by no means the best for every purpose."[26] But it is also true that Bohr's model of the atom— even when it was updated by the Pauli Exclusion Principle—could not adequately explain (1) the length of chemical periodicity and (2) the place in the periodic table where periodicity occurs.[27]

Although it is popularly believed (and stated in most introductory textbooks) that atomic theory affords the complete reduction of the periodic law to some deeper law of physics, this idea has remained problematic. Throughout his scholarship, Eric Scerri has offered a masterful account of the inadequacies of such reductionism.[28] This is especially important since the reduction of chemistry to

---

[25] Ibid., 203.
[26] Ibid., 121.
[27] Ibid., 203.
[28] See in particular his "Electronic Configurations, Quantum Mechanics, and Reduction," *British Journal for the Philosophy of Science*, 42 (1991), 309–325 and also his "Has Chemistry Been at Least Approximately Reduced to Quantum Mechanics?" in D. Hull, M. Forbes, and R. Burian (eds.), *PSA 1994*, Vo. 1 (East Lansing, Mich.: Philosophy of Science Association, 1994), 160–170. This issue is also discussed throughout *The Periodic Table*, especially at 242–248 and 266–275.

physics is taken by many to be a paradigm case of the successful reduction of one scientific explanation to another. In a telling example (that is centrally relevant to our current point about the periodic law), however, Scerri points out that although quantum mechanics can predict from first principles the fact that the electron capacity of successive *shells* is 2, 8, 18, and 32, the lengths of successive *periods*—namely 1, 8, 8, 18, 32, and 32—have yet to be derived from first principles.

Suffice it to say that the general problem here is that some chemical *properties*, which are part of the descriptive referents of the periodic law, have no conceptual correlate at the physical level, nor can our best current physical theory explain every chemical behavior that we witness at the secondary level. One interesting example of this is the "diagonal behavior" of some elements, whereby—in violation of group trends—"two elements from adjacent groups show greater similarity than is observed between the elements and the members of their own respective groups."[29] In considering one such trend—the relationship between lithium and magnesium—Scerri highlights seven separate examples of diagonal behavior that "undermine the simplistic physicist's notion that chemical behavior is governed just by the electronic configuration of atoms."[30] Instead, the explanation relies on irreducibly chemical concepts.

The periodic law is a supervenient law of chemistry. It cannot be completely expressed (or fully explained) at the physical level. Yes, it depends on underlying physical relationships and, as materialists, advocates of supervenience would be loath to make any claims about a break in ontology. Supervenience is not supernaturalism. But this does not detract from the conclusion that the properties that are ensconced in the periodic law—such as density, reactivity, malleability, and color—are chemical properties, and therefore should be explained at the chemical level. Attempting to reduce such phenomena to the underlying physical level, even where possible, may eclipse the very thing that chemical explanation is all about.

Conclusion

In this chapter I have argued that the concept of supervenient laws has great value, not only in the field of chemistry, but also in the other "secondary" disciplines as well. Indeed, more than this, our analysis has revealed that even the idea of whether there are fundamental laws of physics could be enriched by considering the concept of supervenience.

Many who object to the concept of supervenient laws balk at the idea that it would require some sort of metaphysical commitment over and above materialism. But this is not true. What is required instead is realization that scientific explanation is about more than just trying to recreate some Adamic picture of ontology. Scientific explanations are a dialogue between ontology and the descrip-

---

[29] Scerri, 2007, 265.
[30] Ibid., 266.

tions and theories that we use to capture this ontology. But the latter are potentially infinite and, depending on which descriptive terms we use, regularities may emerge given some ways of looking at the world that will elude us using others. There may be one and only one reality but there are an infinite number of ways of describing it. And, since our laws and theories are based not just on reality, but also on these descriptions, there may be infinitely many possible laws as well.

Scientists sometimes talk as if the goal of all scientific explanation were to reduce our understanding to a foundational set of laws. This goal led to some of the greatest breakthroughs in modern science, as it has motivated us to employ redescriptions and seek better theories, when concepts at the secondary level are not working or seem incomplete. But this view is a myth. There are no foundational laws of science because there are no foundational descriptions. Any description, any theory, any way of looking at the world is always subject to revision and, even if we think that there is only one way that the world could be and that there is a truth about it, we still face the problem that there are many different ways of capturing that truth.

Scientific explanation depends not just on nature, but also on us. If we seek to reduce all secondary explanations to their most basic level, then we must ask to *which* base, by *what* set of bridge laws, for it is always possible that alternative descriptions will lead to different foundations. And what will we do then, even if we are successful, when we find that the phenomena we sought to originally explain are not represented in the descriptive terms of our foundational theory? Some laws are reducible and some are not. We need not just a compatible ontology but also compatible concepts and descriptions at each level for any reduction to go through. But a single ontology may support many alternative descriptions, theories, and laws at many different levels, not all of which may be reducible. Given this, it is perhaps best to remember that *all* scientific laws are in some sense supervenient, for they depend not just on ontology, but on our *descriptions* of ontology, which are potentially infinite in number.

The problem that many have had in accepting the idea of supervenient laws is perhaps not that they do not understand (or accept) the idea of supervenience, but that they really have misunderstood what it means to be a scientific law. Laws exist not just in nature, but are dependent on us too. Laws are an epistemological entity that are used to explain and thus they are always developed within the context of a theory, which in turn also depends on a description. Perhaps it is useful, then, to remember that all laws of nature are in some sense secondary. There is no such thing, I maintain, as the discovery of the "true" laws of nature. Some laws are false, of course, because reality rules them out. But this does not mean that there is some other set of laws that is "ideally" true, because it takes more than concordance with reality to make a law true. It also depends on the preferences that we have for a particular description of reality. And, as we have seen, there are many of those and ontology alone cannot choose between them for us.

Sometimes supervenient laws are the only laws in town.

References

Achinstein, Peter. (1971). *Law and Explanation: An Essay in the Philosophy of Science.* Oxford: Clarendon Press.

Armstrong, David. (1983). *What Is a Law of Nature?* Cambridge: Cambridge University Press.

Bird, Alexander. (2002). "Laws and criteria," *Canadian Journal of Philosophy*, 32, 4 (December), 511–542.

Bunge, Mario. (1961). "Kinds and criteria of scientific laws," *Philosophy of Science*, 28, 3 (July), 260–281.

Dretske, Fred. (2004). "Laws of Nature." Reprinted in J. Carroll (ed.) *Readings on Laws of Nature*. Pittsburgh, PA: University of Pittsburgh Press, 16–37.

Hoffmann, Roald. (1995). *The Same and Not the Same.* New York: Columbia University Press.

Kim, Jaegwon. (1984). "Concepts of supervenience," *Philosophy and Phenomenological Research*, 45, 2 (December), 153–176.

Kim, Jaegwon. (1989). "Mechanism, Purpose, and Explanatory Exclusion." In *Philosophical Perspectives*, Vol. 3. (J. Tomberlin, ed.). Atascadero, CA: Ridgeview, 77–108.

Kim, Jaegwon. (1989). "The myth of non-reductive materialism," *Proceedings and Addresses of the American Philosophical Association*, 63, 3 (November), 31–47.

Kincaid, Harold. (1987). "Supervenience doesn't entail reducibility," *The Southern Journal of Philosophy*, 25, 343–356.

Le Poidevin, R. (2005). "Missing elements and missing premises: A combinatorial argument for the ontological reduction of chemistry," *British Journal for the Philosophy of Science*, 56, 117–134.

Lombardi, O. and Labarca, M. (2005). "The ontological autonomy of the chemical world," *Foundations of Chemistry*, 7, 2, 125–148.

McIntyre, Lee. (1996). *Laws and Explanation in the Social Sciences: Defending a Science of Human Behavior.* Boulder, CO: Westview Press.

McIntyre, Lee. (1997). "Gould on laws in biological science," *Biology and Philosophy*, 12, 3, 357–367.

McIntyre, Lee. (2002). "Supervenience and explanatory exclusion," *Critica*, 34, 100, 87–101.

McIntyre, Lee. (2003). "Taking underdetermination seriously," *SATS: Nordic Journal of Philosophy*, 4, 1, 59–72.

Scerri, Eric. (1991). "Electronic configurations, quantum mechanics, and reduction," *British Journal for the Philosophy of Science*, 42, 309–325.

Scerri, Eric. (1994). "Has Chemistry Been at Least Approximately Reduced to Quantum Mechanics?" In *PSA 1994*, Vol. 1 (D. Hull, M. Forbes, and R. Burian, eds.). East Lansing, MI: Philosophy of Science Association, 160–170.

Scerri, Eric. (2007). *The Periodic Table: Its Story and Its Significance.* New York: Oxford University Press.

Teller, Paul. (1985). "Is supervenience just disguised reductionism?" *The Southern Journal of Philosophy*, 23, 93–99.

Tooley, Michael. (2004). "The Nature of Laws." reprinted in J. Carroll (ed.). *Readings on Laws of Nature*. Pittsburgh, PA: University of Pittsburgh Press, 38–70.

| # The Changing Views of a Philosopher of Chemistry on the Question of Reduction

ERIC R. SCERRI

1 The Question of Reduction

The question of the reduction of chemistry to quantum mechanics has been inextricably linked with the development of the philosophy of chemistry since the field began to develop in the early 1990s. In the present chapter I would like to describe how my own views on the subject have developed over a period of roughly 30 years.

A good place to begin might be the frequently cited reductionist dictum that was penned in 1929 by Paul Dirac, one of the founders of quantum mechanics.

> The underlying laws necessary for the mathematical theory of a larger part of physics and the whole of chemistry are thus completely known, and the difficulty is only that exact applications of these laws lead to equations, which are too complicated to be soluble.
>
> (DIRAC 1929)

These days most chemists would probably comment that Dirac had things backward. It is clear that nothing like "the whole of chemistry" has been mathematically understood. At the same time most would argue that the approximate solutions that are afforded by modern computers are so good as to overcome the fact that one cannot obtain exact or analytical solutions to the Schrödinger equation for many-electron systems. Be that as it may, Dirac's famous quotation, coming from one of the creators of quantum mechanics, has convinced many people that chemistry has been more or less completely reduced to quantum mechanics. Another quotation of this sort (and one using more metaphorical language) comes from Walter Heitler who together with Fritz London was the first to give a quantum mechanical description of the chemical bond.

Let us assume for the moment that the two atomic systems ↑↑↑↑... and ↓↓↓↓... are always attracted in a homopolar manner. We can, then, eat Chemistry with a spoon.

(HEITLER 1927)[1]

Philosophers of science eventually caught up with this climate of reductionism and chose to illustrate their views with the relationship with chemistry and quantum mechanics. It must also be said that such a view agreed entirely with the prevailing notion of the unity of science as developed by the Logical Positivist school of philosophy. Here are some examples of what these philosophers wrote on the subject:

> the possibility that science may one day be reduced to microphysics (in the sense in which chemistry seems today to be reduced to it...).
>
> (OPPENHEIM, PUTNAM 1958)

> Certain parts of 19th century chemistry (and perhaps the whole of this science) is reducible to post-1925 physics.
>
> (NAGEL 1961)

> Today it is possible to say that chemistry is a part of physics, just as much as thermodynamics or the theory of electricity.
>
> (REICHENBACH 1978)

Then in the 1960s and '70s logical positivism came under increasing criticism by the likes of Popper, Kuhn, Lakatos, and Feyerabend, many of whom appealed to the history of science to challenge the prevailing view that the various special sciences could be reduced to physics. It became almost politically incorrect to hold reductionist views about any of the special sciences such as chemistry. Not surprisingly, therefore, as a graduate student beginning a PhD thesis on the reduction of chemistry, I duly climbed onto the bandwagon of anti-reductionism and produced a number of articles in which I gave specific instances of what I took to be the failure of reductionism.[2]

Moreover this position seemed to coincide with the fact that the philosophy of chemistry had been almost completely neglected up to that point. If chemistry were indeed reduced to quantum mechanics it would justify the belief that chemistry was very much an applied field with no particularly deep questions or big ideas. One might therefore say that in the early 1990s there were "political reasons" for claiming that chemistry had not been reduced, especially among people like myself who were campaigning for the growth of a philosophy of chemistry. In this chapter I want to take stock of the situation some 20 or more years later.

---

[1] Heitler's program to explain all of chemistry got him in trouble more than once. Wigner used to tease him. He would ask: "what chemical compounds would you predict between nitrogen and hydrogen? And of course, since he did not know any chemistry he couldn't tell me." Heitler confessed as much. I am indebted to Gavroglu and Simoes for this episode.

[2] At a recent workshop on reduction and emergence at the Sorbonne University, Paul Humphreys jokingly suggested that part of the attraction for the concept of emergence comes from New Age thinking. I think the same can be said about anti-reductionism. As much as academics think they look down of such cultural movements I think they still pervade the Zeitgeist of any particular era. The popularity of New Age thinking contributes to the attraction that philosophers have for such notions as emergence and anti-reductionism.

First of all it has to be said that the field has not developed in the way that some of us believed that it might. I find it remarkable that even after 20 years there are still fewer than 10 books devoted specifically to the philosophy of chemistry. In addition, as I will be arguing, I think that our anti-reductionist claims may have been over exaggerated, due perhaps to a form of pioneering zeal on our part.[3]

To take another example from a recent meeting on reduction and emergence in the sciences held in Paris, only 2 of the 15 speakers were people who would describe themselves as philosophers of chemistry. The vast majority of philosophers of science still prefer to consider the philosophy of physics or of biology while conveniently skipping the field of chemistry, which lies between physics and biology in many obvious respects.

I hardly think that the unpleasant reputation that chemistry enjoys because of its messy, smelly, and often dangerous nature can be entirely to blame for the continued neglect among philosophers of science for the field of chemistry. I rather believe that it is because of a popular misconception that there are no "big ideas" in chemistry. In this chapter I will be discussing one undeniably big chemical idea, namely the existence of the periodic system.

Before moving on to the more specific part of this chapter let me also pause to mention the fact that in terms of numbers of practitioners, chemistry outnumbers all other scientific disciplines. In fact according to some measures the number of chemistry practitioners may even outnumber the sum total of combined practitioners from all other scientific fields, with the possible exception of computer science (Schummer 1997).

In very broad terms, the complexity found in the biological world ensures that the non-reducibility of biology is generally upheld. At the same time physics is physics and as such is the supreme reducing discipline while chemistry, which is a closer neighbor than biology is popularly believed to be reducible, and indeed to have been reduced. Let me quickly say that I now believe that this view is essentially correct and that chemistry is more or less a reduced science. There is no denying that chemistry has been living somewhat in the shadow of physics since the beginning of the twentieth century. In this chapter I will be looking at more specific aspects, and especially at parts of chemistry which I personally claimed as examples of the breakdown of reduction but on which I have now changed my opinion as a result of more recent research.

In any case I have always tried to stress that much of the philosophical discussion is far too general to be of much use. To simply ask whether chemistry has been reduced to physics without qualification is rather meaningless.

---

[3] In spite of the lack of monographs on the subject there are two international journals devoted to the philosophy of chemistry. They are *Foundations of Chemistry* (published by Springer) and *Hyle* (published independently of any major publishing house). The parent association for *Foundations of Chemistry* is the International Society for the Philosophy of Chemistry that has held regular meetings in different locations around the world for the past consecutive 18 years. There have also been frequent sessions devoted to the philosophy of chemistry at the biennial meetings of the Philosophy of Science Association. However, as I have suggested in some earlier articles these sessions have tended to be somewhat inward looking and arranged by speakers who have agreed in advance to advance a similar agenda rather than opposing each other's work where this might be a more appropriate course of action.

Similarly, to expect there to be a yes or no answer to such a complex question is, I believe, a gross oversimplification. On this theme I think I have been fairly consistent (Scerri 1994a, Scerri 1994b).

In addition I have rejected the use of the classical criteria for reduction that have been proposed by Logical Positivist philosophers of science in favor of more naturalistic criteria. In examining the extent to which a specific area of chemistry has been reduced to quantum mechanics I have used the criteria that chemists and physicists themselves would regard as constitutive of a reduction. In simple terms this means the extent to which the contents of the secondary science can be calculated in a completely ab initio manner from the principles of quantum mechanics, in very much the same spirit as the quotes from Dirac and London mentioned at the outset.

Let me make one final preliminary remark before presenting my more specific arguments concerning reduction. As in most of my work I will be discussing what is generally described as the epistemological reduction of chemistry to quantum mechanics. This is sometimes also termed theoretical reduction as it concerns the reduction of theories of a special science to the theory or theories of physics. I have not devoted much attention to the more general question of the ontological reduction of chemistry to quantum mechanics since I see no way in which this task can genuinely be conducted without smuggling back epistemological aspects of both the two sciences in question. In saying this I take epistemological reduction to mean whether chemistry for example is nothing but quantum physics irrespective of our laws and theories about the two domains. I disagree with the philosophers that seem to believe that merely switching to talk of entities or laws of the various sciences rather than their theories entitles them to suddenly enter the realm of ontology.

Returning to my own contributions in the area of so-called ontological reduction, as I will call it, my work has consisted of criticizing the views of other authors rather than putting up positive arguments either for or against the ontological reduction of chemistry (Scerri 2007, 2012).

2 The Periodic System—Easily the Biggest Idea in Modern Chemistry

Much has been written on the periodic system of the elements (Van Spronsen 1969, Gordin 2004, Scerri 2007). It is beyond any doubt one of the major discoveries in all of modern science and certainly the most influential discovery in the field of chemistry. The existence of the periodic system serves to unify the whole of inorganic chemistry and upholds the essential unity of all chemical substances, in spite of the tremendous variety possessed by individual elements and their compounds. In addition to being of theoretical value it serves an extremely useful didactic purpose and has permeated the teaching of chemistry since very soon after it was initially discovered in the 1860s (Kaji, Kragh, and Pallo 2015).

To most observers this seems like an obvious case of the reduction of chemistry to quantum mechanics, if not the supreme example of this kind of relationship. As we all learn in high school and college chemistry the wonderful edifice of the periodic table (see figure 6.1) that was gradually discovered on purely chemical grounds became fully understood and explained following the development of quantum mechanics and its talk of electron shells and orbitals. What may originally have seemed like a mysterious system of classification at the turn of the twentieth century scientists became regarded as a simple outcome of the underlying physics of the atom. In order to discuss this subject in greater depth I will begin with a very brief historical tour of the discovery of chemical periodicity and will take as my somewhat arbitrary starting point the publication of a set of atomic weights by John Dalton in 1808.

One can gain a simple understanding of the periodic system by considering what has been called the element line, as shown in figure 6.2. Here the elements have been ordered according to their atomic numbers, meaning the number of protons in the nuclei of their atoms. At the time of the discovery of the periodic system in the late 1860s the quantity used to bring about this ordering was that of atomic weight, such as the values published by Dalton. In any case the

Part of an early table of atomic and
molecular weights published by Dalton

Element	Weight
Hydrogen	1
Azot	4.2
Carbon (charcoal)	4.3
Ammonia	5.2
Oxygen	5.5
Water	6.5
Phosphorus	7.2
Nitrous gas	9.3
Ether	9.6
Nitrous oxide	13.7
Sulphur	14.4
Nitric acid	15.2

FIGURE 6.1 Dalton's atomic weights list of 1808.

Adapted from J. Dalton, *Memoirs of the Literary and Philosophical Society of Manchester*, 2(1), 207, 1805, table on p. 287.

H	He	Li	Be	B	C	N	O	F	Ne	Na	Mg	Al	Si	P	S	Cl	Ar	K	Ca	Sc	Ti	V	Cr	Mn
1	2	3	4	5	6	7	8	9	10	11	12	13	14	15	16	17	18	19	20	21	22	23	24	25

FIGURE 6.2 The element line.

differences in the ordering depending on whether one used atomic weight or atomic number are minimal and this issue need not detain us here (Scerri 2014).

Once the number line is established one then looks for chemical analogies among the listed elements. The next step is to cut the number line in order to obtain certain sequences of elements such that they reflect the chemical analogies when they are stacked on top of one another. In other words chemically analogous elements should fall together into vertical columns or groups as shown in figure 6.3.

We now turn to the reason why the sequence of elements has been terminated in figure 6.2 at the 25th element, manganese. This has been done because a complication occurs with the very next element, iron. Whereas the previous few elements such as potassium, calcium, scandium, titanium, vanadium, chromium, and manganese show increasing maximum oxidations states or combining powers of 1 to 7 inclusive, the element iron shows a value of +3, therefore breaking the previous sequence. What Mendeleev and other early discoverers of the periodic system did was to expel iron and the next few elements from the main body of the periodic table, placing them instead into an additional column on the right-hand side of the table. The sequence of increasing oxidation numbers resumes again with the element rubidium, which displays a maximum oxidation state of +1, and is therefore placed in the first column of the table (figure 6.4).[4]

H							He	
Li	Be	B	C	N	O	F	Ne	
Na	Mg	Al	Si	P	S	Cl	Ar	
K	Ca	Sc	Ti	V	Cr	Mn		

FIGURE 6.3 A short-form periodic table obtained by cutting the element line 3 times and stacking the shorter sequences on top of each other to reflect chemical analogies.

Tabelle II.

Reihen	Gruppe I. — R^2O	Gruppe II. — RO	Gruppe III. — R^2O^3	Gruppe IV. RH^4 RO^2	Gruppe V. RH^3 R^2O^5	Gruppe VI. RH^2 RO^3	Gruppe VII. RH R^2O^7	Gruppe VIII. RO^4
1	H=1							
2	Li=7	Be=9,4	B=11	C=12	N=14	O=16	F=19	
3	Na=23	Mg=24	Al=27,3	Si=28	P=31	S=32	Cl=35,5	
4	K=39	Ca=40	—=44	Ti=48	V=51	Cr=52	Mn=55	Fe=56, Co=59, Ni=59, Cu=63.
5	(Cu=63)	Zn=65	—=68	—=72	As=75	Se=78	Br=80	
6	Rb=85	Sr=87	?Yt=88	Zr=90	Nb=94	Mo=96	—=100	Ru=104, Rh=104, Pd=106, Ag=108.
7	(Ag=108)	Cd=112	In=113	Sn=118	Sb=122	Te=125	J=127	
8	Cs=133	Ba=137	?Di=138	?Ce=140	—	:	—	— — — —
9	(—)				—	—	—	
10	—	—	?Er=178	?La=180	Ta=182	W=184	—	Os=195, Ir=197, Pt=198, Au=199.
11	(Au=199)	Hg=200	Tl=204	Pb=207	Bi=208	—	—	
12	—	—	—	Th=231	—	U=240	—	— — — —

FIGURE 6.4 Mendeleev's short form periodic table including all known elements from hydrogen to uranium.

---

[4] One rather odd feature is that the element copper and a few others appear in two places at once. This was corrected in later versions of Mendeleev's periodic tables.

Although this short form of the periodic table is perfectly adequate in many respects it was eventually replaced by an 18-column or medium-long form table. In this representation the transition elements beginning with scandium, or element 21, are removed from the main body of the short form table and are placed as a central block between the first two and subsequent six columns of the short-form table. The advantage of this format is that it reflects chemical periodicities more accurately than the short form does. For example there is a greater similarity among titanium, zirconium, and hafnium than there is between these elements and carbon, silicon, tin, and lead—all of which find themselves classified together in the short-form table. The modern 18-column or medium-long format table is shown below in figure 6.5.

One final adjustment can be made in order to improve the periodic table further. As the medium-long form table stands it features the elements called the lanthanides and actinides rather awkwardly as a disconnected footnote. In

H																	He
Li	Be											B	C	N	O	F	Ne
Na	Mg											Al	Si	P	S	Cl	Ar
K	Ca	Sc	Ti	V	Cr	Mn	Fe	Co	Ni	Cu	Zn	Ga	Ge	As	Se	Br	Kr
Rb	Sr	Y	Zr	Nb	Mo	Tc	Ru	Rh	Pd	Ag	Cd	In	Sn	Sb	Te	I	Xe
Cs	Ba	Lu	Hf	Ta	W	Re	Os	Ir	Pt	Au	Hg	Tl	Pb	Bi	Po	At	Rn
Fr	Ra	Lr	Rf	Db	Sg	Bh	Hs	Mt	Ds	Rg	Cn		Fl		Lv		

La	Ce	Pr	Nd	Pm	Sm	Eu	Gd	Tb	Dy	Ho	Er	Tm	Yb
Ac	Th	Pa	U	Np	Pu	Am	Cm	Bk	Cf	Es	Fm	Md	No

FIGURE 6.5 18-column, medium-long form periodic table. The lanthanide and actinide series appear as a disconnected footnote below the main body of the table.

H																															He
Li	Be																									B	C	N	O	F	Ne
Na	Mg																									Al	Si	P	S	Cl	Ar
K	Ca															Sc	Ti	V	Cr	Mn	Fe	Co	Ni	Cu	Zn	Ga	Ge	As	Se	Br	Kr
Rb	Sr															Y	Zr	Nb	Mo	Tc	Ru	Rh	Pd	Ag	Cd	In	Sn	Sb	Te	I	Xe
Cs	Ba	La	Ce	Pr	Nd	Pm	Sm	Eu	Gd	Tb	Dy	Ho	Er	Tm	Yb	Lu	Hf	Ta	W	Re	Os	Ir	Pt	Au	Hg	Tl	Pb	Bi	Po	At	Rn
Fr	Ra	Ac	Th	Pa	U	Np	Pu	Am	Cm	Bk	Cf	Es	Fm	Md	No	Lr	Rf	Db	Sg	Bh	Hs	Mt	Ds	Rg	Cn	113	Fl	115	Lv	117	118

FIGURE 6.6 32-column or long-form periodic system.

reality these elements are as much a part of the periodic system as are all the others. In order to encapsulate this fact the periodic system can be expanded further to produce the 32-column or long-form periodic table as displayed in figure 6.6.

3 A Little More History

Returning to the historical sequence of events, let me mention just a few of many momentous discoveries that occurred in physics at the turn of the twentieth century and which contributed to the physical explanation of the periodic system. In a period of three consecutive years starting in 1895, Wilhelm Röntgen discovered X-rays, Becquerel discovered radioactivity, and J. J. Thomson discovered the electron. Each of these achievements was to open up new vistas in the

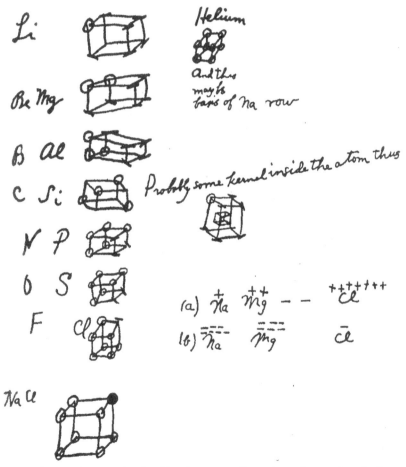

FIGURE 6.7 Lewis' 1902 sketch of the electron configurations of atoms from hydrogen to chlorine to explain chemical periodicity.

study of radiation and matter. Soon afterward Ernest Rutherford and many others pioneered the study of the structure of the atom.

A mere six years after the discovery of the electron, the chemist G. N. Lewis began to suggest how this particle could be used to explain the periodic system.

Lewis's idea is simple and ingenious. The number eight governs chemical periodicity in the short-form table that was prevalent at the time. In other words the properties of elements recur approximately every eight elements. Meanwhile each cube has eight corners. Lewis therefore suggested that on moving through the elements one adds an additional electron onto a corner of a cube until an octet is reached, whereupon the next cube begins to fill its corners with electrons. Elements that fall in the same group of the periodic table share the same number of outer cube electrons—for example, beryllium and magnesium, which each have two outer cube electrons, or boron and phosphorus that each have three outer cube electrons as shown in his sketch (see figure 6.7).

The notion of static electrons was soon demolished by the work of Ernest Rutherford. But Ernest Rutherford's model had little to say about the distribution of electrons around the nucleus, a task that was taken up by Bohr in his famous trilogy paper of 1913. Bohr's model located electrons in successive circular orbits, each specified by just one quantum number. By working backward from chemical behavior and spectral data Bohr succeeded in accommodating chemical periodicity and produced the table of configurations shown in figure 6.8.

We can stop to ask whether this represents a reduction of the periodic system. The answer would have to be very crudely yes, but the fact that Bohr accommodated already known experimental data and used that data to fix the configurations somewhat weakens the claim to any serious form of reduction of chemistry to quantum mechanics.

Subsequent developments over a period of about 10 years saw the introduction of three further quantum numbers to describe further degrees of freedom for each electron in an atom. The fourth quantum number, usually referred to as electron spin, was introduced by Pauli in 1924. By this point the reductive case became considerably stronger. The derivation of the quantization of various properties of the electron—such as its angular momentum and the possible values of the quantum numbers, as well as the manner in which the quantum numbers are all related to each other—were all derived from first principles of quantum mechanics, unlike what Bohr had carried out in the first incarnation of the theory in 1913. At this point it became possible to explain the sequence of numbers that governs the lengths of periods even in the most sophisticated representation of the periodic system, that is to say the long-form table.

The sequence of numbers to be explained is 2, 8, 18, 32, 50—or in general $2n^2$. These numbers drop out of quantum mechanics quite simply and represent periods involving the filling of s orbitals only for number 2; the filling of s and p orbitals for the number 8; filling of s, p, and d orbitals for the number 18; and the filling of s, p, d, and f orbitals for the number 32. Looked at from this perspective the reduction of the periodic system to quantum mechanics represents a tremendously successful enterprise.

1	H	1			
2	He	2			
3	Li	2	1		
4	Be	2	2		
5	B	2	3		
6	C	2	4		
7	N	4	3		
8	O	4	2	2	
9	F	4	4	1	
10	Ne	8	2		
11	Na	8	2	1	
12	Mg	8	2	2	
13	Al	8	2	3	
14	Si	8	2	4	
15	P	8	4	3	
16	S	8	4	2	2
17	Cl	8	4	4	1
18	Ar	8	8	2	
19	K	8	8	2	1
20	Ca	8	8	2	2
21	Sc	8	8	2	3
22	Ti	8	8	2	4
23	V	8	8	4	3
24	Cr	8	8	2	2	2

FIGURE 6.8 Bohr's electronic configurations of the first 24 atoms.
N. Bohr, On the Constitution of Atoms and Molecules *Philosophical Magazine*, 26, 476–502, 1913, 497

4 But Not Everything Is Derived from Quantum Mechanics

In spite of the apparent success offered by the reduction of the periodic system that has just been described, there is a remaining problem that caused authors such as myself to claim that the periodic system had not in fact been fully reduced. This concerns the simple fact that the lengths of all periods, apart from the first short period of two elements, repeat in terms of their length. The correct sequence of successive periods of elements in the medium, or as better displayed in the long-form, table, is 2, 8, 8, 18, 18, 32, 32,…

Nobody has yet succeeded in deducing this sequence from quantum mechanics although there are certain claims to having done so (Löwdin 1969, Allen and Knight 2002, Ostrovsky 2001). In terms of reduction I would now interpret the situation as an example of the fact that each successful step toward reduction raises further questions and new aspects that have yet to be reduced fully. But let me return to the facts of the periodic table and the above sequences of elements. The sequence of elements that denotes successive period lengths can be summarized by the following relatively simple mathematical expression,

$$L_n = \frac{1}{8}\left[2n + 3 + (-1)^n\right]^2$$

where L_n is the number of elements in any period with period number n (Kryachko 2007). However this is merely a mathematical trick, since there is no underlying physical theory that leads us to why this particular equation is required. More specifically the expression has not been derived from quantum mechanics, which is of course the putative reducing theory in this context.

The correct sequence is also generated by the Madelung, or $n + l$, rule, which states that the order of filling of atomic orbitals proceeds with increasing values of the $n + l$ quantum numbers for any particular atomic orbital (Scerri, 2009). Stated in other words, it predicts the following order of orbital filling:

1s < 2s < 2p < 3s < 3p < 4s < 3d < 4p < ...   (1)

The reductive claim on behalf of quantum mechanics therefore appeared to hinge on whether or not this expression had been derived from quantum mechanics. Some authors claim that indeed it had but I believe I showed that this was not in fact the case. This finding strengthened my former anti-reductionist stance regarding the periodic system. And this is where matters stood until 2012 at which time I began to read and understand the work of the theoretical chemist Eugen Schwarz, who coincidentally has been a frequent participant at our summer conferences of the International Society for the Philosophy of Chemistry.

5 The Current Situation: Anti-reductionism No Longer Seems So Clear-Cut

A simple way to express the gist of Schwarz's contribution to the debate would be to say that contrary to what is stated in 99 percent of all chemistry and physics textbooks, the sequence of orbitals shown above in expression (1) does *not* in fact represent the correct order of filling of atomic orbitals, except in the case of metals in the s-block of the periodic table, meaning the two columns on the far left of the periodic table (figure 6.9).

Although the above Madelung rule succeeds in giving the overall configuration of the transition metals beginning with scandium, or element 21, the order of filling is not provided by this rule. There is clear-cut experimental evidence to support this view, a fact that makes the persistence of the incorrectly used Madelung rule somewhat intriguing.[5]

------

[5] This is not something that I will pursue here but I have done so in other publications, such as in Scerri (2013).

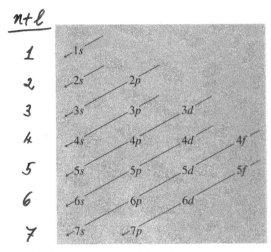

$$\frac{n+\ell}{}$$

FIGURE 6.9 A graphical representation of the Madelung or n + l rule. The order of filling of atomic orbitals is generally supposed to be provided by following the diagonal arrows starting at the top f the diagram with the 1s orbital and then moving downwards to give 2s, followed by 2p, then 3s etc.

Returning to the question of the reduction of the periodic system we must now ask a new question. Given the lack of fundamental status of the Madelung rule, is it still essential for this rule to be derived from quantum mechanics in order to consider that chemical periodicity has been fully reduced? The situation would appear to have changed. Admittedly, the use of the rule is still valid in trying to obtain the expected configuration of any atom but the fact that it fails to predict the precise order of filling of orbitals in any particular atom is a rather serious issue. For example, in the case of the atom of scandium, which has atomic number 21, the order of orbital filling is such that 3d orbital fills before a 4s orbital. This subsequently explains the fact that ionization of a scandium atom to form a Sc$^+$ ion invariably involves the removal of a 4s orbital electron. According to the usual (and incorrect) account whereby the 4s orbital fills preferentially in scandium, the preferential ionization of a 4s electron appears to be a complete mystery, which most books attempt to cover over with all manner of ad hoc maneuvers.

My previous ardent claims that the periodic system has not been fully reduced to quantum mechanics because of a lack of a derivation of the n + l rule has therefore fallen by the wayside. Mea culpa. I was wrong.

6 Another Twist: Those Anomalous Configurations— The Good Side of Reduction

There is another issue concerning reduction and electronic configurations, apart from the question of whether the n + l rule has been deduced from first principles. The anti-reductionist can appeal to the fact that even if the Madelung

1	2	3	4	5	6	7	8	9	10	11	12	13	14	15	16	17	18
H																	He
Li	Be											B	C	N	O	F	Ne
Na	Mg											Al	Si	P	S	Cl	Ar
K	Ca	Sc	Ti	V	*Cr*	Mn	Fe	Co	Ni	*Cu*	Zn	Ga	Ge	As	Se	Br	Kr
Rb	Sr	Y	Zr	Nb	*Mo*	Tc	*Ru*	*Rh*	*Pd*	*Ag*	Cd	In	Sn	Sb	Te	I	Xe
Cs	Ba	Lu	Hf	Ta	W	Re	Os	Ir	*Pt*	*Au*	Hg	Tl	Pb	Bi	Po	At	Rn
Fr	Ra	Lr	Rf	Db	Sg	Bh	Hs	Mt	Ds	Rg	Cn	113	Fl	115	Lv	117	118

FIGURE 6.10 Periodic Table showing gas phase anomalous configurations among d-block elements shown in smaller symbols. These seem to occur almost randomly throughout the d-block. The f-block is not displayed.

rule were derived in a fully convincing fashion there remains one important aspect that suggests that not everything has been reduced. There are approximately 20 elements whose atoms do not follow the Madelung rule in that their configurations are anomalous. For example, the atom of chromium would be expected to have a configuration of [Ar] $3d^4 4s^2$ according to the Madelung rule; yet experimental evidence points to its being [Ar] $3d^5 4s^1$. Figure 6.10 shows all the atoms, which behave in this anomalous manner in the sense that their outermost s orbital does not possess an s^2 configuration. The anti-reductionist can appeal to this behavior in order to maintain an anti-reductionist position. Until recently I would also have supported this view as a further argument against those who would claim that the periodic table has been reduced to quantum mechanics.

More recently I have become persuaded otherwise. It now appears that there is an intriguing explanation for these anomalous configurations, which is provided by quantum mechanics as well as an appeal to experimental data on the spectra of atoms.

7 How Is an Electronic Configuration Obtained from Experimental Data?

In this section we will examine the way in which the electronic configuration of any atom is obtained from spectral data and we will also examine an alternative way of obtaining electronic configurations.

First there is the traditional approach. This consists of examining the spectrum of the gas phase atoms of any particular element and looking for the spectroscopic term of lowest energy. One then tries to identify the electronic configuration which gives rise to this spectroscopic term and one takes this configuration to represent the ground state configuration of the atom in question. Consider for example the spectrum of neutral scandium. Figure 6.11 is a

Sc I

	Config.	Desig.	J	Level	Interval	Obs. g
	$3d\,4s^2$	$a\,^2D$	1½ 2½	0.00 168.34	168.34	0.79 1.20
1	$3d^2(a\,^3F)4s$	$a\,^4F$	1½ 2½ 3½ 4½	11520.15 11557.64 11610.24 11677.31	37.49 52.60 67.07	
2	$3d^2(a\,^3F)4s$	$a\,^2F$	2½ 3½	14926.24 15041.98	115.74	
	$3d\,4s(a\,^3D)4p$	$z\,^4F°$	1½ 2½ 3½ 4½	15672.55 15756.51 15881.76 16026.52	83.96 125.25 144.76	
	$3d\,4s(a\,^3D)4p$	$z\,^4D°$	½ 1½ 2½ 3½	16009.71 16021.78 16141.04 16210.80	12.07 119.26 69.76	
	$3d\,4s(a\,^1D)4p$	$z\,^2D°$	2½ 1½	16022.72 16096.86	−74.14	
3	$3d^2(b\,^1D)4s$	$b\,^2D$	2½ 1½	17012.98 17025.36	−12.38	
4	$3d^2(a\,^3P)4s$	$a\,^4P$	½ 1½ 2½	17918.85 17947.98 18000.25	29.13 52.27	
	$3d\,4s(a\,^3D)4p$	$z\,^4P°$	½ 1½ 2½	18504.06 18515.77 18571.40	11.72 55.63	
	$3d\,4s(a\,^1D)4p$	$z\,^2P°$	½ 1½	18711.08 18855.76	144.73	
5	$3d^2(a\,^1G)4s$	$a\,^2G$	4½ 3½	20237.10 20239.92	−2.82	
	$3d\,4s(a\,^1D)4p$	$z\,^2F°$	2½ 3½	21032.78 21085.84	53.06	
6	$3d^2(a\,^3P)4s$	$a\,^2P$	½ 1½	[21400]+y 21480.40+y	80.40	
	$3d\,4s(a\,^3D)4p$	$y\,^2P°$	{½ 1½}	24656.80		
	$3d\,4s(a\,^3D)4p$	$y\,^2D°$	1½ 2½	24866.18 25014.15	147.97	0.82 1.17

Config.	Desig.	J	Level	Interval
$3d^2(a\,^3F)4p$	$y\,^4D°$	½ 1½ 2½ 3½	32637.40 32659.21 32696.84 32751.54	21.81 37.63 54.70
$3d^2(a\,^3F)4p$	$z\,^2G°$	3½ 4½	33056.19 33151.40	95.21
$3d^2(a\,^3F)4p$	$y\,^2F°$	2½ 3½	33154.01 33278.64	124.63
$3d^2(a\,^3F)4p$	$x\,^2D°$	1½ 2½	33615.06 33707.25	92.19
$3d^3$	$e\,^4F$	1½ 2½ 3½ 4½	33763.57 33798.68 33846.62 33906.40	35.11 47.94 59.78
$3d\,4s(a\,^3D)5s$	$e\,^4D$	½ 1½ 2½ 3½	34390.25 34422.85 34480.05 34567.10	32.60 57.20 87.05
$3d\,4s(a\,^3D)5s$	$e\,^2D$	1½ 2½	35671.00 35745.57	74.57
$3d^3$:	$f\,^2D$	1½ 2½	36276.76 36330.49	53.73
$3d^3$	$e\,^4P$	½ 1½ 2½	36492.82 36515.76 36572.80	22.94 57.04
$3d^2(b\,^1D)4p$	$w\,^2D°$	1½ 2½	36984.15 37089.77	105.62
$3d\,4s(a\,^3D)4d$	$e\,^2P$	½ 1½	37085.72 37148.25	62.53
$3d^2(b\,^1D)4p$	$w\,^2P°$	1½ ½	37086.81 37125.72	−39.41
$3d^2(a\,^3P)4p$	$x\,^4D°$	½ 1½ 2½ 3½	 37486.48 37553.34 37717.11	66.86 163.77
$3d\,4s(a\,^3D)4d$	$g\,^2D$	1½ 2½	37780.83 37855.50	74.67
$3d^2(a\,^3P)4p$	$z\,^4S°$	1½	38179.92	

FIGURE 6.11 An extract from Charlotte Moore's tables of atomic energy levels for the neutral scandium atom. The electronic configuration is generally taken to be whichever configuration gives rise to the spectroscopic term with lowest energy. In this case it is the $3d^1\,4s^2$ configuration (Moore, 1970).

copy of part of the spectrum of neutral scandium for which the spectroscopic term of lowest energy, shown as .00 in the fourth column, originates from the configuration $3d^1\,4s^2$ as shown in the first column. In most treatments of electronic configurations this is the end of the story.

However in more accurate work one seeks the average configuration which is obtained by taking an average of the energies of all the spectroscopic terms arising from each of the lowest lying electronic configurations of any atom (Wang et al. 2006).

In the spectrum shown in figure 6.11 this involves taking an average of all the terms originating from the $3d^1\,4s^2$ configuration and comparing this energy with the average value for all the spectroscopic terms arising from the $3d^2\,4s^1$ configuration—of which there are 15 terms in this case. Moreover the manner in which this averaging is carried out requires making use of the J, or overall quantum, number that is the result of coupling the total orbital angular momentum of the atom L with its total spin angular momentum or S.

In physical terms this represents a move from considering gas phase atoms to atoms in condensed phases, meaning in the liquid or solid states. It also represents a move away from isolated atoms to atoms that are in chemical combination. Broadly speaking both of these changes mean that one is dealing

Configuration averages in scandium

d¹s²

Term	J	E / cm⁻¹	2J+1	E_weighted	Degeneracy	E_average / cm⁻¹	E_average / eV
²D	1.5	0.00	4	0.00	10	101.00	0.012523
	2.5	168.34	6	1010.04			

E_average rel. s¹ / eV
-1.99

d²s¹

Term	J	E / cm⁻¹	2J+1	E_weighted	Degeneracy	E_average / cm⁻¹	E_average / eV
4F	1.5	11519.99	4	46079.96	90	16165.20	2.00
	2.5	11557.69	6	69346.14			
	3.5	11610.28	8	92882.24			
	4.5	11677.38	10	116773.80			
2F	2.5	14926.07	6	89556.42			
	3.5	15041.92	8	120335.36			
2D	2.5	17012.76	6	102076.56			
	1.5	17025.14	4	68100.56			
4P	0.5	17226.04	2	34452.08			
	1.5	17255.07	4	69020.28			
	2.5	17307.08	6	103842.48			
2G	4.5	20236.86	10	202368.60			
	3.5	20239.66	8	161917.28			
2P	0.5	20681.43	2	41362.86			
	1.5	20719.86	4	82879.44			
2S	0.5	26936.98	2	53873.96			

E_average rel. s¹ / eV
0.00

FIGURE 6.12 Calculation of average configuration energies arising from various possible configurations in scandium atom.

with more physically and chemically relevant species than gas phase atoms of isolated atoms, thus providing further motivation for taking these alternative configurations more seriously.

Figure 6.12 presents the results of calculating the lowest lying configuration of the scandium atom, when carried out via this averaging procedure. In the case of this atom the ground state configuration is the same, namely $3d^1 4s^2$, regardless of whether it is obtained in the traditional manner or by this averaging procedure. In fact the $3d^1 4s^2$ configuration is found to be almost exactly 2 eV more stable than the next configuration of $3d^2 4s^1$.

But this is not true in all cases. Figure 6.13 shows the variation in the energies of the s^2, s^1 and s^0 configurations for each atom beginning with calcium and ending with copper. Clearly, as atomic number increases the energy of the s^2 configuration shows an increase relative to that of the s^1 configuration. Whereas the s^2 configuration is considerably more stable for elements such as scandium, the energies of these configurations cross over each other once the atom of iron has been reached. In the case of the nickel atom the s^1 configuration is found to be approximately 1 eV lower than the s^2 configuration. These results imply that the ground state configurations for several atoms are different from what they are generally regarded as according to the traditional approach (Figure 6.14).

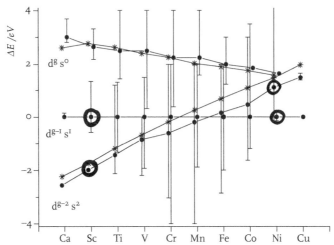

FIGURE 6.13 Variation in energies of s^2, s^1 and s^0 configurations across the first transition series. For iron and elements beyond, the s^1 configuration is more stable than s^2. The atoms of scandium and nickel have been highlighted with circles. In the case of scandium the average s^2 configuration lies lower than the s^1 configuration. In the case of the nickel atom the situation is the other way round. Bold circles represent configurations obtained experimentally from spectral evidence. The crosses represent configuration energies obtained theoretically via the Hartree-Fock method. (Wang et al, 2006).

H																	He
Li	Be											B	C	N	O	F	Ne
Na	Mg											Al	Si	P	S	Cl	Ar
K	Ca	Sc	Ti	V	Cr	Mn	*Fe*	*Co*	*Ni*	*Cu*	Zn	Ga	Ge	As	Se	Br	Kr
Rb	Sr	Y	Zr	Nb	*Mo*	Tc	*Ru*	*Rh*	*Pd*	*Ag*	Cd	In	Sn	Sb	Te	I	Xe
Cs	Ba	Lu	Hf	Ta	W	Re	*Os*	*Ir*	*Pt*	*Au*	Hg	Tl	Pb	Bi	Po	At	Rn
Fr	Ra	Lr	Rf	Db	Sg	Bh	Hs	Mt	Ds	Rg	Cn	113	Fl	115	Lv	117	118

FIGURE 6.14 Periodic Table showing anomalous configurations for condensed phase atoms (italicized symbols). Again the f-block is not displayed.

Returning to the theme of this chapter, this alternative method for calculating electronic configurations of atoms provides a perfectly natural explanation for the so-called anomalous configurations. It could be argued that there are in fact no anomalies since one is merely observing the result of the variation of two energies, those of the s^2 and s^1 configurations, which happen to cross at a certain point along the first transition series. Moreover, and perhaps more pertinent to the present project, the energies of these configurations can be computed from first principles via the Hartree-Fock method and they too show very similar trends, including a crossing of energies at more of less the same point along the transition series.[6]

Conclusions

This study illustrates the fact that the philosophers of chemistry, especially those influenced by the prevailing anti-reductionist zeitgeist, can too easily conclude from episodes in the history of chemistry and physics that reductionism fails. This is in fact how my own work until recently can be characterized, although I like to think I have been consistent in adopting a cautious approach and have stayed close to the scientific facts.

In the case under discussion one can only conclude that the periodic table has not been reduced to quantum mechanics if one restricts oneself to gas phase atoms rather than atoms in condensed phases. But this represents a serious omission since the condensed phases are overwhelmingly more relevant to most of chemistry and physics. Second, there is the related fact that by restricting oneself to gas phase atoms one is only considering isolated atoms of the elements rather than atoms that are bonded and present in compounds.

Third, let me make a general comment about philosophy of science which has been made many times before but which is especially pertinent in the present case. There is a tendency for philosophers of science to obtain their knowledge of science from textbooks, which inevitably present impoverished accounts of the particular fields that they are describing. Of course it is not difficult to understand this tendency, which stems from the fact that philosophers are generally not sufficiently technically proficient to cope with the latest research on the subject and prefer to fall back on the version of the science that they themselves learned during their earlier scientific education.

In the case under discussion in this chapter, experts in chemistry and physics are well aware of the limitations of focusing on the configurations of gas phase atoms. Meanwhile philosophers of science build their anti-reductionist cases on idealist textbook accounts of what electronic configurations really consist of. Neither the n + l rule nor the so-called anomalous configurations of atoms apply in general. The n + l rule applies strictly to just the s-block elements which constitute about 10 percent of all the known elements.[7]

---

[6] The only disagreement would seem to be over the atom of manganese. According to spectral data the s configuration has a lower average energy than the s configuration while according to the theoretical prediction the energy order is reversed.

[7] The precise value depends on how one performs such a calculation. I have calculated the proportion by considering the 13 s-block elements as a percentage of the 118 currently known elements. Others might want to exclude the super-heavy elements in which case the percentage of s-block elements will be a little higher.

The electronic configurations that are generally believed to be anomalous because they feature an incomplete outer s-orbital are seen in a completely new light when one turns to considering the average configuration of atoms taken over all spectroscopic terms that emerge from any particular configuration (figure 6.14). Any claims to the failure of reductionism that are based on the n + l rule or the anomalous configurations are thus rendered invalid. The reduction of the periodic table to quantum mechanics is far more successful than contemporary philosophers of chemistry have been willing to admit.

Perhaps the best way to think of reduction might be as a "direction" rather than as a goal. Each attempt to explain chemical and physical phenomena such as the anomalous electronic configurations generally results in a deeper understanding of the phenomena. The goal of complete reduction may never be reached but scientific knowledge continues to advance, whatever some uninformed philosophers might believe.

References

Allen, L.C., Knight, E.T. (2002). "The Lowdin challenge: Origin of the n + l, n (Madelung) rule for filling the orbital configurations of the periodic table," *International Journal of Quantum Chemistry*, 90, 80–88.

Dirac, P.A.M. (1929). "Quantum mechanics of many-electron systems," *Proceedings of the Royal Society of London A* 123 714–733; quote on p. 714.

Gordin, M. (2004). *A Well-Ordered Thing, Dimitrii Mendeleev and the Shadow of the Periodic Table*. New York: Basic Books.

Heitler, W. (1927). Letter from Heitler to London, September (Fritz London Archives, Duke University).

Kaji, M., Kragh, H., Pallo, G. (eds.). (2015). *Early Responses to the Periodic System*. New York: Oxford University Press.

Kryachko, E.S. (2007). *International Journal of Quantum Chemistry*, 107, 372–373.

Löwdin, P.-O. (1969). "Some comments on the periodic system of the elements," *International Journal of Quantum Chemistry*, IIIS, 331–334.

Moore, C. (1970). *Selected Tables of Atomic Spectra*. Washington, DC: National Bureau of Standards, National Standard Reference Series.

Nagel, E. (1961). *The Structure of Science. Problems in* The Logic of Explanation. New York: Harcourt, Brace & World.

Oppenheim, P., Putnam, H. (1958). "Unity of science as a working hypothesis," *Minnesota Studies in the Philosophy of Science*, 2, 3–36.

Ostrovsky, V.N. (2001). "What and how physics contributes to understanding the periodic law," *Foundations of Chemistry*, 3, 145–181.

Reichenbach, H. (1978). *Selected Writings: 1909–1953*, M. Reichenbach and R.S. Cohen (eds.). Dordrecht: Reidel.

Scerri, E.R. (1994a). "Has Chemistry Been at Least Approximately Reduced to Quantum Mechanics?" In *Philosophy of Science Association Proceedings*, volume 1 (D. Hull, M. Forbes, and R. Burian, eds.). East Lansing, MI: The Philosophy of Science Association, 160–170.

Scerri, E.R. (1994b). "It All Depends What You Mean By Reduction." In *From Simplicity to Complexity, Information, Interaction, Emergence*. Proceedings of the 1994 ZiF Meeting in Bielefeld (K. Mainzer, A. Müller, and W. Saltzer, eds.). Bielefeld, Germany: Vieweg-Verlag, 77–93.

Scerri, E.R. (2007). "Reduction and emergence in chemistry, proceedings of the PSA 2006 Meeting," *Philosophy of Science*, 74, 920–931.

Scerri, E.R. (2007). *The Periodic Table, Its Story and Its Significance*. New York: Oxford University Press.

Scerri, E.R. (2009). "The dual sense of the term 'element,' attempts to derive the Madelung rule and the optimal form of the periodic table, if any," *International Journal of Quantum Chemistry*, 109, 959–971.

Scerri, E.R. (2012). "Top down causation regarding the chemistry—physics interface—A skeptical view," *Interface Focus,* Royal Society Publications, 2, 20–25.

Scerri, E.R. (2013). "The trouble with the Aufbau Principle," *Education in Chemistry*, 50 (November), 24–26.

Scerri, E.R. (2014). "Master of missing elements," *American Scientist*, 102, 358–365.

Schummer, J. (1997). "Scientometric studies on chemistry I: The exponential growth of chemical substances, 1800–1995," *Scientometrics*, 39(1), 107–123.

Van Spronsen, J. (1969). *The Periodic System of the Chemical Elements, A History of the First Hundred Years*. Amsterdam: Elsevier.

Wang, S.G., Qiu, Y.X., Fang, H., Schwarz, W.H.E. (2006). "The challenge of the so-called electron configurations of the transition metals," *Chemistry A European Journal*, 12, 4101–4114.

| Contingencies in Chemical
Explanation

NORETTA KOERTGE

Introduction

"Chemistry has a position in the center of the sciences, bordering onto physics, which provides its theoretical foundation, on one side, and onto biology on the other, living organisms being the most complex of all chemical systems" (Malmström et al.). Thus begins a recent essay on the development of modern chemistry.[1] Philosophers have long wrestled with how best to describe the exact relationship between chemistry and physics. Is it an example of a classic reduction? But before we ask whether chemistry could in principle be derived from physics, there is a prior question: How well integrated is the science of chemistry itself? This chapter argues that although there is a coherent explanatory core within chemical theory, contingency plays a larger role than is usually recognized. Furthermore, these phenomena at the boundaries of traditional chemistry education are where some of the most important current research is occurring. I will first adopt a quasi-historical approach in this essay, including anecdotes from my own educational trajectory. I then briefly discuss how our current understanding of the explanatory structure of chemistry should be reflected in education today.

1 The Explanatory Core of Traditional Chemistry

The professor of quantum chemistry at the University of Illinois in the 1950s told us a story from his PhD defense. His director, Linus Pauling, walked into the room and said something to this effect: "Well, Karplus, you've done a bunch of calculations on the hydrogen molecule ion (H_2^+). Very nice. But you claim to

---

[1] Many thanks to Nicholas Best for suggesting useful references and also his remarks about how what he called contingent chemistry figured in the curriculum in New South Wales, Australia.

be a *chemist*. So please write the Periodic Table on the board for us." Who knows exactly what point Pauling was actually trying to make, but it reminds us of this basic point. The periodic table with its horizontal and vertical trends is still the basis of the classification of enormous amounts of information about the formulae and properties of chemical compounds. Mendeleev would not have understood talk of strontium-90, but he would have realized immediately that this product of nuclear testing would enter the body in a manner similar to calcium.

In retrospect, the transition from a classification system based on atomic weight to one based on atomic number may *look* completely seamless, although Paneth reminds us of early debates about whether each isotope should now be considered to be a different chemical element (Paneth 1962). Mendeleev's concern about the three reversed pairs was instantly resolved. More importantly, Mendeleev had puzzled over the question of how one could ever give a deep explanation of chemical regularities using only the concept of weight and Newton's laws (Mendeleev, 1905). But now there was a way forward to explain the myriad empirical generalizations connected with the periodic table in terms of charged subatomic particles. One could now account for the similarities between members of the same vertical column, such as Ca and Sr, and horizontal trends in composition, such as CH_4, NH_3, H_2O, and HF. Eventually, there was even a non–ad hoc explanation of the strange family of rare earths, which had led Mendeleev to despair that they had "broken" his periodic table (Mendeleev 1905).

Although Pauling may have poked a little fun at quantum mechanical studies of super-simple arrays, such as H_2^+ or the bifluoride ion, his monumental textbook *The Nature of the Chemical Bond and the Structure of Molecules and Crystals* used the quantum mechanical description of electron orbitals to add a whole new dimension to the periodic table. We could now explain why an H_2O molecule was angular while a CO_2 molecule was linear. At first the existence of compounds such as $Cr(NH_3)_6Cl_3$ where precise numbers of neutral ligands surrounded metallic ions was a surprising addition to the simple patterns canonized in the periodic table. But with the development of Crystal Field Theory, which also relied on quantum chemistry, one could make sense of both the geometry of coordination compounds and their spectra. Throw in a little thermodynamics and as my professor of inorganic chemistry put it, "With the periodic table, you don't need to make a handbook out of your head!"

But then I took organic chemistry and we students were confronted with the multiplicity of possible ways to synthesize secondary amines with little guidance as to which method might work in a particular case. (This is an ongoing area of research according to Salvatore 2001.) Although one might create a reaction mechanism once a synthesis was known to work, there seemed few ways to predict ahead of time how to make even relatively simple organic compounds.

Chemists have always been interested in synthesizing new compounds with unusual properties (such as the mineral acids) as well as preparing pure samples in bulk of naturally occurring materials. Johann Juncker and Stahl also emphasized the epistemic value of always following analysis by synthesis (Koertge 1969). They took the idea from mathematics, where the analysis of, say, a geometric figure was a method of discovery, which then had to be followed by synthesis, the derivation from axioms. Another influential example was Newton; after analyzing white light with a prism, he recombined the separate colors to produce white (Koertge, Analysis, 1980). Lavoisier's famous experiments on oxygen illustrate what Stahl would have called "completing the proof" by not only decomposing mercuric oxide but also synthesizing it (Koertge 2010). (One is also reminded of Friedrich Wöhler's synthesis of urea, although the reasoning there is more complicated. When his experiment was used as an argument against vitalism it was significant that a *synthesis* from inorganic materials was required. Failure to find a vital component upon analysis was not sufficient (Cohen 1996)). With today's sophisticated analytical techniques, the ability to synthesize a material is less important as a double check. However, the project of predicting the path of reactions remains an important theoretical challenge, especially in industrial operations such as the "cracking" of petroleum to make gasoline, where catalysts play such an important role.

The above anecdotes/examples remind us of how a myriad of facts about thousands of individual chemicals can be organized into families and trends, using the periodic table supplemented with quantum chemistry. Furthermore, the explanatory core of modern chemistry alluded to so far applies universally. The chemical properties of methane found in ancient ice cores are the same as those of methane isolated from the burps of cows today and would be the same as samples brought back from an exoplanet. Contrast the situation in biology or geology, where knowing the origins of a particular organism or formation is essential to understanding it. That is why philosophers of science often contrast the types of explanation offered in the physical sciences, said to involve only universal laws, from explanations in the historical sciences, said to be contingent on initial conditions. But as we will now see, some properties of chemical materials also depend on their history!

2 Recognizing the Role of Contingency

We begin again with an example from carbon chemistry, but we will soon see that similar issues arise with inorganic materials. As students we were fascinated to learn about the amazing physical differences between graphite and diamonds and it was very gratifying to relate those differences to the underlying molecular structures—the familiar tetrahedral bonds in the case of diamonds and the somewhat more complicated hexagonal layers of graphite. But what was not explained at all was how those differing structures arose—other than vague allusions to geological processes involving high temperatures and pressure. And why was it so difficult to make synthetic diamonds of a respectable size and transparency? A recent paper provides a theory about the intermediate steps ("Trouble" 2011). But so far the honest answer seems to be that the process is very sensitive to complicated boundary conditions and although we may eventually succeed in replicating those conditions, we don't know how to describe them. There has recently been much more practical success in synthesizing other allotropic forms of carbon, popularly known as buckyballs and nanotubes. However, once again the product you get is contingent on very specific boundary conditions ("Fullerenes and 'Buckyballs'" 2014).

Not all allotropes are so mysterious. Red and white phosphorus have been known for centuries and in this case we know not only their molecular structure, but also the conditions for their formation. But at the other extreme are the allotropes of sulfur. They are of less practical importance, but from a theoretical point of view they are more numerous and even more difficult to characterize than is the case with carbon ("Allotropes of Sulfur" 2014).

Chemists have known of the existence of allotropes since the discovery of oxygen and ozone in the early nineteenth century. Allotropes provide an excellent way to teach students that sometimes knowing the chemical composition of a material is not enough to explain all of its properties; sometimes we also need to describe its atomic or molecular structure. (Other familiar examples are Cis-trans isomers and even optical isomers, although in the latter example fewer properties depend on structural differences.) None of this is news, but what we sometimes forget is how these phenomena complicate the explanatory structure of chemistry. The original periodic table dream was to be able to give at least rough predictions of the properties of new materials simply based on their chemical composition, similar to the way Mendeleev predicted the existence and properties of the three so-called missing elements. But allotropes and other instances of polymorphism remind us that the path by which a chemical is synthesized can have dramatic effects on the resultant product.

Another class of materials not easily subsumed under the traditional explanatory structure centered on the periodic table are the clathrates. Methane clathrate, which now plays an important role in climate science, has been assigned a molecular weight of 957, corresponding to a formula of $(CH_4)_8(H_2O)_{46}$, often simplified as $CH_4 \cdot 5.75H_2O$. Methane molecules are trapped within cages formed from dodecahedral and tetrahedral arrays of water molecules ("Methane Hydrate-I Mineral Data" 2014). Methane clathrates are formed in the sea under cold conditions and high pressure. They can also solidify in pipelines carrying natural gas if water vapor is present.

In the preceding cases, the substances in question all have a fixed composition and structure, which is typically known. Gaps in our understanding of them have to do with the conditions and processes by which they are formed. In the case of polymers, however, there is also tremendous variability in the formula weights of the products themselves and theory is often of little help in either characterizing them or discovering methods of producing materials with the desired properties.

Alloys present another interesting challenge to chemical theory. One might naively suppose that you could just mix metals in any desired proportion and that the properties of the product, such as hardness, would tend to vary linearly.

In actuality, not every composition of the constituents leads to a stable alloy. Too much or too little of a constituent may lead to the precipitation of so-called impurities, which may in fact strengthen the alloy. Metallurgists describe the crystal and lattice structures of alloys and even describe the phase changes they undergo as they cool. See figure 7.1 below (Chakrabarti and Laughlin 2004).

Alloys do not fit neatly into our usual categories of mixture versus pure substance or homogeneous versus heterogeneous solids. Their physical properties depend on their "molecular" structure, which depends on their composition, which can vary, but only within limits.

So-called non-stoichiometric compounds such as the mineral wüstite (mainly ferrous oxide) also have a composition that varies, but over a small range. Despite its variable composition, it has a definite crystal structure; it crystallizes in the isometric-hexoctahedral crystal system. It is not just a mixture of ferrous and ferric oxides. Berthollet argued that such compounds violated Dalton's Law of Definite Proportions. Dalton won as far as introductory textbooks are concerned, but Berthollet gets a nod in solid state chemistry today. It is also interesting to note that petroleum engineers characterize the size of keragen particles before and after fracking in terms of Daltonian atomic mass units, which can run into the tens of thousands (Kleinberg 2011). Unlike the wüstite example, these variable huge numbers are just the result of combining the molecular weights of the variety of hydrocarbons in petroleum.

Colloids are another example of familiar materials that are very difficult to integrate into the explanatory structure of traditional chemistry. Colloids occur

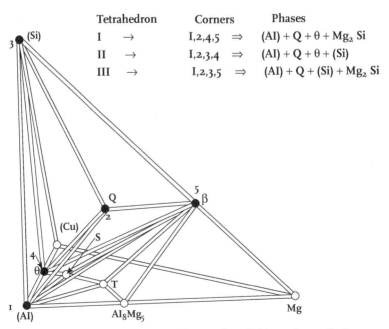

FIGURE 7.1 Line diagram of stable equilibrium phase fields in Al–Mg–Si–Cu system at room temperature.

naturally (e.g., in milk, blood, and smoke) and also have important commercial uses (e.g., paints, gelatin, Styrofoam). A historically important example is so-called cranberry glass, in which colloidal particles of gold are deposited in molten glass resulting in a red color. The basic idea in colloid formation is simple—small particles must somehow be suspended in a supportive medium. Chemical knowledge can be relevant to the prediction of the stability of the colloid and its properties. For example, is the material to be suspended in water hydrophilic? Once again, trial-and-error craftsmanship is necessary to discover the contingencies of preparation that result in the desired product.

Recent developments in climate science have focused attention on the formation conditions and properties of aerosols. For example, soot sent up into the atmosphere from charcoal stoves and older coal-powered power plants will sometimes act as a greenhouse gas, thus contributing to global warming, but it can also have a cooling albedo effect, as do clouds. A recent article studying the catalytic effect of transition metal ions in the clouds on the role of sulfur dioxide emissions raises the following provocative question:

> But should we consider aerosols such as smoke and soot to lie within the domain of chemical phenomena? Typically it is physicists or engineers who study the dynamics of particles in a fluid. Yet so-called sulfate aerosols are formed by a variety of chemical reactions within the atmosphere, chemical information that is relevant to their effects on climate.
>
> (HARRIS ET AL. 2013)

Some of the above examples of familiar chemical materials, whose structure and properties depend on the contingent conditions under which they are formed, pose interesting research problems. In other cases the details of these processes are so idiosyncratic that it seems unlikely that there will ever be a general explanatory theory of their chemistry. (A somewhat similar prognosis undoubtedly obtains at the interface of chemistry and biology, but we have not surveyed that territory.) So why is this situation of philosophical interest? Even the most optimistic defenders of the project of reducing chemistry to physics would have conceded that all they were envisaging was a successful reduction *in principle*—no one said that it would be easy in practice. But what the above argument claims is that there are chemical phenomena that cannot be fully explained within traditional chemical theory. So even if systematic chemistry *were* reduced to physics, important chemical knowledge would be left behind. More important, however, is the realization that there are important research opportunities at the boundary between chemistry and physics, areas that are now being studied by geologists, metallurgists and engineers. And we need to ensure that we prepare future scientists to explore these borderlands.

3 Implications for Chemistry Education

To improve science education at the K-12 level in American schools, a consortium of scientists and science educators has recently proposed a New Generation of Science Standards (NGSS). The NGSS are based on a framework developed by the National Research Council and have received favorable reviews in *Science*, so they provide at least some indication of new directions in science education. What are the implications for chemistry? Two general observations: First, standards for chemistry are blended into a category called physical science, which is treated on a par with biological science and earth science. Perhaps not surprisingly a disproportionate number of the concepts that students are expected to learn in physical science come from introductory physics. Some critics are concerned that high schools will be increasingly likely to steer students into an integrated physical science class and perhaps even stop offering separate chemistry classes.

The second general feature of the NGSS may have a more interesting impact on chemical education. Although we call the new curriculum plans *science* standards, they are explicitly designed to promote proficiency in STEM (Science, Technology, Engineering, Math) subjects. Not surprisingly, the progression of so-called Disciplinary Core Ideas (DCI) is intended to be aligned with the new national standards for mathematics education known as the Common Core. The emphasis to be placed on engineering and technology, however, is quite new. In fact, engineering concerns or activities are recommended for every single one of the DCI. The inevitable result of this new conception of science education when applied to chemistry will be an increased emphasis on the sort of contingency circumstances discussed earlier. For example, a reviewer of a draft of the new standards objected when the DCI for molecular weight included table salt as a typical simple compound whose molecular weight students should be able to calculate. He pointed out that in this case NaCl represents the *formula* weight—there are no separate molecules. Insisting on this distinction is becoming more common (Molecules and moles 2014). It was also suggested that from the beginning students should learn the difference between ionic solids (e.g., salt) and networked solids (e.g., quartz). Whether or not one believes all of these distinctions should be introduced in high school, they reflect the growing emphasis on topics that used to be relegated to chemical engineering.

A similar broadening of the chemistry curriculum at the college level is illustrated by the undergraduate program at my university. In the Indiana state university system, agriculture and engineering are only taught at Purdue, but Indiana University has strong programs in the sciences. Our chemistry department has six divisions: In addition to analytic, organic, inorganic, and physical, there are chemical biology and materials science. The partition between chemical biology (within Chemistry) and the Department of Biochemistry is historical in origin and we need not pursue it, but the recognition of materials science as a field of concentration for chemistry undergraduates reflects the growing importance of this kind of research.

An essay surveying a century of Nobel Prizes in chemistry points out that although a majority of the prizes have been given for research in organic chemistry, many recent prizes have been awarded for contributions that "lie close to industrial applications, for example, those in polymer chemistry." Prizes in the first decade of the twenty-first century show a similar pattern (Malmström et al. 2014).

A challenge for chemical education today is to prepare students to do research in areas that border on physics, biology, and geology while retaining a focus on and appreciation of the traditional explanatory core of chemistry.

4 Implications of Contingent Chemistry for Philosophy of Science

Most the examples of what might be called *contingent chemistry* discussed above are not new discoveries. So why have they been largely ignored? One answer is obvious: All of us, but especially theoreticians and philosophers of science, like tidy, unified systems of knowledge. Who could not be impressed when Euclid with only five axioms rationalized the myriad geometrical truths that carpenters and surveyors had discovered empirically? But perhaps admirers of chemistry are especially prone to be defensive about the explanatory power of modern chemistry, given its inevitable comparisons with physics.

Consider the following potted history of theories of chemical elements. The *Oxford English Dictionary* definition of chemistry as "the branch of science concerned with the substances of which matter is composed, the investigation of their properties and reactions, and the use of such reactions to form new substances" would have been heartily endorsed by chemists and alchemists for millennia. Aristotle proposed four elements (earth, air, fire, water) that he hoped could account for the properties and natural motions of terrestrial materials, such as smoke and steam, blood and bile. (A fifth element was invoked for celestial bodies.) Alchemists introduced three additional principles (salt, sulfur, mercury) because they were interested in explaining phenomena such as acidity, combustion, and the striking differences between metals and other solids. All seven of these fundamentals of matter were "philosophical" elements, which were not expected to ever be isolated. The properties of different ordinary materials were to be explained in terms of different proportions of the elements, which could vary continuously. Lavoisier (following Boyle) proposed an operational definition of element as the last product of analysis and proposed a table of 23 elements, most of which we still recognize today. By the time of Mendeleev, roughly a century later, the list was three times as long.

Let's pause for a moment to note what theoretical havoc would have resulted if chemists of Mendeleev's time had tried to predict and explain the properties of materials by assuming, as Aristotle and earlier chemists may well have, that all elements could combine in all proportions. That was *not* the historical situation, of course, for two important reasons. First, Dalton with his Atomic Theory and Law of Definite Proportions had essentially quantized the elements. They could not combine in continuously varying proportions. Second, experience showed that no element could combine with all the others. Many chemists started to propose low-level generalizations about families of elements with similar properties and Mendeleev's periodic table incorporated all these regularities and more. As we have discussed, when quantum mechanics started providing an account of the nature of the chemical bond, it's not surprising that chemists would not place high priority on trying to gain a theoretical understanding of phenomena at the boundaries of chemistry. (Industrial chemists, of course, had different agendas.)

As quantum mechanics started providing a foundation for the periodic table philosophers of science became fixated on the question of whether all of chemistry could be reduced to physics. There was broad agreement that this was possible in principle, but wide-ranging disagreement about exactly what concept of *reduction* was applicable. Soon philosophers of biology were also enticed by the reductionist program (Brigand et al. 2012). If genetics could be reduced to biochemistry and chemistry could be reduced to physics, the result would be a beautiful, multilayered cognitive structure with physics as the foundation. However, enthusiasm for reducing molecular biology to chemistry has waned. The biology of organisms alive now or in the past is obviously contingent not only on unpredictable large-scale geological changes (e.g., the Cambrian extinction and subsequent explosion) but also on small-scale mutations to the DNA. Recently, biologists interested in what is sometimes called "evo-devo" have shown how the DNA of offspring can be methylated by contingencies in the environment both in the womb and after birth so even the backbone of the "Central Dogma" is not invariant. As Mitchell notes, even "Mendel's classical law of 50:50 segregation will hold just as long as the genetic and environmental conditions persist which render 50:50 segregation adaptive" (Mitchell 2000, 260). And Cat shows how even the supposedly pristine domain of physics is sprinkled with phenomena that aren't easily explained with the usual natural laws because of complications due to contingent conditions (Cat 2005).

I have argued in this chapter that we chemists and philosophers of chemistry could learn from biology to pay more attention to contingencies in our field. Past geological conditions have affected the structure of gems and minerals in interesting ways reminiscent of the situation in biology. There are regularities to be described and explained, but they don't fit neatly into our core chemical conceptual scheme. And even in laboratory or industrial settings we cannot always control, predict, or even explain after the fact exactly what is happening—these reactions are contingent on details of the environmental conditions.

References

http://www.nextgenscience.org/ (2014) (accessed August 8, 2014).

"Allotropes of Sulfur." (2014). http://www.nuffieldfoundation.org/practical-chemistry/allotropes-sulfur (accessed August 6, 2014).

Brigand, Ingo, and Love, Alan. (2012). *Reductionism in Biology*. http://plato.stanford.edu/entries/reduction-biology/#3.1 (accessed August 8, 2014).

Cat, Jordi. (2005). "Modeling cracks and cracking models: Structures, mechanisms, boundary conditions, constraints, inconsistencies and the proper domains of natural laws," *Synthese*, 146, 447–487.

Chakrabarti, D. J., Laughlin,David E. (2004). "Phase relations and precipitation in Al–Mg–Si alloys with Cu additions," *Progress in Materials Science* (Elsevier), 49, 389–410.

Cohen, P. S., Cohen, S. M. (1996). "Wöhler's synthesis of urea: How do the textbooks report it?" *Journal of Chemical Education*, 73, 9, 883–886.

"Fullerenes and 'Buckyballs'." (2014). http://www.statesymbolsusa.org/Texas/Buckyball_Molecule.html (accessed August 6, 2014).

Harris, Eliza; Sinha, Bärbel; van Pinxteren, Dominik; Tilgner, Andreas; Wadinga Fomba, Khanneh; Schneider, Johannes, et al. (2013). "Enhanced role of transition metal ion catalysis during in-cloud oxidation of SO2," *Science*, 340, 727–730.

Kleinberg, Robert. (2011). *The Marcellus Shale: Science and Technology.* February 15, 2011. http://www.cornell.edu/video/the-marcellus-shale-science-and-technology (accessed August 8, 2014).

Koertge, Noretta. (1969). "The General Correspondence Theory." PhD dissertation. University of London.

Koertge, Noretta. (1980). "Analysis as a Method of Discovery." In *Scientific Discovery: Case Studies* (Thomas Nickles, ed.). Dordrecht: Reidel, 21–49.

Koertge, Noretta. (2010). "How Should We Describe Scientific Change?" In *Discourse on a New Method* (Mary Domski and Michael Dickson, eds.). Chicago: Open Court Press, 511–522.

Malmström, Bo. G., and Andersson, Bertil. (2001). *The Nobel Prize in Chemistry: The Development of Modern Chemistry.* http://www.nobelprize.org/nobel_prizes/themes/chemistry/malmstrom/index.html (accessed August 8, 2014).

Mendeleev, Dimitri. (1905). *Principles of Chemistry.* 7th Russian Edition, 3rd English Edition. Translated by G. Kamensky. London: Longmans, Green and Co.

"Methane Hydrate-I Mineral Data." (2014). http://webmineral.com/data/Methane%20hydrate-I.shtml#.U-Oi_ohgOq4 (accessed August 7, 2014).

Mitchell, Sandra D. (2000). "Dimensions of scientific law," *Philosophy of Science*, 67(2), 242–265.

"Molecules and Moles." (2014). http://chemistry.about.com/od/atomicmolecular structure/a/moleculesmoles.htm (accessed August 8, 2014).

Paneth, F. A. (1962). "The Epistemological Status of the Concept of Element." *British Journal for the Philosophy of Science*, 13, 1–14, 144–160.

Pauling, Linus. (1960). *The Nature of the Chemical Bond and the Structure of Molecules and Crystals.* Ithaca, NY: Cornell University Press.

Salvatore, R. N., Yoon, C. H., Jung, K. W. (2001). "Synthesis of secondary amines," *Tetrahedron*, 57 (37, 10 September), 7785–7811.

"The Trouble with Turning Graphite into Diamond." (2011). http://www.technology review.com/view/422350/the-trouble-with-turning-graphite-into-diamond/ (accessed August 6, 2014).

CHAPTER 8 | Reaction Mechanisms

RICHARD M. PAGNI

AFTER I HAD BEEN in graduate school for several months, I decided to work in a group that studied photochemical reactions, those that are initiated by ultraviolet or visible light. My research advisor was interested in discovering new reactions and deducing experimentally how they occurred—the reaction mechanism. I remember my first group meeting where the topic of discussion was the ring-opening of cyclobutenes (compounds with four carbon atoms in a ring opening up to form compounds with no rings). I still recall people describing the potential ways in which the ring-openings occurred as domino and antidomino (today called conrotatory and disrotatory). Thermally induced reactions, that is, those initiated by heat, occurred one way and photochemically induced reactions, the other. Even though these reactions had been studied thoroughly, the reaction mechanisms were considered incomplete because nobody could explain the dichotomy between the thermally- and photochemically induced reactions. Why these reactions occurred in the manner they did was unknown. When quantum mechanical explanations were later proposed to explain the ring-opening reactions, the reaction mechanisms might be said to be complete, although the related question of the cause of the reactions—in other words, why they happen at all—still had to be addressed. Reaction mechanisms consist in more than merely knowing the pathways by which reactants are converted into products. Until all related questions are answered satisfactorily, a mechanism may be considered incomplete.

1 Historical Background

The evolution of modern chemistry from its origin in the late eighteenth century to its present day power and sophistication is remarkable (Brock 1992, Greenberg 2000, Bensaude-Vincent and Simon 2008, Chalmers 2011). Space limitations preclude more than a very brief summer of this history; references are included for the interested reader.

Oxidation was weaned from its alchemical origins and the first elements and gases synthesized during the early decades of modern chemistry (Smartt

Bell 2005, Thorpe 2007, Holmes 2008, Jay 2009). Additional elements were then discovered and their properties measured and compared (Scerri 2007). Methodology, laboratory technique, and apparatus were developed to carry out these new tasks, primarily in the nineteenth century (Faraday 1960, Buckingham 2004). After compounds began to appear during this period, structural theory began slowly to develop (Rocke 2010), including that of stereochemistry (Eliel et al. 1994, Debrè 1994, Wagnière 2007, Buchwald and Josefowicz 2010). Reactions began to appear in the middle of the nineteenth century and have continued to blossom ever since. Physical chemistry, which included the new fields of thermodynamics and kinetics, came of age in the latter part of the nineteenth century (Laidler 1993, Masel 2001). At the dawn of the twentieth century the existence of atoms, molecules, and reactions was well established. When this degree of sophistication had been attained, it was possible to study reaction mechanisms in detail; this began in earnest during the 1920s. These burgeoning studies were facilitated by developments in theory (Nye 1993), instrumentation (Reinhardt 2006), spectroscopy (Harris and Bertolucci 1978), crystallography (Giacovazzo 2002), and computers. Model building and paper chemistry, which originated in the nineteenth century, became powerful tools for drawing structures and their reactions simply (Hoffmann and Laszlo 1991, Klein 2001). There is hardly a chemical publication, lecture, or informal discussion today that does not contain simplified pictorial representations of molecules. It is now possible to study virtually any reaction mechanisms in detail (Lowry and Richardson 1987, Maskill 1985, Reichardt 2003, Anslyn and Dougherty 2006).

2 Philosophical Considerations

2.1 Scientific Realism

Even though most chemists cannot articulate a philosophy of chemistry, I believe that a large majority of them are realists (Carpenter 2000, Harré 1970). This is due to the fact that chemists make and manipulate chemical substances. What does a realist believe? There are many types of realism, but the following definition will suffice. Realism posits the view that there is a real and orderly universe, one independent of human beings and minds, and one that is amenable to objective and unbiased study and understanding by the interplay of observation, experimentation, hypothesis, and *theory*. Scholars in many disciplines have voiced their opinions about scientific realism. Some scholars are concerned with the human element and others the scientific method and its results and conclusions including questions related to ontology, epistemology, and semantics (Chakravartty 2007).

There are those such as Steven Weinberg, the American theoretical physicist, who are strongly committed to realism, including its most arcane, yet successful, theories, and believe that the "theory of everything" is within our grasp (Weinberg 1993), while other physicists are more sanguine about this prospect (Gleiser 2010, Rosen 2010). There are many individuals who do not believe that the scientific method yields genuine knowledge (Brown 2001, Feyerabend 2010, Giere 2010, Haack 2003, Kagan 2009, Marks 2009). Science in some views is socially constructed, or it is the purview of white males, or it is subjective rather than objective, or it is a product of the West, or it is perspectival rather than objective, or it is too metaphysical. There are still other points of view. Anti-realists do not believe that unobservable entities have a place in science, especially those that are undetectable by our senses or instrumentation (van Fraassen 1980, Chakravartty 2007). In their view electrons *may* exist because they can be detected in cloud chambers or galvanometers, whereas the undetectable virtual photons, believed to be the carriers of electrical and magnetic forces, do not. Other philosophers of science believe that theories, being metaphysical statements about the universe, should play no part of science; only empirically derived knowledge can legitimately have a place in science. People with this perspective are called positivists or empiricists, the conservatives of the scientific realm.

There are many reasons why some philosophers of science and scientists find theorizing invalid (Leplin 2001), but most chemists are not among them. Chemists, being realists, believe in the existence of atoms, molecules, gases, and reactions and their pathways or mechanisms. In what follows the realist position will be presented. Nonetheless, it is doubtful if there are many chemists who share Weinberg's belief that a simple "theory of everything" will be found. How can we fully understand the universe in which we are embedded? How can our limited senses, even including instruments that enhance them, explain in detail how the universe functions? How can we be certain that a perfectly functioning "theory of everything" is correct?

Before proceeding further, there are two additional things that the reader should keep in mind. First, the following discussion deals with organic reaction mechanisms although inorganic or organometallic reaction mechanisms could have been considered. Second, the narrative which follows describes one way in which a chemist might create a philosophy of reaction mechanisms (Berson 2003, Hoffmann 2007). There are undoubtedly others.

2.2 Reaction Mechanisms

There are two words in "reaction mechanism." The first word—reaction—is easy to explain. The definition of a chemical reaction is not problematical because various sources define it similarly as a chemical transformation or change. A chemical reaction might also be described as the conversion of one or more substances (reactants) under a given set of conditions including pressure, temperature, phase, and so forth into one or more other substances (products), with a substance made up of atoms, molecules or ions (species with positive or negative charge). A more formal definition of a reaction may be found in Schummer (1997). The chemist, of course, must identify the reactants and products to be certain that a reaction has occurred.

The definition of mechanism is more difficult because it depends on which scientific discipline one is considering. An anthology on the philosophy of science describes mechanism as "the view that all phenomena can ultimately be explained in terms of cause and effect" (Boyd et al. 1997). Cause-and-effect has historically been a thorny philosophical problem, but it can be argued that it is an important, but incomplete, descriptor of chemical mechanism. The role of cause and effect will be discussed later; for now the discussion will concentrate on mechanism phenomenologically which involves identifying the reactants, intermediates, transition states (see the next section on the S_N2 reaction and mechanism), and products, and numerous other features. Explanations for these features also come later.

The idea of mechanism has received attention in recent years from a number of philosophers of science who seek a universal definition of the concept even though their examples come almost exclusively from physics and the biological sciences (Glennan 1996, 2002a, 2002b, 2008, 2010; Machamer et al. 2000; Tabery 2004; Woodward 2002). Whether a universal definition is possible is an open question, however. Little attention has been devoted to reaction mechanisms in spite of the fact that they have been topics of interest for over a century (Carpenter 2000; Del Re 2003; Goodwin 2003; Ramsey 2004, 2008; Stemwedel 2006). Let me begin by seeing whether the ideas of philosophers of science can be applied to a simple reaction mechanism and then to other examples from chemistry. Most of the universal attributes of mechanism, as you will see, can be applied to reaction mechanisms.

2.3 The S_N2 Reaction and Mechanism

The reaction of the iodide anion (I^-) with methyl bromide (CH_3Br; CH_3 = methyl) in water (solvent) yields the bromide anion (Br^-) and methyl iodide (CH_3I), an example of an S_N2 reaction (Substitution, Nucleophilic, 2nd order kinetics), first promulgated in its modern form during the 1930s by Edward D. Hughes and Christopher K. Ingold (Akeroyd 2000a, 2000b). In this reaction the bromine atom in methyl bromide is replaced by an iodine atom—the substitution. The electron-rich iodide anion is the nucleophile or nucleus seeker; the leaving group, the bromide anion in this instance, is the nucleofuge. Because the

$$I^- + CH_3Br \xrightarrow{\text{water}} CH_3I + Br^-$$

reaction kinetics or rate is 2nd order overall, first order in each reactant, the reaction is believed to occur in one step. One could imagine a multi-step mechanism, and chemists have (Sneen and Larsen 1966, 1969), but the alternative is more complex and less believable than the one-step mechanism (Hoffmann et al. 1996). Based on the stereochemistry of similar reactions (Walden 1896), the reaction occurs by backside attack of the iodide anion on the carbon atom of the methyl bromide and the departure of the bromide anion from the other side of the carbon atom. A single energy barrier separates reactants from products, with the highest energy point between reactants and products called the (very short-lived) transition state. The name is apropos because the highest energy species is transitional between reactants and products. Note that in the transition state the carbon atom is bonded or partially bonded to five other atoms (3 H's, 1 I, and 1 Br). Based on the Hammond Principle (Hammond 1955), a chemist arrives at a picture of the transition state even though this entity is never seen. The transition state has a structure which is a weighted average of the entities which generate it, the reactants, and the entities it produces, the products. The transition state structure will resemble more closely the species to which it is closer in energy, which in this case are the reactants. Chemists believe that the structure of the transition state is reasonably accurate; based on this picture, researchers can make accurate predictions of future chemical behavior of this reaction, for example by changing the solvent from water to acetone.

From this evidence, the reaction mechanism for this S_N2 reaction is written in the following manner (Figure 1):

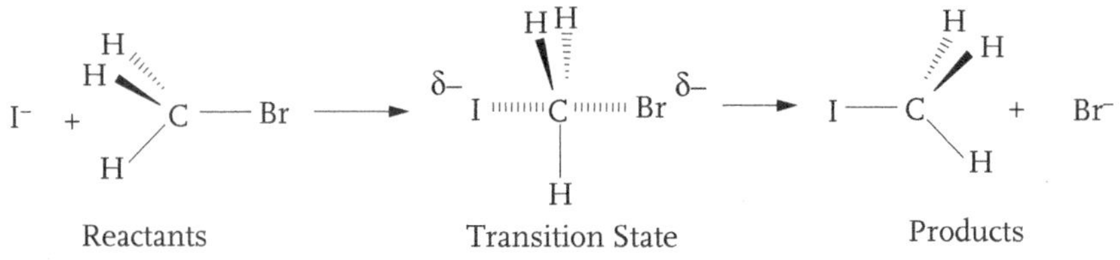

Reactants   Transition State   Products

FIGURE I

In the pictorial representation of the reaction, the straight lines represent single bonds in-plane from carbon to hydrogen, iodine, and brome; the filled triangles represent single carbon-hydrogen bonds projecting above the plane; dashed lines single carbon-hydrogen bonds to the rear of the plane; and straight dashed lines as partial single bonds from carbon to iodine and bromine in the transition state. The δ-symbols indicate that the transition state is negatively charged and the incoming iodide and outgoing bromide are negatively charged. The resulting picture is a two-dimensional representation of a three-dimensional reaction; the picture tells the chemist that methyl bromide and methyl iodide have tetrahedral structures, while that of the transition state trigonal bipyramid.

Some features of the reaction are not shown in the mechanism. The reactants and products are stable compounds (the bromide and iodide anions as salts). The transition state is not stable, but must be considered an entity because it is an integral part of the reaction. A moment's thought shows that there must be an essentially infinite number of species between reactants and products based on the degree of bond-making (carbon to iodine) and bond-breaking (carbon to bromine). It is not clear how a philosopher of science would categorize these ephemeral species which the chemist believes exist by inference. Every entity in the reaction interacts with solvent, especially the bromide and iodide anions; this is called solvation. The iodide and bromide anions are partially bonded to five or six water molecules in solution. Some water must be removed from the vicinity of I⁻ for it to undergo reaction with the methyl bromide, for example. The reaction mechanism shown above is static, unlike what exists in the reaction medium. The species are rapidly translating, rotating, and vibrating, all of which may influence the course of the reaction. The iodide anion, for instance, must collide with methyl bromide with the proper orientation for reaction to occur, a low probability event. Even if the reactants collide with the proper orientation, they must do so with enough energy so that the transition state of higher energy is reached. Most correctly oriented collisions do not have sufficient energy for reaction to occur, another low probability event. Although it is not necessary in this case, oxygen and water from the atmosphere are ordinarily excluded from reaction media because they often interfere with the reaction being studied. Reactions are usually carried out in a reaction vessel, often while being stirred, and perhaps with external heating. Do these external features have to be considered for a reaction mechanism to be fully described?

Another way to describe a reaction and its mechanism is with a reaction coordinate diagram (Figure 2). Here potential energy or Gibbs free energy, neither of which will be defined here, is plotted on the y axis and a reaction in progress is plotted on the x axis. It is important to know that energy increases as one goes up the y axis. The reaction coordinate under consideration, the x axis, shows the forming carbon-iodine bond and breaking carbon-bromine bond which are co-linear with one another. As we look from left to right along the x axis, we are looking at the reaction as it occurs as a function of energy. There are numerous other coordinates that could have been chosen for the x axis, but only the one involved in bond making and bond breaking is appropriately chosen. This process is often called a one-dimensional trajectory. You can see from the diagram below that energy must be provided to the iodide anion and/or methyl bromide, most likely by collisions with solvent molecules, in order for the transition state to be attained. When the transition state yields products, this energy—and more—is returned to the solution. This is due to the fact that the products are lower in energy than that for the reactants.

Keep in mind that this is a powerful, but simplified, view of a multidimensional process, and other processes such as quantum mechanical tunneling and bifurcation may be important (Carpenter 1992, 2000; Collins 2013, 2014).

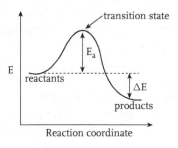

FIGURE 2

Based on the voluminous literature on the S_N2 reaction of the last 160 years, it is possible to write a generalized mechanism for this reaction, as shown (Figure 3):

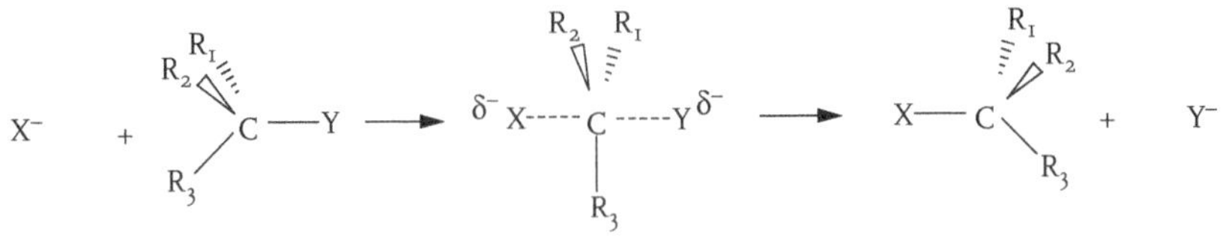

FIGURE 3

where X^- is the nucleophile, Y^- is the nucleofuge, and R_1, R_2, and R_3 are atoms or groups of atoms on one of the reactants, the transition state and one of the products. The nucleophile and nucleofuge need not have a negative charge; ammonia (NH_3) may serve as a nucleophile, for example. The Rs may be hydrogen, alkyl groups such as methyl (CH_3) or ethyl (CH_3CH_2), or other things. Although not shown in the above model, the choice of solvent will influence the speed with which the reaction occurs. In the reaction of iodide with methyl bromide, where the nucleophile is negative charged, the reaction is faster in acetone than in water. Chemists have also spent considerable time deducing which species will function as nucleophile and nucleofuge. The iodide anion is both a good nucleophile and nucleofuge, while the hydroxide anion (OH^-) behaves as a nucleophile but not as a nucleofuge. Chemists have proposed various explanations for the behavior of nucleophiles and nucleofuges. Chemists have also examined how varying the number and structure of the R groups affects the reaction. When all three R groups are alkyl groups, for example, a substitution reaction may occur but will do so by a different mechanism called the S_N1 reaction in which case the kinetics and stereochemistry are different than that observed for the S_N2 reaction. Explanations have been proposed on why the number of alkyl groups attached to the substrate influences whether an S_N2 or S_N1 reaction occurs. Furthermore, when the structures of the substrate are sufficiently complex, competing reactions with names such as the E1 and E2 reactions may also occur. Chemists want to explain all of these features. When they have done so, they can use the reaction in synthesis with confidence. Chemists believe that genuine progress has been made.

Unlike physics, chemistry rarely deals with laws except in situations that are very physical in nature such as the ideal gas law. (The Hammond Principle might be such a chemical law.) The above picture of the S_N2 reaction does not represent a chemical law, probably because it is not sufficiently general or inclusive. Not all reactions occur by the S_N2 mechanism; in fact very few of them do. All gases, on the other hand, obey the ideal gas law, at least under conditions of low temperature and pressure. The S_N2 reaction is a reaction type, one of hundreds, whose pictorial representation might be described as a model of the reaction. The generalized mechanism might be described as the theory of the reaction.

Before moving forward, there are several others issues to keep in mind. A reaction mechanism is never seen and its features only deduced by inference based on a series of experiments. Just as chemical structures are represented by a simple pictorial notation, the overall mechanism is as well. Most features of a reaction mechanism are deduced from the domain of physical chemistry and its kinetics, thermodynamics, isotope effects, and quantum mechanical calculations.

2.4 Philosophical Views of Mechanism

Mechanism has been a topic of concern to philosophers of science for some time. Glennan has discussed this topic in two recent review articles (Glennan 2008, 2010). Two contemporary ideas of potential universal applicability seem of relevance to reaction mechanisms, "parts and their interactions" (Glennan 1996, 2002a, 2002b) and "entities and activities" (Machamer et al. 2000). A third idea is that mechanism requires both interactions and activities (Tabery 2004).

Glennan, the author of the "parts and interactions" concept, has described a mechanism for a float valve which regulates the water level in a toilet tank. The valve consists of a hollow ball which, owing to its buoyancy, floats on the water in the tank. The hollow ball has an arm that is connected to an on-off switch behind which there is water under pressure. When the water level in the tank is at the fill point, the float and its arm are in a position which keeps the switch at the off position. When the toilet is flushed, a flap at the bottom of the tank opens releasing the water in the tank into the bowl at which point the water in the tank and the float valve drop. The arm attached to the hollow ball now turns the switch to its on position which then discharges water into the tank. Because the flap at the bottom of the tank is now closed, the tank fills with water until the fill point is reached at which time the float arm returns the water switch to its off position. There are thus several parts that work in concert to make the device function. Note the importance of time and volition in the operation of the float valve. This mechanism and in fact all mechanisms operate through time. Without someone flushing the toilet, the act of volition, the float valve is a non-functioning mechanical device. Note in addition that the float valve is a product of human invention that has no counterpart in the natural world. Each part of the valve operates in a way consistent with laws of physics such buoyancy, density, fluid dynamics, and gravity. These laws in turn are described in terms of simpler, more fundamental laws. When the hierarchical process is carried back to nature's fundamental laws, the reductive process stops because we cannot explain from whence these laws come. Glennan also shows that a voltage switch with no moving parts nonetheless operates analogously to the float valve, that is, through the interaction of its parts. Glennan believes that his causal mechanism is universally applicable to all domains of science including chemistry. Mechanism is intimately connected to causality in this picture and operates via well-established laws of physics.

How does this view of mechanism fit into chemistry and its mechanisms? If one considers the dissolution of a solute in a solvent, the picture works well, with the solute and solvent being the parts and the solvation of the solute by the solvent the interaction. Chemists in general will have no difficulty in seeing atoms, molecules, transients, and even transition states as parts but may have difficulty in seeing reactions as interactions. In the S_N2 reaction the interaction would presumably be between the two reactants colliding *prior* to their reacting with one another. In a unimolecular reaction, on the other hand, what is interacting with what? In a gas phase reaction, is it the collision of reactant with the reaction vessel, or in solution, is it the collision of reactant with solvent molecules? Perhaps the interaction would have to be one part of a molecule interacting, that is, reacting, with another part of the molecule such as in a sigmatropic shift (see the first reaction directly below; Figure 4). Bond breaking is not necessary for a reaction to occur, however. The interconversion of gauche and anti-conformers of butane (see the second reaction directly below; Figure 5) is a unimolecular reaction that does not break or form bonds. Can one thus equate interaction with reaction? I do not believe so. Even more troublesome is the idea that physical laws are the basis for all mechanisms. This may be true in the world of physics, but not necessarily so in chemistry. There are very few laws in chemistry that do not arise from physics. Even though Glennan's idea works well in physical situations, it does not work as well for chemical phenomena. Further clarification is in order.

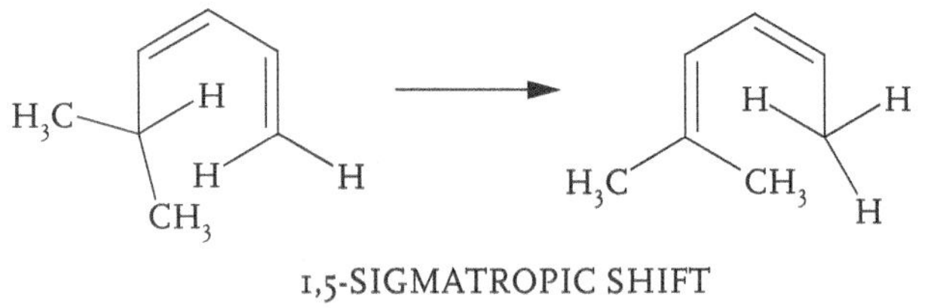

1,5-SIGMATROPIC SHIFT

FIGURE 4

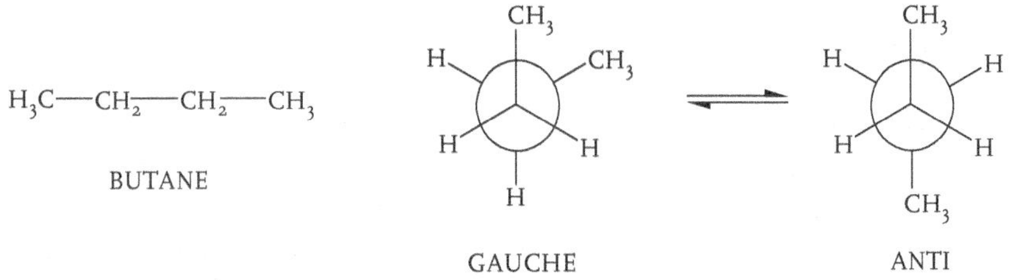

BUTANE

GAUCHE   ANTI

FIGURE 5

Machamer, Darden, and Craver (MDC), who are interested in protein synthesis and electrical transmission between neurons, both complex chemical phenomena and each with its own mechanism, describe mechanisms in terms of entities and activities. They define mechanism as "entities and activities organized such that they are productive of regular changes from start or set-up to finish or terminal conditions." In terms of reaction mechanism, the entities would constitute the reactants, products, transients, transition states, and products and their respective properties; the activities, the producers of change, would be the individual reactions. According to MDC a mechanism must be regular (reproducible), productive (conversion of reactants to products), and temporal, all concepts consistent with a chemist's view of reaction mechanisms. The MDC concept of mechanism is well suited for describing reaction mechanisms including the S_N2 reaction, the sigmatropic shift, and the anti-gauche interconversion of butane conformers. The reason for this may be the fact that, if their definition of mechanism works on things as complex as neurotransmission and protein synthesis, it will work on systems of less complexity, especially if the less complex systems are also chemical in nature. This is a top-down approach based on complexity. Glennan's mechanism, on the other hand, is a bottom-up approach which works well for physical systems, but less well for chemistry, a less fundamental and more complex discipline. Shortly, however, Glennan's concept will be important in explaining the mechanism of a very complex biochemical system.

Tabery believes that a complete description of mechanism requires a blend of the two previously described mechanisms. A system where he believes that both ideas are required is in the initial steps in photosynthesis. As is well known, the first step is the absorption of photons by the chloroplast of the plant which is followed by a series of exquisite electron transfer reactions whose ultimate outcome is to make chemical energy in the form of ATP. Electromagnetic energy has been converted into chemical energy. A complex series of beautiful chemical reactions occurs here. There are also many three-dimensional chemical interactions in the overall chemistry as well. The three-dimensional order that one encounters in the photosynthetic apparatus is not ordinarily encountered for simple reactions in solution. There are features here where one concept is preferable and other features where the other is preferred. Tabery concludes that this system requires both interpretations of mechanism. The chemical reactions are best described by MCD's entities and activities, while the overall process from photon to chemical bond may be imagined as a sophisticated naturally occurring mechanical device, best described by Glennan's parts and interactions.

In the fundamental discipline of physics, I contend, "parts and interactions" is all that is needed to describe a physical mechanism. In chemical reactions, be they in the gas phase, solution, the crystalline phase, amorphous regimes, and on surfaces, "entities and activities" is better descriptor of what is occurring. In more complex biochemical reactions both concepts may be required to adequately describe a mechanism.

There are situations in chemistry where both ideas are also needed, for instance, in the anodic oxidation of a compound at an electrode surface. The electrochemical apparatus, which consists of an anode, cathode, standard electrode, salt bridge and various kinds of electronic devices, "parts and interactions" is the best descriptor of the physical apparatus. The chemistry brought about by the apparatus is still best described in terms of "entities and activities." Photochemical reactions behave in the same manner. Here there is a physical device to generate the photons which impinges on a chemical and initiates a photochemical reaction.

There are other situations in which it is not clear what role "parts and interactions" plays in chemistry. Reactions are usually run in glass vessels, often with stirring, in the absence of oxygen, and at a specific temperature. Chemicals are often slowly added to the vessel via dropping funnels. Do these addends constitute part of the chemical system? If so, should one deem their contribution to the overall chemistry in terms of "parts and interactions"?

2.5 Reaction Mechanisms, Chemical Concepts, and Explanatory Power

Most chemists are not interested in how reaction mechanisms fit in with mechanisms in other disciplines. This is not to say that they are not interested in generality, but only within their own field. What chemists seek is a (minimum) number of fundamental concepts that can be applied to all known reactions. When new chemical phenomena are discovered, new concepts may have to be created if the older concepts are inadequate. Application of these concepts to reactions and their mechanisms yields explanatory power. This power can be used to carry out new examples of old reactions and create entirely new reactions. Explanatory power is perhaps the most important thing that a chemist wishes to have (Goodwin 2003, Del Re 2003, Ramsey 2008). Experienced chemists can propose a sequence of reactions that is almost certain to give the desired product, be it a compound of theoretical interest, a natural product with a very complex structure, or a prescription drug. Experienced chemists create new and useful reactions or discover new ways of carrying out old reactions. Experience may be equated to an intimate knowledge of both fundamental principles and the repertoire of known reactions. Because the fundamental concepts do not apply in every case, or are applied in a different order, there never will be a unified theory of reaction mechanisms. There will always be new reactions, new compounds, and new functional groups to be discovered, and new concepts to be created.

How do these concepts arise? Consider the S_N2 reaction. Primary alkyl halides, those containing a single group such as methyl, generally undergo the reaction in the presence of good nucleophiles (and nucleofuges). It is possible, however, to hinder the nucleophile from attacking the backside of the primary carbon by appropriately substituted "large" groups. If the group is made sufficiently large, the S_N2 reaction will not occur. This is due to a steric effect. The concepts of nucleophile, nucleofuge and steric effect are concepts applicable to many reactions other than the S_N2 reaction.

New reactions are often discovered when a desirable outcome cannot be attained with available methods. OH^- is a poor nucleofuge in the S_N2 reaction. There are many circumstances when a synthetic pathway requires that OH^- function as a nucleofuge. An old scheme will carry out this chemistry in two distinct steps. A new scheme involving the Mitsunobu reaction will carry out the chemistry in one step. The mechanism of the Mitsunobu reaction is readily understood in terms of a few fundamental concepts and reactions.

Let me further illustrate how concepts are used to understand a reaction mechanism by examining an example of electrophilic aromatic substitution (EAS), representing a vast number of reactions many of which were discovered in the nineteenth century. The terms *electrophilic, aromatic,* and *substitution* are fundamental concepts in chemistry. The example of EAS that I wish to examine in a little detail is the nitration of toluene (methyl-substituted benzene), a so-called aromatic compound, using a mixture a nitric acid (HNO_3) and sulfuric acid (H_2SO_4) to form three organic products, *ortho-, meta-* and *para*-nitrobenzene, and water (Figure 6).

In this reaction hydrogen on the aromatic ring is replaced (substituted) with nitro (NO_2), thus the origin of the term nitration. Nitration of aromatic compounds has been studied extensively for over a century because the reactions are easy to carry out in numerous ways and many of the products have military and societal utility.

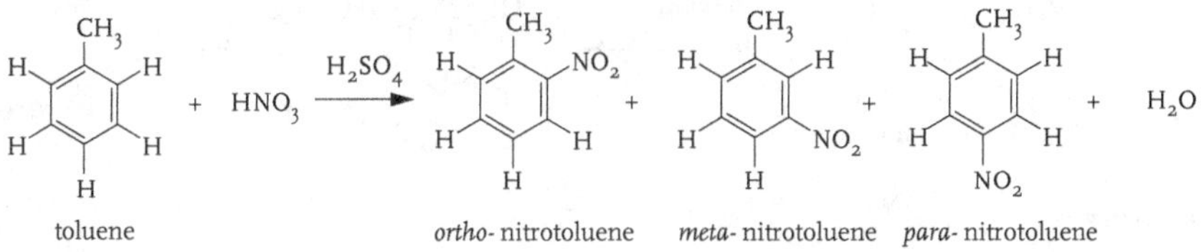

toluene   ortho- nitrotoluene   meta- nitrotoluene   para- nitrotoluene

FIGURE 6

What follows is the well-established reaction mechanism for the formation of the para product; the ortho and meta products are formed similarly (Figure 7). Four reactions are required to go from reactants to product: (1) the reversible protonation of nitric acid by sulfuric acid, the catalyst in the reaction; (2) the reversible heterolytic cleavage of protonated nitric acid to form water and the nitronium ion, NO_2^+, a very reactive electrophile; (3) the irreversible attack of the nitronium ion at the para position of toluene to form an intermediate called a sigma complex; (4) the irreversible removal of the proton at the para position of the sigma complex by a base to form the product which now also has an aromatic ring. Below each reaction is a short description of what is occurring and how the chemicals are behaving. I will not explain the terminology in detail as this will divert the discussion away from the point I wish to make, in other words, there are many universally applicable concepts that can be applied to reaction mechanisms. The interested reader is referred to Orchin et al. (2005) for definitions of the various terms and concepts mentioned below. Let me state again that the terminology is quite general and has been applied to many other reactions. The behavior of the reactants is also general and has also been applied to other reactions. Sulfuric acid typically behaves as a proton donor, for example. Exceptions do occur, however. Nitric acid normally functions as an acid, but here behaves as a base.

$$HONO_2 + H_2SO_4 \rightleftharpoons H_2ONO_2^+ + HSO_4^-$$

(1) reversible reaction; (2) acid-base reaction; (3) sulfuric acid – catalyst;

(4) sulfuric acid - Brønsted-Lowry acid; (4) nitric acid – base

$$H_2ONO_2^+ \rightleftharpoons H_2O + NO_2^+$$

(1) reversible reaction; (2) acid-base reaction; (3) heterolytic cleavage; (4) NO_2^+ - reactive intermediate, transient; (5) NO_2^+ - Lewis acid; (6) NO_2^+ - electrophile; (7) NO_2^+ - isoelectronic with carbon dioxide

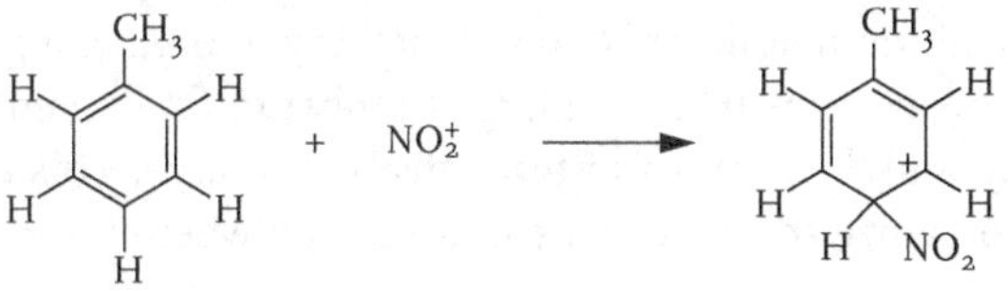

(1) para attack; (2) irreversible reaction; (3) rate-determining step; (4) Lewis acid-base reaction; (5) sigma complex; (6) resonance structures; (7) resonance effect; (8) inductive effect; (9) electron-donating methyl group

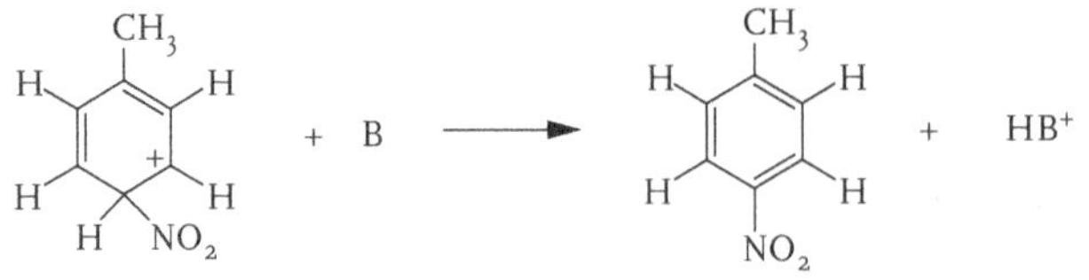

FIGURE 7

(1)B – base; (2) irreversible reaction; (3) Brønsted-Lowry acid-base reaction; (4) regenerate an aromatic compound

Two other features of this reaction illustrate other observed chemical effects: (1) *regiochemistry* where NO_2^+ preferentially attacks the ortho and para positions of toluene over the meta; (2) *relative reactivity* where toluene is more reactive in the nitration reaction than is benzene (C_6H_6), the parent aromatic compound, under the same reaction conditions. Both the regiochemistry and relative reactivity can be rationalized on the basis of what chemists call resonance and inductive effects in the three sigma complexes derived from toluene.

Terms such as *acid, base, Bronsted-Lowry, Lewis, electrophile, nucleophile, reversible, irreversible, rate-determining, resonance, inductive, intermediate, transient, steric*, and many others occur repeatedly in chemists' explanations of why reactions occur the way they do. In other chemical disciplines such as organometallic chemistry other concepts such as oxidation addition and reductive elimination are useful in understanding their reaction mechanisms.

Unlike these terms which are mainly qualitative, some terms have numerical values. Solutions of acids can be described by their pH or H_0, the acidity function. Acid strengths are often compared by compounds' pK_a values. In the case of the nitration of an aromatic compound, its kinetics is often measured as a function of acid strength of the reaction medium, that is, pH, which can be varied by combining various proportions of sulfuric acid and water.

Explanatory power also entails explaining why reactions do not occur in certain ways. Why is less *meta*-nitrotoluene formed in the above nitration reaction than the ortho and para nitro products? When the chemist has rationalized what has occurred and what has not, s/he is satisfied. Nonetheless, chemists delight in finding exceptions to the rule. The mechanism for the nitration of aromatic compounds is quite secure because it has been studied so extensively and yet there are many exceptions.

2.6 Organization of Reactions and Reaction Mechanisms

The reader may think that there is no rational way of organizing reactions and their mechanisms because there is a large number of fundamental concepts to know and apply, and the number of reactions is very large. This is not the case. Chemists can organize their reactions in a scheme where the mechanisms go from the most general characterization to more and more specific characterizations. Chemists do not ordinarily write out schemes such as the one shown below (Figure 8), but they are there in their minds. There are many such schemes. A chemist may also organize reactions based on whether the mechanisms involve reactive intermediates such as cations, anions, and uncharged free radicals.

SUBSTITUTION

Free Radical

Thermally-Induced

Photochemically-Induced

Nucleophilic

S_N2

S_N1

Aromatic

Direct

Via Benzyne

Via Aryl Cation

Electrophilic

Aromatic

Friedel-Crafts

Alkylation

Acylation

Nitration

Halogenation

Sulfonation

FIGURE 8

There is still another method for organizing reactions and their mechanisms, one that is particularly important in synthesis. This is via the chemical connection between different functional groups, where a functional group is a group of atoms with distinctive structural and chemical properties. These include alcohols, carboxylic acids, ketones, and aromatic compounds; the carbonyl group (C=O) figures prominently in many of these groups. The object is to discover methods for converting one functional group into another, preferably in one step. As there are dozens of functional groups, there are thousands of linkages between one functional group and another. Imagine that there are three functional groups: A, B, and C. The chemist wants to find ways to carry out the reactions A →B, B→A, A→C, C→A, B→C, and C→B, six reactions in total. There are some transformations that are easily accomplished, and in many ways. Consider the conversion of a ketone such as 2-butanone into 2-butanol (Figure 9). There are dozens of ways to carry out the transformation such as the use of lithium aluminum hydride or sodium borohydride, both commercially available reducing agents.

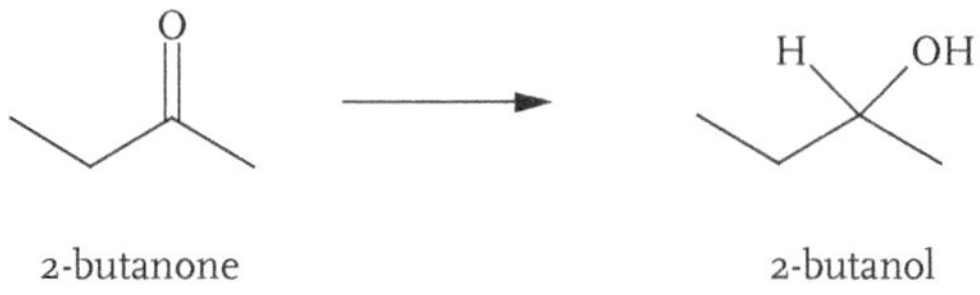

2-butanone   2-butanol

FIGURE 9

Why would a synthetic chemist want so many methods for carrying out the same reaction? In the synthesis of a compound possessing several functional groups one reagent may be selective and reduce the carbonyl group of a ketone without reacting with the other functional groups. Both lithium aluminum hydride and sodium borohydride will reduce ketones to alcohols but lithium aluminum hydride will reduce many other functional groups that sodium borohydride will not. Chemists say that sodium borohydride is more selective than lithium aluminum hydride. On the other hand there are functional groups in a molecule that are reduced by lithium aluminum hydride but not by sodium borohydride.

Most syntheses require more than one step. As the distance between reactants and desired product becomes larger, as gauged by the number of reactions involved, the number of plausible schemes for carrying out the synthesis also becomes larger. How does the chemist decide which scheme to use? Elegance and beauty are important considerations in choosing a scheme. Aesthetics becomes an important criterion in evaluating a successful synthetic strategy. Nonetheless, mechanistic criteria must always be taken into account in order to attain a synthetic goal.

There is perhaps no better way of appreciating these organizational principles than by examining their use in introductory texts in organic chemistry such as the classic Morrison and Boyd (1959) and the current Loudon (2002).

2.7 Cause and Effect

Even after two millennia of reflection "cause and effect" continues to be a topic of interest and contention among philosophers. What is of interest here is its connection to mechanism. If causation is the manner in which we get from one place to another, mechanism and causation in essence are the same thing. Glennan in fact refers to his mechanism of parts and interactions as a causal mechanism. This makes sense because this mechanism is caused by the fundamental laws of physics which have no cause; they just are. The "entities and activities" mechanism of Machamer, Darden, and Craver can be viewed in the same light. Reaction mechanism may also be seen in the same light.

Chemists are not satisfied with comingling mechanism with cause and effect in the manner described above. Yes, reactants yield products and in certain reproducible ways, but what makes reactants behave as they do? What makes them reactive? Is not the answer to these questions the cause? Chemists wish to use more fundamental concepts in their attempts to understand how reactions occur much as Glennan wants to describe mechanism in general with the most fundamental physical laws.

Chemists associate causality with two things: energy and what I shall call property. Energy is required for any reaction to occur (Fair 1979). Reactions, for example, are activated by heat, light, microwaves, and so forth. Even reactions, which are carried out at a very low temperature, still require the input of energy for them to occur. Thermally activated reactions do not occur at temperatures a few degrees above absolute zero, but reactions may occur at these temperatures if the reactants are exposed to photons, a source of energy.

Property refers to the characteristics that atoms and molecules possess that make them react the way they do. In the S_N2 reaction a nucleophile is a species possessing a pair of non-bonding electrons having the property of nucleophilicity. The S_N2 reaction occurs because certain species behave as nucleophiles. Likewise the species being attacked reacts because it contains a nucleofuge. In the electrophilic aromatic nitration reaction, sulfuric acid *causes* the reaction to occur because it behaves as a catalyst.

Other things related to property influence causality as well. Steric effects may cause a reaction to alter its regiochemistry or prevent a reaction from occurring at all. Orbital overlap may cause the stereochemistry of a reaction to occur one way over another. In other examples, the orbital characteristics of the transition states dictate how these reactions occur.

We have seen previously that chemical concepts are related to explanation. We have now seen that chemical concepts are also related to causation. It follows that causality is intimately connected to explanation as well.

2.8 Falsification

Every budding chemist has been told that a reaction mechanism is never proven and one purpose of studying a mechanism is to falsify it. This is psychologically odd. Why would anybody spend time falsifying something? Falsification is just one tool in the scientist's arsenal for studying nature. The concept of falsification probably originated with Karl Popper (Popper 1968), the eminent and controversial twentieth-century philosopher of politics and science. He believed that a theory was only valid scientifically if it were potentially falsifiable. Physics and chemistry are valid scientific disciplines in this context because their theories can be tested and potentially invalidated, whereas Freudian psychology is not a valid science. The Ptolemaic picture of our solar system, on the other hand, is a valid theory because it can and ultimately was falsified by Copernicus, Johannes Kepler, and Newton. A reaction mechanism is a theory of how a reaction occurs, and in principle can be shown to be incorrect. As a reaction is studied in more detail, its mechanism becomes more secure because there is more evidence; falsification thus becomes less probable. Nonetheless, it will never be possible to probe every facet of a reaction, nor will it be possible to run every example of a particular reaction; some things will always lie outside of our purview. Furthermore, it is not possible to know things as they really are. What is an electron? Its numerous physical characteristics are known, but not its essence. The quantum nature of elementary particles weakens what can be known with certainty even more.

The interested reader is referred to a series of lively articles in the *Journal of Chemical Education* on whether a reaction mechanism can ever be proven (Buskirk and Baradaran 2009, Brown 2009, Lewis 2009, Wade 2009, Yoon 2009). It is a reasonable hypothesis to believe that the mechanism for a reaction that has been studied for an extended period of time is secure even though there are always new experiments which could be carried out on it. Induction favors the well-studied mechanism, whereas probability tells us to be cautious in our belief. No exceptions have ever been found for the mechanism of the S_N2 after over a century of study. As discussed earlier, Sneen and Larsen proposed an alternate mechanism for this reaction (Sneen and Larsen 1966, 1969), which took considerable effort and ingenuity to refute. It could have turned out differently, of course, because no one has control over how the universe operates. The mechanism of the S_N1 reaction, one superficially similar to the S_N2 reaction, also seems secure, but owing to the fact that it is far more complex than that for the S_N2 reaction, there is a greater likelihood that certain features of this reaction will be shown to be false in the future.

Even though a mechanism may be identical for a large number of examples, this does not imply that there are never any exceptions. There is hardly a reaction known for which there is not an example that does not occur by the expected mechanism. These exceptions enrich chemistry, and most chemists would be happy to discover one. These exceptions, which are ordinarily rare, do not falsify the standard mechanism but enrich understanding of the material world. Two examples will illustrate exceptions to the rule.

The nitration of aromatic substrates such as toluene using sulfuric acid and nitric acid has been studied since the nineteenth century and its mechanism well-established, as described earlier in the chapter. Any chemist wishing to carry out a new example of this reaction will assume the standard mechanism to be operating. Nonetheless, there are five cases in which the standard mechanism has to be modified or a new mechanism proposed (Coombes and Russell 1971; Ridd 1971, 1978; Taylor 1972; Myhre 1972; Clemens et al. 1983; Johnston et al. 1989). All of these exceptions falsify the standard mechanism.

The photochemistry of anthracene, a polycyclic aromatic hydrocarbon, has been studied in solution since the middle of the nineteenth century. When carried out in the absence of oxygen, the reaction occurs in exactly the same way in countless solvents (Figure 10). The reaction yields the same product in every instance and proceeds through the same intermediates. Changing solvents only alters the lifetimes of the various intermediates. When the reaction is carried out in water, however, where the solubility of the hydrocarbon is extremely low (circa 10^7 molar), a different set of products is produced by a completely different mechanism (Figure 10) (Sigman et al. 1991). Two reasons have been offered in this change in behavior: (1) the very low concentration of the substrate in the solvent, and (2) the ability of water to accept an electron from the excited state of the substrate. Water even becomes incorporated into the products. There are far more examples where the standard mechanism operates than the one exception found to date. Chemists use induction—and probability—when carrying out new examples of well-established reactions.

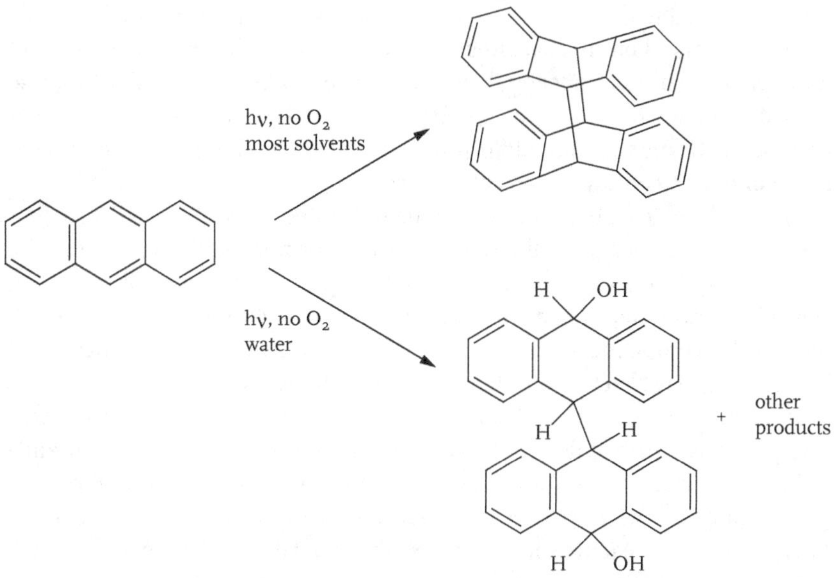

FIGURE 10

Does falsification have any bearing on how a chemist synthesizes molecules? When a chemist develops a scheme for synthesizing a target molecule, s/he assumes that each reaction in the scheme will proceed by the standard mechanism. The chemist assumes a 100 percent probability that each reaction will go according to plan. Reactions, however, do not always react according to plan, not because a reaction occurs in an aberrant manner, but owing to the complexity of the reactant, an undesirable functional group reacts in preference to the reaction of a desirable functional group. If the reactant contains, for example, two non-equivalent aldehyde groups (CHO), the chemist may have reason to believe that aldehyde 1 will undergo reaction in preference to aldehyde 2, but the opposite occurs. The chemist's plan for making the target molecule has been thwarted and s/he must develop a new synthetic strategy. What makes a synthesis hard is the difficulty of accurately predicting where a reaction will occur on a reactant. Popper's falsifiability plays no role in synthesis.

Concluding Remarks

It took more than a century for chemists and physicists to establish the existence of atoms and molecules. Synchronous with this evolution was the development of structural theory and the discovery and development of stereochemistry and chemical kinetics. Beginning in the last decades of the eighteenth century and picking up speed during the nineteenth century chemists discovered more and more reactions. When a critical mass of structures and reactions was attained, chemists began to wonder how this chemistry took place. Wonderment led to experiment which in turn led to mechanism. As reactions became more complicated, and thus more difficult to study, more sophisticated methods were developed to deduce how these reactions occurred. Over time hundreds of reactions and their mechanisms have been discovered. A surprisingly small number of chemical concepts have evolved to explain how and why these myriad reactions occur as they do. Explanatory power has reached a stage of great sophistication. Tremendous progress has occurred over the last two centuries in understanding how reactions occur.

Although philosophers continue to be interested in mechanism, their purview has not included reaction mechanisms. For notable exceptions see the articles by Weisberg et al. (2011) and Goodwin (2010). An especially appealing view of reaction mechanisms is that they consist of entities and activities although the competing view that reaction mechanisms consist of parts and interactions may also be important in certain situations.

A mechanism may be the cause and effect by which reactions are converted into products, what philosophers call a causal mechanism. This is certainly valid, but chemists want more. As Glennan has suggested for physical mechanisms, chemists want to look at causality in a more fundamental manner. Energy and the reactive properties of the reactants must be considered. Property is related to chemical concepts and explanatory power. As a result, causality and explanatory power are intimately related.

Can reaction mechanisms be proven? No more so than anything else. It is not in the nature of universe that we know things as they really are. Quantum mechanics also fundamentally delimits what it is possible to know with certainty. Nonetheless, chemists believe that significant progress has been made in understanding how reactions occur. That such is the case is the extraordinary success of synthesis. Synthesis of new materials would not be possible without an intimate knowledge of how the reactions occur. Our civilization would not be possible without this success.

Acknowledgments

I thank Michael Weisberg, Mark Goodwin, and Stuart Glennan for their help in preparing this chapter.

References

Ackeroyd, F. M. (2000a). "Why was a Fuzzy Model so successful in physical organic chemistry?" *Hyle*, 6, 161–173.

Ackeroyd, F. M. (2000b). "The foundations of modern organic chemistry: The rise of the Hughes and Ingold theory from 1930–1940," *Foundations of Chemistry*, 2, 99–125.

Anslyn, E. V., Dougherty, D. A. (2006). *Modern Physical Organic Chemistry*. Sausalito, CA: University Science Books.

Bensaude-Vincent, B., Simon, J. (2008). *Chemistry—The Impure Science*. London: Imperial College.

Berson, J. (2003). *Chemical Discovery and the Logicians' Program*. New York: Wiley-VCH.

Boyd, R., Gasper, P., Trout, J. D. (1997). *The Philosophy of Science*. Cambridge, MA: MIT Press.

Brock, W. H. (1992). *The Norton History of Chemistry*. New York: Norton.

Brown, J. R. (2001). *Who Rules in Science—an Opinionated Guide to the Wars*. Cambridge, MA: Harvard University Press.

Brown, T. L. (2009). "A discussion of 'Can reaction mechanisms be proven?'," *Journal of Chemical Education*, 86, 55.

Buchwald, J. Z., Josefowicz, O. G. (2010). *The Zodiac of Paris*. Princeton, NJ: Princeton University Press.

Buckingham, J. (2004). *Chasing the Molecule*. Thrupp, UK: Sutton.

Buskirk, A., Baradaran, H. (2009). "Can reaction mechanisms be proven?," *Journal of Chemical Education*, 86, 551–554.

Carpenter, B. (2000). "Models and Explanations: Understanding Chemical Reaction Mechanisms." In *Of Minds and Molecules* (Bhushan, N., Rosenfeld, S., eds.). Oxford: Oxford University Press, 211–229.

Carpenter, B. K. (1992). "Intramolecular dynamics for the organic chemist," *Accounts of Chemical Research*, 25, 520–528.

Chakravartty, A. (2007). *A Metaphysics for Scientific Realism*. Cambridge: Cambridge University Press.

Chalmers, A. (2011). *The Scientist's Atom and the Philosopher's Stone*. Dortrecht: Springer.

Clemens, A. H., Ridd, J. H., Sandall, J. P. B. (1983). "*Ipso*-Attack at the 4-position in the nitration of 4-nitrophenol," *Chemical Communications*, 343–344.

Collins, P., Carpenter, B. K., Ezra, G. S. Wiggins, S. (2013). "Nonstatistical dynamics on potentials exhibiting bifurcations and valley-ridge inflection points," *Journal of Chemical Physics*, 139, 154108/1-154108/12.

Collins, P., Kramer, Z. B., Carpenter, B. K., Ezra, G. S., Wiggins, S. (2014). "Nonstatistical dynamics on the caldera." *Journal of Chemical Physics*, 141, 034111/1 034111/17.

Coombes, R. J., Russell, L. W. (1971). "Electrophilic aromatic substitution. Part VIII. Isomer ratio and assignment of mechanism in aromatic nitration. o-Xylene and biphenyl," *Journal of the Chemical Society B*, 2443–2447.

Debré, P. (1994). *Louis Pasteur*. Baltimore: Johns Hopkins University Press.

Del Re, G. (2003). "Reaction mechanisms and chemical explanation," *Annals of the New York Academy of Sciences*, 988, 133–140.

Eliel, E., Wilen, S. H., Mander, L. N. (1994). *Stereochemistry of Organic Compounds*. New York: Wiley.

Fair, D. (1979). "Causation and the flow of energy," *Erkenntnis*, 14, 219–250.

Faraday, M. (1960). *The Chemical History of a Candle.* Viking, New York, 1960

Feyerabend, P. (2010). *Against Method* (4th ed.). London: Verso.

Giacovazzo, C., Monaco, H. L., Artioli, G., Viterbo, D., Ferraris, G., Gilli, G., Zanotti, G., Catti, M. (2002). *Fundamentals of Crystallography* (2nd ed.). Oxford: Oxford University Press.

Giere, R. N. (2010). *Scientific Perspectivism.* Chicago: University of Chicago Press.

Gleiser, M. (2010). *A Tear at the Edge of Creation.* New York: Free Press.

Glennan, S. (1996). "Mechanism and the nature of causation," *Erkenntnis*, 44, 49–71.

Glennan, S. (2002a). "Rethinking mechanistic explanations," *Philosophy of Science*, 69, 8342–8353.

Glennan, S. (2002b). "Contextual unanimity and the units of selection problem," *Philosophy of Science*, 69, 118–137.

Glennan, S. (2008). "Mechanisms." In *The Routledge Companion to the Philosophy of Science* (Psillos, S. and Card, M. eds.). New York: Routledge.

Glennan, S. (2010). "Mechanisms." In *The Oxford Handbook of Causation* (Beebee, H., Hitchcock, C., and Menzies, P., eds.). Oxford: Oxford University Press.

Goodwin, W. (2003). "Explanation in organic chemistry," *Annals of the New York Academy of Sciences*, 988, 141–153.

Goodwin, W. (2010). *Mechanisms and Chemical Explanation. Handbook of the Philosophy of Science 6.* Amsterdam: Elsevier.

Greenberg, A. (2000). *A Chemical History Tour.* New York: Wiley-Interscience.

Haack, S. (2003). *Defending Science—Within Reason.* Amherst, NY: Prometheus

Hammond, G. (1955). "A Correlation of Reaction Rates," *Journal of the American Chemical Society*, 77, 337–338.

Harré, R. (1970). *The Principles of Scientific Thinking.* Chicago: University of Chicago Press.

Harris, D. C., Bertolucci M. D. (1978). *Symmetry and Spectroscopy.* New York: Dover.

Hoffmann, R., Laszlo, P. (1991). "Representation in Chemistry," *Angewandte Chemie International Edition*, 30, 1–16.

Hoffmann, R., Minkin, V. I., Carpenter, B. K. (1996). "Ockham's razor and chemistry," *Bulletin de la Société chimique de France*, 133, 117–130.

Hoffmann, R. (2007). "What might philosophy of science look like if chemists built it?," *Synthèse*, 155, 321–336.

Holmes, R. (2008). *The Age of Wonder.* New York: Pantheon

Jay, M. (2009). *The Atmosphere of Heaven: The Unnatural Experiments of Dr Beddoes and His Sons of Genius.* New Haven, CT: Yale University Press.

Johnston, J. F., Ridd, J. H., Sandall, J. P. B. (1989). "The borderline between the classical and the electron transfer process in nitration by the nitronium ion," *Chemical Communications*, 244–246.

Kagan, J. (2009). *The Three Cultures—Natural Sciences, Social Sciences, and the Humanities in the 21st Century.* Cambridge: Cambridge University Press.

Klein, U. (ed.). (2001). *Tools and Modes of Representation in the Laboratory Sciences.* Dordrecht: Springer Netherlands.

Klein, U. (2003). *Experiments, Models, Paper Tools: Cultures of Organic Chemistry in the Nineteenth Century.* Stanford: Stanford University Press.

Laidler, K. J. (1993). *The World of Physical Chemistry*. Oxford: Oxford University Press.

Leplin, J. (2001). "Realism and Instrumentalism." In *A Companion to the Philosophy of Science* (Newton-Smith, W. H., ed.). Malden, MA: Blackwell.

Lewis, D. E. (2009). "A discussion of 'Can reaction mechanisms be proven?'," *Journal of Chemical Education*, 86, 554.

Loudon, G. M. (2002). *Organic Chemistry* (4th ed.). Oxford: Oxford University Press.

Lowry, T. H., Richardson, K. S. (1987). *Mechanism and Theory in Organic Chemistry*. New York: Harper and Row.

Machamer, P., Darden, L., Craver, C. F. (2000). "Thinking about mechanisms," *Philosophy of Science*, 67, 1–25.

Marks, J. (2009). *Why I Am Not a Scientist—Anthropology and Modern Knowledge*. Berkeley: University of California Press.

Masel, R. I. (2001). *Chemical Kinetics and Catalysis*. New York: Wiley-Interscience.

Maskill, H. (1985). *The Physical Basis of Organic Chemistry*. Oxford: Oxford University Press.

Morrison, R. T., Boyd R. N. (1959). *Organic Chemistry*. Boston: Allyn and Bacon.

Myhre, P. C. (1972). "Nitro group migrations during aromatic nitration," *Journal of the American Chemical Society*, 94, 7921–7923.

Nye, M. J. (1993). *From Chemical Philosophy to Theoretical Chemistry: Dynamics of Matter and Dynamics of Disciplines 1800–1950*. Berkeley: University of California Press.

Orchin, M., Macomber, R. S., Pinhas, A. R., Wilson, R. M. (2005). *The Vocabulary and Concepts of Organic Chemistry* (2nd ed.). New York: Wiley-Interscience.

Popper, K. (1968). *The Logic of Scientific Discovery* (2nd ed.). New York: Harper and Row.

Ramsey, J. L. (2004). "Straining to explain strain and synthesis," *Foundations of Chemistry*, 6, 81–91.Ramsey, J. L. (2008). "Mechanisms and their explanatory challenges in organic chemistry," *Philosophy of Science*, 75, 970–982.

Reichardt, C. (2003). *Solvents and Solvent Effects in Organic Chemistry*. Weinheim: Wiley-VCH.

Reinhardt, C. (2006). *Shifting and Rearranging: Physical Methods and the Transformation of Modern Chemistry*. Sagamore Beach, MA: Science History Publications.

Ridd, J. H. (1971). "Mechanism of aromatic nitration," *Accounts of Chemical Research*, 4, 248–253.

Ridd, J. H. (1978). "Diffusion control and preassociation in nitrosation, nitration, and halogenation," *Advances in Physical Organic Chemistry*, 16, 1–49.

Rocke A. (2010). *Image & Reality: Kekulé, Kopp, and the Scientific Imagination*. Chicago: University of Chicago Press.

Rosen, J. (2010). *Lawless Universe—Science and the Hunt for Reality*. Baltimore: Johns Hopkins University Press.

Scerri, E. (2007). *The Periodic Table: Its Story and Its Significance*. New York: Oxford University Press.

Schummer, J. (1997). "Towards a philosophy of chemistry," *Journal for General Philosophy of Science*, 28, 307–336.

Sigman, M. E., Zingg, S. P., Pagni, R. M., Burns, J. H. (1991). "Photochemistry of anthracene in water," *Tetrahedron Letters*, 32, 5737–5740.

Smartt Bell, M. (2005). *Lavoisier in the Year One*. New York: Norton.

Sneen, R. A., Larsen, J. W. (1966). "Identification of an ion-pair intermediate in an S_N2 reaction," *Journal of the American Chemical Society*, 88, 2593–2594.

Sneen, R. A., Larsen, J. W. (1969). "Substitution at a saturated carbon. X. Unification of mechanisms of S_N1 and S_N2," *Journal of the American Chemical Society*, 91, 362–366.

Stemwedel, J. D. (2006). "Getting more with less: Experimental constraints and stringent tests of model mechanisms of chemical oscillations," *Philosophy of Science*, 73, 743–754.

Tabery, J. J. (2004). "Synthesizing activities and interactions in the concept of a mechanism," *Philosophy of Science*, 71, 1–15.

Taylor, R. (1972). "Electrophilic aromatic substitution. VII. Effect of nitrosation on the ortho: para ratio in nitration of biphenyl," *Tetrahedron Letters.*, 18, 1755–1757.

Thorpe, T. E. (2007). *Humphry Davy*. Chalford, UK: Nonsuch.

Van Fraassen, B. (1980). *The Scientific Image*. New York: Oxford University.

Wade, P. A. (2009). "A discussion of 'Can reaction mechanisms be proven?'," *Journal of Chemical Education*, 86, 558.

Wagnière, G. H. (2007). *On Chirality and the Universal Asymmetry: Reflections on Image and Mirror Image*. Zürich: VHCA.

Walden, P. (1993). "Über die gegenseitige Umwandlung optischer Antipoden," *Berichte*, 29, 133–138.

Weinberg, S. (1993). *Dreams of a Final Theory*. New York: Pantheon.

Weisberg, M., Needham, P., Hendry, R. (2011). "Philosophy of Chemistry." *Stanford Encyclopedia of Philosophy* (http://plato.stanford.edu/archives/march2011/entries/weisberg.

Woodward, J. (2002). "What is a mechanism? A counterfactual account," *Philosophy of Science*, 69, 8366–8377 (2002)

Yoon, T. (2009). "A discussion of 'Can reaction mechanisms be proven?'," *Journal of Chemical Education*, 86, 556.

| Metaphysical Issues

CHAPTER 9

Causality in Chemistry
Regularities and Agencies

ROM HARRÉ

THE MEANING OF CAUSAL statements and claims has been and continues to be a topic of great interest to philosophers.[1] Before we turn to the role of causal concepts in chemical discourse a survey of the main lines of the philosophical debates on the meaning of causal language will provide the necessary groundwork for this study. It will be useful to maintain the following distinction: The word "causation" will be used as the name for the relation between causes and effects; the word "causality" will be used for the corresponding concept and its many synonyms as they appear in causal discourses.

Since antiquity causal discourses have mostly been taken to be species of explanatory discourses. The analysis of explanatory discourses begins, for our purposes, with Aristotle's writings on the subject (Aristotle 1984). For him, an explanation requires four components: a material *aitia*, what stuff is involved; a formal *aitia*, what structures are to be taken into account; an efficient *aitia*, what brings about the change to be explained; and a final *aitia*, to what end does the thing to be explained tend.[2] Subsequent philosophical discussion of the nature of causation has led to two main proposals According to one popular view "causation" refers to the production or generation of effects by material and human agents, often allied to the Aristotelian view of efficient causes and in recent writing to the revival of the notion of "causal power" (Kistler and Gnassanou 2007). The alternative builds on the principle that "causation" refers to an observed regular concomitance between similar pairs of events leading to an expectation of the occurrence of the second event on the occasion of the occurrence of the first, as Hume argued. The philosophy of chemistry has inherited this problem "space." Should we follow the lead of Aristotle or take our analytical tools from Hume? Or should we look for some kind of hybrid?

---

[1] Some parts of the last two sections of this text have been drawn from Harré (2012), and are reproduced by kind permission of the editor.

[2] Translations of Aristotle's writings into Latin used *"causa"* for the Greek *"aitia."* Only one of the above four *aitia* resembles the modern vernacular sense of "cause," namely efficient cause.

Recent attempts by followers of the Aristotelian plan to revive the concept of causal power have included conceptual studies of what this concept and others allied in use mean (Cartwright 1987) and analyses of the logical form of attributions of causal powers (Cheng 1997; Hiddleston 2005). The traditional alternative to causal powers has been the regularity analysis pioneered by David Hume (1739). Instead of causation as a process in which *active agents* exercise their productive and generative powers, the regularity analysis is based on the presumption that causation is a relation between *pairs of events*. The alleged efficacy of causes is a psychological phenomenon, an expectation induced in us by repeated experience of similar correlated pairs of phenomena. Can the use of causal concepts in chemistry be accounted for in the Humean way as referring to regular sequences of events, or does it need to include the role of causal agents in producing those sequences?

The analyses of causal efficacy by Cheng (1997) and Hiddleston (2005) seem to be based on two interrelated conceptual contrasts:

1. Causes as ephemeral events, echoing Hume (1739) contrasted with causes as persisting efficacious agents.
2. Causality as referring to observed repeated correlation between kinds of events (passivity) contrasted with causality as referring to the exercise of causal powers (activity).

There should be a simple one-to-one mapping between these pairs of concepts. Thus we would have *pairs of correlated events* inducing expectations of their future repetition, contrasting with *persisting powerful particulars*, causal agents.

Cheng introduces a third contrast—that between observables, empirical regularities, and unobservables, theoretical entities—that maps neatly on to both 1 and 2 above. She uses the contrast between an empirical regularity and a theoretical explanation to locate the place for causal powers in her analysis. Causal *powers* are unobservables, while exercises of causal powers by powerful particulars are routinely observable, at least in principle and relative to the available technology.

In chemistry not only is there the question of what brings about changes in materials, but also what accounts for the stability of chemical elements and compounds. There seems to be the need to recognize two kinds of causality—dynamic (or occasional) and static (or continuous). The former covers cases in which causes and effects are events in time and space, the latter covers questions of what causes components of a complex system to maintain interrelations stable enough for the system to be taken to be an individual. A lattice of $Na+$ and Cl^- ions can be a long lasting salt crystal. A benzene ring can pass from molecule to molecule as a single unit, though it is formed from a group of carbon atoms. The philosophical story of the interatomic bonds that maintain the structure of such organic entities as benzene rings is not strictly deterministic. Benzene displays "resonance." It can be imagined to take two forms—with alternating single and double bonds. It is a conceptual "hybrid" (McMurry 2003, 126).[3]

---

[3] An anonymous reviewer saved me from projecting this *picture* on to whatever the molecule structure might be *in rerum natura*.

Probability can enter causal discourses with respect to causes or with respect to effects. What is the probability that this effect was caused by this event? What is the probability this event will or did follow from this cause? In one case we know what happened, say the house burned down, and we want to know the probable cause. In the other case we know that something has happened and we want to know what the probable effect will be. Studies of probabilistic causal discourses have been initiated by Judea Pearl (2000). Using acyclic graphs to map the relation between cause events and effect events and Bayesian probability principles to calculate probabilities, he has created a method for representing probabilistic causality and calculating the probability of outcomes from initial conditions. It is difficult to find cases of indeterministic chemical phenomena in which this kind of probabilistic language seems to fit. In the case of the transformation of one group of substances into another very few reactions go through completely. In the reaction between silver nitrate and sodium chloride

$$AgNO_3 + NaCl \rightarrow AgCl + NaNO_3$$

there will always be small quantities of the original reactants that have not been transformed. But this is not probabilistic, except for individual molecules. The way reactions occur requires the introduction of another distinction, that between complete causation and partial causation. The sense of causality in descriptions of everyday events is deterministic. The brick does not partly break the window (though it may break only part of the window). Closer to the chemical case might be the failed attempt of the murderer in which the arsenic only makes the victim very ill. This does not seem to lend itself to a probabilistic description. The assailant thinks the victim will probably die, but nature may determine otherwise.

Chemistry forces us to consider both the causes of change, sequential phenomena, and the causes of stability among the things in the world with which chemistry traditionally deals.

1 The Grammar of Event Causality

Let us briefly return to Hume's ingenious and powerfully reductive analysis and one of its modern descendants, John Mackie's reduction of the concept of causality to the satisfaction of what he calls the INUS conditions on the antecedents of an effect. According to Hume (1739: Sect XIV) the causal relation as it is thought to hold between a pair of events, one the cause and some subsequent event the effect, includes three components. There must be regular concomitance in *contiguity* and *succession* between the types of events of which the cause event and the effect event are instances. Though Hume uses the word "object" in the earliest formulations of his analysis, it seems clear that by this word he means events rather than things. According to Hume the elements of experience are impressions and ideas are their mental shadows. As well as regular contiguity and succession the idea of causality includes the idea of *necessity*—that is, the idea that the effect event *must* follow the cause event, ceteris paribus. After pointing out that natural necessity is not the same as logical necessity, Hume offers his ingenious reduction of necessity to regularity. The meaning of an idea derives from a corresponding impression, so to understand what the idea of necessity means in the context of natural science we must locate the impression that corresponds to it. According to Hume it is the psychological state of expectation brought about by repeated experience of the correlations of instances of types of events. "Necessity" as a feature of the flow of events disappears in favor of the psychological effect of observing regularity in contiguity and succession.

There must also be a second order regularity between first order regularities of contiguity and succession and the psychological state of expectation, since the former, according to Hume, causes the latter. There must be contiguity and regularity between instances of continuity and regularity and coming to expect the effect-event to occur after the cause-event.

Two-and-a-half centuries later Mackie (1974, 62) offers something similar, the INUS conditions on which he bases his analysis of causality in *The Cement of the Universe* (1974, 36). He argues that an event is to be taken as causal only in certain circumstances, a field of relevant material conditions.

To deal with the problem of specifying the link between causes and effects he makes do with the idea of "the world runs on" (1974, 55). Causal statements are generally taken to support or even entail corresponding counterfactuals. If it is true that the freezing of water causes it to expand, then it should also be true of some sample of liquid water that if it were to be (had been) frozen then it would (would have) expanded. Such inferences seem to require that the possibilities in the situation are taken seriously, perhaps in planning some action. If causal statements are nothing but summaries of what actually happens, how can their extension to possibilities which may never happen be supported? We will return to this conundrum later.

In defining his version of causality as a feature of the conditions under which an event occurs, Mackie writes that a state or event A is the cause or some part of the cause, if it is "an *insufficient* but *non-redundant* part of an *unnecessary* but *sufficient condition*" (Mackie 1974, 62). This yields the neat acronym "INUS" conditions. There may be several sets of conditions sufficient to produce an effect, so no one of them is necessary—that is without which the effect would not occur. However, among these conditions are some which are more relevant to a causal story than others. These are candidate causes. How the relative relevance of conditions is established in the absence of attention to causal mechanisms and natural agents is left to the insights of scientists and other engaged observers, such as bakers and blacksmiths.

For Mackie the ontological categories of causes must be found in what could count as a relevant condition for the coming to be of an effect. The conditions we advert to in science and everyday life include such things as standing states, events, material agents, substances, and anything else natural science, cooking, and gardening might favor. What determines the relevance of conditions could hardly be empirically observed correlations since without a rationale for the relevance of some event to the production of the effect it would not conform to the INUS requirements for identifying causes.

One could read Hume, and perhaps Mackie as well, as outlining the conditions under which we are entitled to surmise that a causal relation exists between an initial state of a system and a subsequent state rather than analyzing the meaning of the causal relation. This would leave room for a distinction between the criteria for picking out a certain pair of events as likely to be causally related and analyzing the content of the concept of causality. Eliding the content of a concept with its truth conditions is the defining mark of positivism. For all sorts of reasons we want to avoid being driven to adopt that stance. Some of the recent writing on causation seems to fall back into the positivist verificationist doctrine of meaning, that is the doctrine that the meaning of a statement is the method of by which it could be verified or refuted.

Despite the fact that it is quite difficult to find Humean explanations in everyday discourse, many philosophers have taken event causality as paradigmatic of the concept of causality and some still do. Instead of saying that it was water that washed away the foundations, that it was aspirin that cured the headache, that it was the brick that broke the window, and so on, we transpose these explanations into the "event" form. Then we have something like this: the impact of the water, the taking of the aspirin, and the striking of the window brought about the effect event or state. However, behind the advice one might give to someone complaining of a headache, "Take an aspirin," it is surely not a belief that the *taking* of the pill will cure the headache. It is the aspirin, the stuff, acetylsalicylic acid, working away on the nervous system that does the trick.

Why does any of this matter? However artificial the paradigm of event causality may seem when imposed on everyday discourse, it serves to support the denial of agentive powers not only to substances but also to people. There are many philosophers who think that the discoveries of neurophysiology and genetics have rendered the idea of human agency redundant. The defense of the role of agent causality in explanations in recent years has been concerned very largely with the metaphysics of human action, with consequences for moral philosophy and philosophical psychology (O'Connor 2002, 45). O'Connor comments that Thomas Reid, for example, held that agent causation was given in human experience, and was extended from that origin to inanimate beings. Agent causation in the human case is offered as the root idea.

The argument I propose is rather the reverse. In passing, I wish to throw doubt on the claim made by some of those who would deny radical agency to people that physics and chemistry do not make use of the concept of causal agency in explanations. Here is a spirited version of this claim:

> How does an *agent* cause an effect without there being an event (in the agent,
> presumably) that is the cause of that effect (and itself the effect of an earlier

cause, and so forth)? Agent causality is a frankly mysterious doctrine, positing something unparalleled by anything we discover in the causal processes of chemical reactions, nuclear fission and fusion, magnetic attraction, hurricanes, volcanoes, or such biological processes as metabolism, growth, immune reactions, and photosynthesis.

(DENNETT 2003, 100)

Admirable though Dennett's approach to cognitive psychology is, this passage betrays a woeful misunderstanding of the ontology of chemistry and physics. What is supposed to be the event internal to a south magnetic pole that brings it about that it attracts a north pole? And what might the internal events be that bring it about that an electron displays a unit negative charge? How could there be photosynthesis without photons?

To see what has gone wrong here reflect on the conditions under which a heavy body falls in a gravitational field. True, an event occurs (the removal of a support) prior to the beginning of the descent; the descent, however, is a process, not an event. But it would be bizarre to cite the removal of the support as the cause of the body's descent, though it might be cited as the cause of the beginning of the descent. The cause of the continuous downward acceleration of the body is the potential energy of the gravitational field. Try removing the support from under a cannon ball in interstellar space. Nothing happens! Why? No gravitational field! Fearing we might want to bring back spirits, souls, and other dubious beings, some philosophers slip back into the old positivist move of eliminating them by eliminating everything agentive from our ontologies by excising all that is unobservable. Of course causal powers are unobservable, just as their particular instantiations, such as charges, field potentials, and the like are unobservable. However their material sources such as magnets, flasks of acids, fires, planets, and so on are easily found. Some bearers of causal powers are difficult to display but their unobservability is usually the result of a technical limit on equipment rather than a matter of metaphysics.

It does not follow that a defense of the fundamental and indispensable role of agentive causality as it appears in the physical sciences provides any ready-made answer to the question of human agency. It does follow that the alleged support from the way those sciences have developed for dispensing with the foundational role of powerful particulars in philosophical psychology and moral philosophy is chimerical. Explanatory patterns in the natural sciences are based on some version of agentive causality. The concept of "agency" in the natural sciences, especially physics and chemistry, includes the root ideas of "spontaneity" and "activity." However, agency in the human domain also includes the root idea of "free choice," which is foreign to the ontology of physics and chemistry.

From Hume's writings of 1739 through Mackie's proposal of INUS conditions in 1974 we can find the same inadequacies when we try to match up their analyses against the intuitions we extract from the conceptual systems in use during that period in the natural sciences. There are two areas of inadequacy:

1. The ontological status of causes and effects is not well studied, eventuating in a ragbag of very different kinds of beings offered as causes and as effects, at least things, substances, conditions and events.
2. The link between cause and effect is not well studied either, with suggestions ranging from Hume's psychological account to positivistic regularities of a variety of kinds, including Mackie's "laws of working" as what lies behind "the world runs on." In everyday and in scientific practice if some event is chosen to play the role of cause it is usually taken to be linked to the effect by the operation of a causal mechanism. Suppose heating mercuric oxide with carbon is picked out as the cause of its becoming mercury and carbon dioxide. The relevant chemical equation describes the mechanism by which the effect is brought about. The mechanism and the process by which the effect is produced are not observable in the same way that the reagents are.

Ehring's (1997) study of causation does make use of the idea of an intervening mechanism which, triggered by a stimulus of a certain type, regularly brings about a subsequent type of event, the relation between the events meeting Hume criteria for causation, that is contiguity, regularity, and natural necessity. However, he leaves no place for causal agents in his account so the activation and efficacy of such mechanisms is left unaccounted for. An agent seems to be required to complete the story of a causal process.

2 The Grammar of Agent Causality

I will try to show that a pair of events meeting the Hume/Mackie conditions can be certified as causally connected only if a plausible account of the production of the effect event citing a causal agent, can be supplied.

The ontology of the natural sciences, as they are actually practiced, rests on the concept of a natural agent—that is, a material being endowed with certain causal powers. The plausibility of this intuition can be demonstrated with examples from chemistry. The generic concept of "power" seems to involve two root ideas, "spontaneity" and "efficacy." The field potentials of an electric charge do not require anything to bring them into being. Their efficacy is displayed in whatever happens when constraints on motion are removed or absent.

The argument will unfold in two phases. In the first phase I offer an account of natural agency as it appears in the way we treat a wide variety of non-human entities or stuffs as causally efficacious. The second phase presents an analysis of causal regresses in chemistry, in modern times reaching through the level of chemical agents into physics.

E. H. Madden and I argued that attributions of event causality made sense only on the presumption of an underlying agent causality (Harré and Madden 1975). There are material beings with powers to bring about various kinds of changes, events, and new states of affairs. Our argument was based on the analysis of the concept of causality as it appeared in everyday thinking, but also in the discourse of the physical sciences, though rarely expressed by the word "cause" in that context.

Causal efficacy is ascribed to a powerful particular in the language of dispositions. "If such and such conditions are fulfilled, then such and such an effect will (is likely to) occur." A dispositional ascription is a truth function of two empirical concepts, one representing the conditions and the other the effect. However, to answer the question "Why does this or that entity or stuff display such and such a disposition?" we need to invoke causal powers. Causal powers are theoretical properties that account for the manifestation of observable phenomena, but they are not themselves observable. Causal powers are attributes of material entities, such as plasmodia (with respect to malaria), magnets (with respect to a compass needles), engines (with respect to the motion of trains), or force fields (with respect to covalent bonding). Some powerful particulars are observable, some are not. But in none of them are their powers observable except in how they are manifested, say in the motion of a body in a field.

The causal powers of natural agents usually persist in time. There are two patterns of persistence. The fact that this stuff in the bottle has the power to alleviate pain is true now, but lasts only until the analgesic is metabolized. The fact that this elementary charge has the power to repel some elementary charges and attract others is true now and it persists through a great many interactions in which this entity plays a part.

Causal powers are individuated in terms of their effects. For example, "Aspirin has the power to reduce pain," that is, it is an analgesic. Codeine too has the power to reduce pain—that is aspirin and codeine display the same dispositional property, just as blood and stop signs are both have the power to induce sensations of red in a conscious observer. Powers are attributes of material entities in just the way that "red" is an attribute of flags and noses. However, the differentia of powerful particulars, the agents which have such powers, include the chemical structure of the molecules of each type of particular and the active route by which each brings about its effect.

Contemporary physics is grounded in charges as powerful particulars and their dispositions as associated fields of spatiotemporally distributed potentials. Insofar as contemporary chemistry is grounded in contemporary physics, it too rests on a metaphysics of charges and fields. Nancy Cartwright has defended a similar metaphysics for physics based on the concept of "natural capacities" (Cartwright 1987).

According to Purvis and Cranefield (1995, 4), there are two basic aspects of a causal agent model:

1. The nature or structure of the agent itself.
2. The architecture of the system in which the agent operates.

The first basic aspect covers two very different cases. A simple agent may possess the power to act in a certain way, as its defining or one of its defining attributes. The question as to the internal structure that endows an electron with a unit charge has no place in the physics that plays a foundational role in chemistry. For chemical purposes, for example using the Lewis theory of bonding, an electron is a simple agent. A complex agent may possess its powers as emergent properties of a more basic structure. Analgesics and antibiotics are agents like this.

3 Causal Laws and Counterfactuals

Events as causes are supposed to necessitate their effects, ceteris paribus, causal agents are supposed to necessitate their effects ceteris paribus too. Should an event of a type that is well known to have a certain kind of effect fail to do so we look for an interfering factor—the powder was damp, the concentrations of the reagents were too low, and so on. Should something well known to have certain causal powers fail to produce the expected effect we adopt the same strategy. Either it has deteriorated or the conditions are not right.

This strategy is reflected in the logician's claim that causal hypotheses are linked to counterfactuals. A causal principle applies not only to actual cases of the production of an effect. It also applies to cases in which we could say had the cause—be it event or activity of a powerful particular—occurred, though in fact it did not, the effect would have been observed. A causal principle or hypothesis is not a mere summary of actual happenings, past, present, or future, but also can be used to reason about what would happen in imagined situations. Some such situations are similar to what really happens, and some differ from what we believe is the realm of reality to various degrees. The principles needed for plausible reasoning in relation to possibility and necessity can be made sense of by imagining them to be used for reasoning about events and things in possible worlds (Lewis 1986).

Proposing events for the ontology of causality renders the link between causal laws and counterfactuals mysterious. Events are ephemeral and actual and so cannot serve to support counterfactuals, since these refer to possibilities rather than actualities. Causal powers endure whether exercised or not. They are well fitted to be an ontological basis for causality that makes the link between causality and the link with counterfactuals intelligible. The emphasis that Purvis and Cranefield put on the structure of the context in relation to the possibility of a power being exercised ties in with the requirement that a power be possessed by a powerful particular whether exercised or not.

An intriguing way of introducing possibility into the analysis of causation that contrasts with analyses that give priority to agency has been proposed by Belnap (2005). He adds branching space-time to the repertoire of analytical tools to make sense of the way possibility serves as a support for the necessity of causation as an essential presupposition of causal necessity. Nevertheless the ontology of causality on which the analysis rests is still Humean. His *causae causantes* are events. The causative events are picked out from all antecedent ephemeral states of the universe by Mackie's INUS conditions. His account inherits the same problem of connection: How are pairs of events, sequences of states of a system, actual or possible, on different branches of space-time, linked by causation? However, there is a more fundamental difficulty in that is very hard to see how an event could be a genuine originating condition, unless it were to appear spontaneously in the flux of time. The agency concept seems to imply that at least some events are not preceded by a regular concomitance. New kinds of mechanisms can appear or be constructed out of the same old stuff and yet generate new kinds of events. There are plenty of examples of this in biochemistry, for example substituting chloride ions for hydrogen on a benzene ring, a process which so far as I know does not occur in nature.

4 The Possession of Powers

The requirement that causal powers can exist in situations in which they are not there and then being exercised leads to the idea of an enduring powerful particular which can properly be said to have a power when not exercising it. In chemistry this quest leads to a hierarchy of structural hypotheses, which are taken to be the ontological basis for the powers of complex beings that appear as emergent properties of complex structures but are not powers of the components of those structures. Acidity is an emergent property of molecules of HCl whether or not the acidity of hydrochloric acid is being exercised on anything— for example, on zinc in the exemplary experiment with which most of us enter the world of chemistry, after we have blown our CO_2-rich exhalations through lime water! These powers are not attributes of hydrogen ions in general, though unless HCl is ionized in aqueous solution the acidity will not be realized. In modern chemistry "acidity" is interpreted as a readiness to share electrons and "basicity" as a readiness to share protons. The causality is still agentive, in that a new realm of powerful particulars enters the story—electrons and protons.

Hiddleston (2005) has made an interesting attempt at a comprehensive analysis of the concept of a causal power. It will prove instructive to test out his analyses of the main features of the concept as intuition reveals them in its use in chemistry, past and present. Chemistry is replete with causal power concepts, such as "elective affinity," "valency," and so on, and so is an ideal test context for this analytical tool and the results of using it.

Hiddleston (2005) begins with a critical examination of David Lewis's claim (Lewis 1986) that the generic concept of causality is transitive. Thus, if A causes B and B causes C then A causes C. Clearly the Lewis claim assumes an event ontology for causality. Hiddleston rightly centers his discussion on counterfactuals. Why should we believe the associated counterfactuals about a powerful particular?

Hiddleston develops the important idea of an active route between cause-event and effect-event. He turns to agent ontology in two striking cases. In the case of the sharpshooter (the firing of whose rifle kills the soldier), the problem of associated counterfactuals is resolved by introducing the moving bullet as the basis of an "active route between the cause-event and the effect-event," or in this case the consequent state, the being dead of the victim. Similarly in the case of someone who ducks out of the way of a falling boulder, but might not have done, the counterfactual is supported by the boulder as persisting causal agent, which status it has whether or not the hiker ducked. The moving boulder is there as an active agent in both scenarios. The hiker is dead, and we say had he ducked he would have survived. The hiker is alive, and we say had he not ducked he would be dead. Of all the goings on in the vicinity of the hiker, why is the boulder so important? It is the powerful particular.

5 A Hybrid Structure

Cheng's (1997) studies seem to suggest that there is no radical disparity between the use of concepts of event-causality and concepts of agent-causality in the discursive practices of actual cases of people reasoning causally. They tend to identify a correlation as causal if they think that there has been a causal power at work. According to scientific realism every reliable observable causal relation between events and conditions antecedent to an event taken as their effect, must be explicable by reference to a generally unobservable generative processes. Cheng summarizes her project as follows:

> I propose that causal power is to covariation [among instances of event-types] as the kinetic theory of gases is to Boyle's law. When ordinary folks induce the causes of events, they innately act like scientists in that they postulate unobservable theoretical entities (in this case the intuitive notion of causal power) that they use to interpret and explain their observable models (in this case their intuitive covariational model). That is, people do not simply treat covariation as equivalent to causal relations; rather, they interpret their observations of covariation as manifestations of the operation of unobservable causal powers, with the tacit goal of estimating the magnitude of these powers.
>
> (CHENG 1997, 369)

Cheng's thesis needs to be "bulked up" to make explicit the concept of "causal agent" or "powerful particular." People surely do not just postulate causal powers as unobservable entities. They presume that there are material beings in the situation with causal powers. So, in addition to satisfying the Hume/Mackie requirements which an observable pattern of events must satisfy to be a candidate causal relation, there must be the presumption of a material, spatiotemporally continuous link between cause-event and effect-event. This is the role for the causal agent or powerful particular. For events to count as causes they must activate agents with causal powers. This is the principle on which the identification of certain covariations in event-types as causal chains is based, whether in physics, chemistry and biology or in everyday life.

6 Versions of Causality in Chemistry

The discourses of chemistry mostly propose hypotheses about the unobservable links between initial states of a mix of material substances of certain kinds, reactions among which lead reliably to new substances which have come into being in the course of the reaction. One conglomerate of material substances is transformed into another, though rarely is the transformation complete.

The analysis of description of chemical phenomena actually poses two questions:

1. How are transformations of initial conglomerations of substances into different consequential conglomerations effected? Sometimes just assembling the ingredients is enough to induce a transformation, but sometimes an external influence is required. Bring together sodium bromate and sulphuric acid only produces bromine vapour when the mixture is heated.
2. Why are the molecules of most compounds stable? There must be continuously active causes maintaining the atomic structure of a compound. For example, "the greater acidity of phenols compared with alcohols stems from the ability of the negative charge in the phenol ion to spread to the benzene ring which serves *to stabilise* the conjugate base" (Atkins and Beran 1992, 893).[4]

Between what "objects" are causal relations supposed to hold in chemistry? Are causal stories in chemistry about events? We can express the story of a chemical reaction pattern in event terms. "The event of the bringing together of a solution of silver nitrate and a solution of sodium chloride in a test tube is followed by the event of the appearance of silver chloride and sodium nitrate appearing subsequently in the same test tube." But causation as a process linking events does not seem to fit the stability story. Phenol is stable because the carbon, hydrogen, and oxygen atoms are interacting with one another continuously.

Viewing chemistry as the study of the transformation of one conglomeration of substances into another or as a sequence of events, as in the paragraph above, the causal story is filled out with hypotheses about causal *mechanisms*. The usual story involves the rearrangement of atoms into new molecular clusters thus bringing about an observable transformation of substances.

There seem to be three cases of this kind of causation.

1. Once assembled the transformation is spontaneous.
2. An external input is needed to drive the transformation, for example, heating the mixture.
3. An additional substance is required that does not appear in the description of the initial and final conglomeration of substances, the catalyst.

---

[4] My emphasis.

The catalyst can be of the same phase as the reactants, for example a liquid, for example, liquid bromine is the catalyst in the production of oxygen from hydrogen peroxide, or in a different phase, for example finely ground vanadium pentoxide is the catalyst in the production of sulphuric acid from sulphur dioxide and oxygen. These processes are intermediate causal mechanisms that intervene between the original assembly of reactants and the products but the catalysts do not appear as among the new products. While the vanadium pentoxide does not take part chemically in an intermediate reaction, the bromine does.

$$Br_2 + H_2O_2 \rightarrow 2Br^- + 2H+ + O_2$$

$$2Br^- + H_2O_2 + 2H+ \rightarrow Br_2 + 2H_2O$$

However the products of the intermediate bromine reaction do not appear among the transformed substances at the end of the reaction, water and oxygen where once there was hydrogen peroxide.

7 Historical Development of the Chemistry of Substance Stability

Newton was reluctant to accept the idea of simple material inter-atomic linkages. Of what stuff would these be made? After several false starts his account of chemical bonding and so of chemical processes in the *Opticks*, Query 31, expresses a view similar to that generally current today:

> Have not the small Particles of Bodies certain Powers, Virtues, or Forces, by which they act at a distance on one another for producing a great Part of the Phaenomena of Nature? For it is well known, that Bodies act one upon another by the Attractions of Gravity, Magnetism, and Electricity; and these Instances shew the Tenor and Course of Nature, and make it not improbable that there may be more attractive Powers then these. How these Attractions may be performed, I do not here consider.... I use [the word] "attraction" here to signify only in general any Force by which bodies tend towards one another whatsoever be the Cause.
>
> (NEWTON 1730 [1952], 376)

By the end of the eighteenth century the Newtonian hypothesis of inter-atomic *forces* won out over material linkages as an explanation of stability of combinations of corpuscles. Chemical forces were identified with electrical forces by Humphry Davy and Jöns Jacob Berzelius. Though the details of the ionic theory have changed greatly the outline story, insofar as it makes use of causal concepts such as the power of attraction between positive and negatively charged bodies, has remained much the same.

Explanations of stability in chemistry involve electrons as powerful particulars but in two ways.

1. Electrovalency uses charges as agencies, with mechanisms of electron and proton *exchanges* between atoms to create anions and cations. The continuous causation is electrostatic attraction. Metal/non-metal compounds such as copper sulphate are usually electrovalent.
2. Covalency is based on the *sharing* of electrons between two or more atom-cores in the molecule. Non-metal/non-metal compounds such as boron halides are usually covalent.

As Llored (2010) has shown, the success of Mulliken's "molecular orbital" concept in interpreting covalency requires a different interpretation of the concept of "electron" from that which it has in either classical or in quantum chemistry. It has become a constituent in a working model of the architecture of atoms rather than a genuine component. How did it come to play that role?

To understand what Mulliken did, a concept generally unfamiliar to chemists but taken for granted by all of them is needed: This is the concept of an "affordance." The word was invented by the psychologist J. J. Gibson (1968) to refer to the features of a physical system that are brought out, particularly as perceptibles, by certain specific actions upon that system. A steel rod affords tensile strength when supporting a heavy weight. A hammer affords shaping power when brought down on a red hot piece of iron at the temperature at which it affords malleability to that blacksmith and that hammer.

7.1 Causes of Transformation and Change

The spontaneity component of the content of the concept of "chemical agency" is displayed in reactions. Certain bodies, when placed in contact, exhibit a proneness to combine with one another, or to undergo decomposition, while others may be most intimately mixed without change. The actual phenomena of combination suggest the idea of peculiar attachments and aversions subsisting between different bodies. A specific attraction between different kinds of matter must be admitted as the cause of combination, and this attraction may be conveniently distinguished as *chemical affinity*.

George Fownes introduces the causal powers terminology as follows:

> The term chemical affinity, or chemical attraction, has been invented to describe that particular power or force, in virtue of which, union, often of a very intimate and permanent nature, takes place between two or more bodies, in such a way as to give rise to a new substance, having for the most part, properties completely in discordance with those of its components.
>
> (FOWNES 1856, 214)

Though this passage was written in mid-century, by 1807 Humphry Davy had already made the connection between chemical affinity and electricity:

> In the present state of our knowledge it would be useless to speculate on the remote causes of electrical energy. Its relation to chemical affinity is, however, sufficiently evident. May it not be identical with it, and an essential property of matter?
>
> (DAVY 1807, 39)

So if a chemist hopes to bring about changes in a substance, for example, to decompose it into its basic constituents, one way would be to use electrical forces to rupture the forces holding the original structure together, since they too are electrical.

The final step in building something like the near-modern conception of chemical reaction came with the idea of "order of affinity."

> The affinity of bodies appears to be of different degrees of intensity. Lead, for instance, has certainly a greater affinity than silver for oxygen. But the order of affinity is often more strikingly exhibited in the decomposition of a compound by another body.

> (GRAHAM 1850, 223)

Here Graham completes the analysis of the concept of chemical attraction in terms of causal powers, adding "efficacy" to "spontaneity." However, we are still within the Newtonian framework in which "forces" do duty as the relevant powerful particulars. What sort of entities are these?

7.2 From Forces to Fields

By the mid nineteenth century the ontology of physics was changing from an essentially Newtonian atomism with inter-atomic forces toward charges and their fields. The concept of "field" in more or less its modern form had been introduced by Gilbert (1600) as the foundations of a working ontology for explaining magnetic phenomena. The orienting power of the "orbis virtutis" accounted for the way a compass needle behaved in the vicinity of a magnet, rather than explaining it as the result of the interaction of polar forces. Gilbert's inspired anticipation of the field concert hung fire for a couple of hundred years until Michael Faraday's elastic lines of force model brought the concept to life again. However, chemists were slow to adopt field concepts. Only in the mid twentieth century do they displace force concepts altogether.

However, this displacement affects only the most recondite research. In practice, most chemists in most situations continue to make use of a theory of valence based on electric polarities that Berzelius would have recognized. This is elementary chemical theory, but instructive in the way it reveals the underlying ontology of the practice of chemistry, even into the third millennium. Discussions with one of the research chemists in my college have shown me how widely the entitative electron model is used in the everyday work of chemists.[5]

Somewhere between the correlations of the Humeans and the powerful particulars of their opponents lies the causal mechanism, an enduring device that is stimulated or runs on to produce a change in the materials or their phase that would be the sought after chemical result. For the most part we can rarely do more than create plausible models of such mechanisms. The chemistry of solutions offers plenty of examples of this way of supporting causal

---

[5] I am grateful to Gareth McGuire for instruction in these matters.

hypotheses. Thinking of the distinction between acids and bases in aqueous chemistry we get the definitions:

Acid: when added to water produces H+ ions.
Base: when added to water produces hydroxide (OH−) ions.

But this is generalized into two complementary definitions in terms of models making use of powerful particulars.

Bronsted model: an acid donates a proton, a base accepts a proton.
Lewis model: an acid accepts a pair of electrons, a base donates a pair of electrons.

Covalency—for example in the union of two chlorine atoms to form the Cl_2 molecule—is explained as the result of the sharing of two electrons, one electron from each atom completing the 8 shell of the other and so ennobling both.[6] Nothing could be more entitative, as we imagine the electron sometimes orbiting one of the chlorine atom-cores and at another time the other.

According to Friend (1915, 170), theories postulating attractive and repulsant forces in explaining molecular structures and the reactions which give rise to them ought to be considered as the ground level of chemistry. Those theories which set out to explain the "actual causes of the chemical affinity," postulated by chemists are properly the province of physics. Even in 2005 something like this is still true, in that the majority of chemists are most likely to go no further than the electron theory leaving the formulation of the wave functions for multiple electronic configurations to the members of another "club."

The situation is actually more interesting ontologically than this division of labor would suggest (Brown 1973). To the ontology of charges and their fields that is realized in the electron theories of valency chemists added a third concept with ontological overtones, namely "energy." For example ionization energy is the energy required to abstract an electron against the attraction of the nucleus, which is complemented by the energy released when an electron is taken up by an atom or ion.

Suppose we explain the existence of NaCl molecules by the Berzelian hypothesis that there is an electrostatic attraction between Na^+ ions and Cl^- ions. Why should this set-up be stable? The solution is due to Albrecht Kessol (1853–1927). By losing one electron to the chlorine atom the sodium atom displays a net positive charge (a cation), and the outer shell of the remaining electrons matches that of the noble gases in having 8 electrons. By acquiring an electron the outer shell of the chlorine also emulates the electron configuration of a noble gas, and moreover is now negatively charged (an anion). So we have an account of the electrostatic forces *and* of the stability of ions within the same picture. This is "ionic" or electrovalency.

---

[6] The expression "noble octet" for the stable electron configuration derives from the determining role this structure has on the properties of the "noble" gases argon, neon, etc.

How would the Hume/Mackie way interpreting causality fare in chemistry? Take the case of a simple double replacement reaction. In symbols:

$$2NaCl + Pb(NO_3)_2 \rightarrow 2NaNO_3 + PbCl_2$$

all in aqueous solution.

By considering electron transfer among the ions a new pair of +/– charged entities comes into being. The key is the transfer of powerful particulars which endow the ions with the Berzelian charge structure that accounts for their stability. Electron transfer leads to new products, either because one of them is insoluble and precipitates, or because the new products are more stable than the reactants were.

Similarly there are many cases of proton transfer, in which a hydrogen ion is relocated. Consider the reaction between ammonia and hydrochloric acid yielding ammonium chloride. The ion equation is something like this:

$$NH_3 + H_3O + Cl^- \rightarrow NH4 + Cl^- + H_2O$$

By deleting the Cl^- on either side of the equation we get a net ion reaction, which reflects the transfer of H^+, a proton, from the complex water ion to the ammonium radical (Atkins and Beran 1992).

These transfers of electrons and protons as powerful particulars endow the ions with the necessary electrostatic charges to sustain Berzelian bonding. Covalency is more complicated, but the principle of referring bonding "forces" to elementary charges is made use of there, too.

To return to Hume's criteria and Mackie's INUS conditions; could their style of causal concepts be applied in the cases I have described, which are no more than routine inorganic chemistry? Well, the test tube must actually contain ammonia and hydrochloric acid brought together in solution. They could perhaps have been poured from separate vessels into a common container. The contiguity and succession conditions are clearly met since reaction occurs almost immediately and the reaction products are readily identifiable afterward. Since this reaction always occurs, ceteris paribus, we might admit the possibility that experience of this would lead to an expectation that it would happen again, Humean "necessity," though most chemists have seen it only once or twice. An event, the mixing as the cause of the process that leads to the reaction products, would be a candidate cause. These are more or less Mackie's INUS conditions, though we might just squeeze the reagents themselves into the story via the relevance requirement. Given the implausibility of Hume's psychological analysis of causal necessity, would a new version of the regularity theory in terms of probability make sense here? Do chemists mean by talking of reagents and products, their favored causal concepts, that adding ammonia to hydrochloric acid would increase the probability that ammonium chloride would be formed? Of course not. But given the causal agents involved, they would, no doubt, conclude that the formation of ammonium chloride was very likely, but in a jokey kind of voice!

In all the test cases that have been discussed the concept of causality appears in the hybrid form, that is expressed in terms of hypotheses as to the unobservable links between the initial and final states of some mix of material substances, when there seems to be a good correlation between them.

In analyzing a case of the application of dynamicist metaphysics, the prime ontological problem is to locate the powerful particulars, the sources of activity. These are the "uncaused causes." Charges satisfy the requirements for having this status in chemistry in the case of the electrovalent bond and the chemistry of ions perfectly. To the question "Why has this ion a net negative charge?" the answer is found in the distribution of elementary charges. But the question "Why has this electron a unit negative charge?" there is no chemical answer. We have reached bedrock and as Ludwig Wittgenstein quipped "my spade is turned."

8 The Chemist as Powerful Particular

In the days before the positivistic influence led to the attempt to present the results of chemical research as if they had been achieved without the touch of human hand, the role of the active experimenter was often made very clear. In his famous study of the successive rewritings of Michael Faraday's discovery of electromotive force David Gooding (1990) showed how. step by step, all indexical features of the original research report (the one Michael Faraday made for himself in his notebooks) were gradually deleted. The final, much shorter, version lacked not only the role of the experimenter but the details of the actual procedure that would make the result easily repeatable. A very similar way for the indexical presentation of experimental results was used by Lavoisier.[7]

His account begins with the active intervention by the experimenter: "Après que tous a été ainsi préparé, *on fait rogui au feu* un fer recourbé…on la passe par-dessous la cloche…on l'approche un petit morceau de phosphore…." However, the story continues with the ascription of causal power to heat: "…légère augmention dans le volume de l'air, en raison de la dilatation *ocassionné de la chaleur…*" (from part 1 chapter 3 of *Traité elementaire de la chimie*, 1864, 41).

Later in the book (p. 654) the activity of the experimenter and of the process are again included in the same description: "*J'ai fait passer* dans un bocal l'air…la volume avait été diminué d'un onzième *par la combustion de phosphor….*" The force concept appears more overtly in the following: "tous les combinations chimique…un certain degree d'adherence plus or moins grand, en raison de la différence d'affinité qu'il avait avec des différentes substances" (488). Here we have chemist as agent and the affinity of material substances all in the same sentence.

Causality in the active sense appears in a static and a dynamic context in Lavoisier's writings—the power that holds acid and base together as a salt,

---

[7] Alchemists used the opposite procedure, inventing or adopting obscure terminology just to prevent anyone else repeating their experiments. In our day "gold" as bling is the reward of fame.

and the power to initiate or drive a process forward. Sometimes the powerful particular at work in the latter context is a human being, the experimenter. But in the former sense the agency is ascribed to material substances. For example in reporting work on combustion by Comte de Saluces (p. 48) Lavoisier uses active verbs (in English translation) "the air disengaged from effervescing substances...*extinguished* flame." And "nitrous acid, when mixed in vacuo, with a fixed alkali *produced* no air."

Conclusion

The two root ideas of causation, regularity of similar sequences of phenomena and the activity of powerful particulars in bringing about change and maintaining the interconnections between the components of stable systems have continued to be studied by philosophers and can be found in use in the characterization and explanation of a wide variety processes and contexts. In chemistry it seems that observations of regularities among phenomena do not end a causal quest but provide the occasion for undertaking a search for hidden causal mechanisms. However, research programs do not end there, but continue in the efforts to identify the powerful particulars that are the source of the capacities to bring about change and to maintain the stability of chemical structures against tendencies to disintegration and decay (Mulliken 1935). Treating the two main concepts of causality in this way allows us to find a place for both in exploring the rationale of humanity's most fascinating and practical science (Harré 2008). With respect to the main varieties of the concept of causality and the processes of causation chemistry is a hybrid science.

References

Aristotle (1984). *Physics*, II 3, 194 b 17–20. In *The Complete Works of Aristotle* (J. Barnes, ed. and trans.). Oxford: Clarendon Press.

Atkins, P. M. & Beran, J. A. (1992). *General Chemistry*. New York: Scientific American Books, Freeman.

Belnap, N. (2005). "A theory of causation: causae causantes (originating causes) as Inus conditions in branching space-times," *British Journal for the Philosophy of Science*, 56, 221–253.

Brown, G. I. (1973). *A New Guide to Modern Valency Theory*. London: Longman.

Cartwright, N. (1987). *Natural Capacities and their Measurement*. Oxford: Oxford University Press.

Cheng, P. (1997). "From covariation to causation: a causal power theory," *Psychological Review*, 104, 367–405.

Davy, H. (1807). "On some Chemical Agencies of Electricity," *Phil. Trans. Roy. Soc.*, 97, 1–56.

Dennett, D. (2003). *Freedom Evolves*. London: Allen Lane.

Ehring, D. (1997). *Causation and Persistence*. New York: Oxford University Press.

Friend, J. N. (1915). *The Theory of Valency*. London: Longmans Green.

Gibson, J. J. (1968). *The Ecological Approach to Visual Perception*. New York: Houghton Mifflin.

Gilbert, W. (1600). *De Magnete*. London.

Gooding, D. (1990). *Experiments and the Making of Meaning*. Dordrecht: Kluwer.

Graham, T. (1850). *On the Diffusion of Liquids*. London: Taylor.

Harré, R. and Madden, E. H. (1975). *Causal Powers*. Oxford: Blackwell.

Harré, R. (2008). "Some presuppositions in the metaphysics of chemical reactions," *Foundations of Chemistry*, 10, 19–38.

Harré, R. (2012). "Do explanation formats in elementary chemistry depend on agent causality?" *Foundations of Chemistry*, 13/3, 187–200.

Hiddleston, E. (2005). "Causal powers," *British Journal for the Philosophy of Science*, 56, 27–59.

Hume, D. (1739). *A Treatise of Human Nature*. London: John Noon, Book One.

Kistler, M. & Gnassanou, B. (2005). *Dispositions and Causal Powers*. Aldershot, UK: Ashgate.

Lewis, D. K. (1986). *Counterfactuals*. Oxford: Blackwell.

Llored, J.-P. (2010). "Mereology and quantum chemistry: The approximation of molecular orbitals," *Foundations of Chemistry*, 12, 203–221.

Mackie, J. R. (1974). *The Cement of the Universe*. Oxford: Clarendon Press.

McMurry. J. (2003). *Fundamentals of Organic Chemistry*. Pacific Grove, CA: Brooks-Cole.

Mulliken, R. S. (1935). "Electronic structures of polyatomic molecules and valence VI: On the method of molecular orbitals," *Journal of Chemical Physics*, 3, 376.

Newton, I. (1730) [1952]. *Opticks*. New York: Dover.

O'Connor, T. (2002). *Persons and Causes*. Oxford: Oxford University Press.

Pearl, J. (2000). *Causality: Models, Reasoning and Inference*. Cambridge: Cambridge University Press.

Purvis, M. K. & Cranefield, S. J. S. (1995). "Causal agent modelling." Unpublished paper, Victoria University, Wellington, New Zealand.

| How Properties Hold Together
in Substances

JOSEPH E. EARLEY, SR.

1 What Has Chemistry to Do with Philosophy?

A main aim of chemical research is to understand how the characteristic properties of specific chemical substances relate to the *composition* and to the *structure* of
those materials. Such investigations assume a broad consensus regarding basic
aspects of chemistry. Philosophers generally regard widespread agreement on
basic principles as a remote goal, not something already achieved. They do not
agree on how properties stay together in ordinary objects. Some follow John
Locke [1632–1704] and maintain that properties of entities *inhere in substrates*. The
item that this approach considers to underlie characteristics is often called "a bare
particular" (Sider 2006). However, others reject this understanding and hold that
substances are *bundles of properties*—an approach advocated by David Hume
[1711–1776]. Some supporters of Hume's theory hold that entities are collections of "tropes" (property-instances) held together in a "compresence relationship" (Simons 1994). Recently several authors have pointed out the importance
of "structures" for the coherence of substances, but serious questions have been
raised about those proposals. Philosophers generally use a time-independent
(synchronic) approach and do not consider how chemists understand properties of chemical substances and of dynamic networks of chemical reactions.

This chapter aims to clarify how current chemical understanding relates to
aspects of contemporary philosophy. The first section introduces philosophical
debates, the second considers properties of chemical systems, the third part
deals with theories of wholes and parts, the fourth segment argues that *closure*
grounds properties of coherences, the fifth section introduces *structural realism*
(SR), the sixth part considers contextual emergence and concludes that *dynamic
structures of processes* may qualify as *determinants* ("causes") of specific outcomes, and the final section suggests that ordinary items are based on *closure
of relationships* among constituents *additionally determined* by selection for
integration into more-extensive coherences.

1.1 Substances

Ruth Garrett Millikan discussed the concept of *substance* in philosophy:

> Substances...are whatever one can learn from given only one or a few encoun-
> ters, various skills or information that will apply to other encounters.... Further,
> this possibility must be grounded in some kind of natural necessity.... The func-
> tion of a substance concept is to make possible this sort of learning and use of
> knowledge for a specific substance.
>
> (MILLIKAN 2000, 33)

Substances necessarily persist through time and thus are distinguished from
events. Chemists, however, define chemical substances as materials of constant composition and definite properties—and also consider all systems to
be composed of smaller items. They hold that macroscopic samples are made
up of molecules and that molecules have atomic nuclei and electrons as components. All of these bits are in incessant motion, and in continual interaction with other items. The long-term stability of composite chemical entities
implies that internal motions are *somehow constrained* so that the composites
retain integrity over time and through interaction. Paul Weiss described this
situation in terms of the philosophers' concept of substance:

> Each actuality is a substance. It maintains a hold on whatever it contains, pro-
> duces, and intrudes upon. It persists and it acts. It has an irreducible, indepen-
> dent core, and receives determinations from insistent, intrusive forces.... If an
> actuality were not a substance, its parts would not belong to it, and it would dis-
> perse itself in the very act of making its presence evident. The very items which
> it dominates, it would not control; nor would it continue to be despite an involve-
> ment in change and motion. It would be inert and solely in itself, or it would be
> a mere event. In either case, it would not be a source of action.
>
> (WEISS 1959, 109)

The well-established dynamic aspect of nature requires attention to how properties of chemical entities are maintained through time, and how they maintain
integrity during interaction. Failure to consider factors involving time (use of
synchronic rather than *diachronic* approaches) is not acceptable (Humphrys
2008, 1997; Earley 2012b).

1.2 Properties and Relations

Natural human languages function *as if* all items fall into one or the other of
two great classes: *subjects* (substances, particulars, or individuals) and *predicates* (attributes or universals—including properties and relations). Aristotle's
early definition: "A substance... is that which is neither said of a subject nor [is]
in a subject" (*Categories* 5 (2^b, 13–14), Barnes 1984, 4) is similar to Bertrand
Russell's characterization: "An 'individual' is anything that can be the subject
of an atomic proposition" (Russell's "Introduction" in Whitehead and Russell
1970, xix).

Russell considered that whatever can be truly asserted concerning an individual (any predicate whatsoever) is a property (attribute) of that particular. Hilary Putnam (1969) pointed out that philosophers who use such broad understandings often run into difficulties that use of narrower property-concepts would avoid. D. H. Mellor (2006) conceded real existence only to those properties and relations that science discovers to be involved in causal laws and argued that many generally-accepted properties (including redness) *do not exist*—because any causal regularity can be dealt with without recognizing them.

Chemists generally use a restricted notion of property which was described by American chemist and philosopher Charles S. Peirce [1839–1914]—that is, a property is *how a thing behaves, or would behave, in a specified operation* (CP 8.208).[1] This usage exemplifies Peirce's Pragmatic Maxim: "Consider what effects, that might conceivably have practical bearings, we conceive the object of our conception to have. Then our conception of these effects is the whole of our conception of the object" (CP 5.402). We follow the usual practice of chemists and use Peirce's concept of property.

Properties filling either Mellor's or Peirce's property-concepts (including powers, capacities, vulnerabilities, and affordances) are classed as dispositional properties. Non-dispositional properties are designated as categorial or substantive properties. *Intrinsic* properties are distinguished from *structural* properties—features of entities that depend on relationship(s). *Relations* involve two or more individuals as *relata*. Some claim that many-place (polyadic) relations can be *reduced to* single-place (monadic) properties. But Russell argued that *unsymmetrical* relations such as *greater than* cannot be reduced to monadic properties.

In the mid twentieth century some held that all problems could be understood in terms of non-composite entities (elementary particles) with intrinsic properties. An alternative view is that every apparently-intrinsic property derives from *interaction* of less extensive components—as the mass of the proton mainly arises from *the combination* of its three component quarks, not from their intrinsic masses (Dürr, Fodor, Frison,Hoelbling, Hoffmann, Katz, et al. 2008). On the former view, intrinsic properties are primary and structural (relational) properties secondary: on the alternative basis structural properties are fundamental.

1.3 Alternatives to the Substance-Attribute Approach

Alfred North Whitehead held that: "All modern philosophy hinges round the difficulty of describing the world in terms of subject and predicate, substance and quality, particular and universal" (Whitehead 1978, 49). Whitehead *rejected* Locke's category of substance and asserted: "'Actual entities'—also termed 'actual occasions'—are the final real things of which the world is made up" (18). *Process* for Whitehead is all of a single sort—self-creation of actual occasions.

---

[1] Paragraph 208 of volume 8 of the electronic version of *The Collected Papers of Charles Sanders Pierce*. Further citations to this work will follow this style.

"Actual entities perish, but do not change; they are what they are" (35). Actual occasions are not substances—they come to be and, in so doing, perish.

Donald W. Mertz (1996, 2003) avoids the substance-property distinction in a different way. Mertz's "Instance Ontology" operates with a single ontological category—called property instance, state of affairs, or fact of relationship. By recognition of intension-types (universals), Mertz's approach qualifies as realism. Each state of affairs corresponds to *unification* of its relata. Gilbert Simondon recommended that we should "seek to know the individual through individuation rather than individuation through the individual" (Simondon 1964, 22). Whitehead made the *achievement* of individuality by each actual occasion a focus of his system. Mertz does not deal with the *process* of individuation.

1.4 Substrate-Bundle Debates

Locke admitted that the concept of substratum was "only a supposition of he knows not what support" of properties (1690, chapter XXIII, section 2): in an Appendix to his *Treatise,* Hume (1739) reported: "I am sensible, that my account is very defective." In the end, Hume was as dissatisfied with his bundle approach as Locke had been with his substratum theory: Present-day advocates of either approach rarely express similar reservations.

Jiri Benovsky (2008) examined several versions of both substrate and bundle theories—some involving tropes and others recognizing universals, some with a single identical unification-relation for all objects, others with variable numbers of relata (*polyadicity*), and still others with distinct unification-relations for each object. Benovsky considered how proponents of each of these versions defended against objections and concluded that both substratum and bundle theories *share a common central postulate*—the concept that each object has a unifying feature. (This is the substrate/bare particular or the compresence relationship.) He identifies these as *theoretical entities* (items "individuated by their theoretical role") and points out that all "play the same role in the same way" in all their applications—therefore they are "identical (*metaphysically equivalent*)" (183). He concludes that substratum and bundle theories are twin brothers not enemies—and that both are seriously deficient, since neither has clarified the unification that they both require. That unification remains a "something, he knows not what" as it was for Locke: both substrate and bundle approaches are still "very defective" as Hume found his own theory to be.

1.5 A Proposed Principle of Unity for Bundles

David Robb (2005, 467) described an explicit search for a *principle of unity* for properties. A certain tennis-ball (called Alpha) figures in these discussions. Robb begins:

> It's hard to deny that there are *natures* or *ways of being*.... How could there be being without a way of being? To be is to be some way or other. I'll call these ways *properties*.... Ordinary objects (chairs, trees, human beings, electrons, stars)...are

merely bundles of properties. Not only must a being be some way or other, it is *exhausted* by ways of being.

Robb holds that an object (such as Alpha) "is a unified, persisting, independent being" (468) but does not examine the basis of such unity, persistence, and independence. Although he considers that objects necessarily persist, Robb specifically states "I will not address diachronic unity here" (476). His treatment is synchronic. Robb states: "Whether something counts as an object may for the purposes of this paper be taken as a primitive fact about it" (468). Elsewhere (Robb 2009) he expands on various aspects of notions of substantiality—but does not propose a criterion of what unifies an object. He holds that parts of objects that are themselves objects are substantial parts: parts that are properties (rather than objects) are qualitative parts. Each object has both substantial and qualitative unity—these are distinct, but "we should expect there to be some systematic relations between them such that our choice of one constrains our choice of the other" (474).

Robb declines to express an opinion as to what the principle of unity of substantial parts might be since: "My concern here in defending the bundle theory is only with the principle of qualitative unity" (474). He uses David Armstrong's concept of a *structural* property: "a kind of complex property, one composed of the properties of and, in most cases, relations among [an] object's parts" (476). Adding: "to say that certain substantial parts are *exhaustive* at a particular mereological[2] level means that those parts are *all* of an object's parts at that mereological level" (477), he proposes a *qualitative principle of unity:*

> (CU) For any substantially complex object **O** and properties F and G, F and G are qualitative parts of **O** iff F and G are both structured on the (exhaustive) substantial parts of **O** at some mereological level.

That is to say, if an object has parts that are themselves objects, then the properties of the composite object are some combination of the properties of the parts—providing that all the parts identified at a particular mereological level are taken into account. Those resultant object-properties may be regarded as parts of the object, but they are parts in a qualitative sense that is different from the usual (substantial) sense of parthood. Robb assumes that the properties of ordinary objects derive directly and exclusively from the properties of their components.

The principle of unity (CU) does not apply to simples—objects that have no substantial parts. Robb presumes that simples exist and invokes the principle that "if no objects exist in their own rights then no objects exist at all" (485). Robb proposes: "A simple object...just *is* a single, simple property" (486) and formalizes that assertion as:

> (SU) For any substantially simple object *O* and properties F and G: F and G are *qualitative parts* of *O* iff F and G are each identical with *O*.

---

[2] Mereology is defined as "the abstract study of the relations between parts and wholes" [*Shorter Oxford English Dictionary, Volume I.* (1993). Oxford: Clarendon Press, 1747]. See section 3.1 of this chapter.

Robb argues vigorously that SU does not involve the category-mistake of confounding (adverbial) ways of being with (substantive) entities, but observes that if SU does involve a category-mistake then "it's one that bundle theorists have been making all along" (486).

2 Properties of Chemical Systems

Some characteristics of chemical systems (called *molecular* properties) depend mainly on the characteristics of components of those systems together with how those constituents are connected. In contrast *spectroscopic properties* involve *transition* between energy states of entities—with concomitant emission or absorbtion of energy. *Chemical properties* involve interaction of substances with like or different others to bring about transition to alternative compositions—thereby producing new *connectivities* (three-dimensional arrangements) of elemental centers ("atoms"; Bader and Matta 2013).

2.1 Molecular Properties

In chemical entities, attractive forces (such as between unlike electrical charges) pull components together; repulsive interactions (as between like charges) drive constituents apart. As parts separate, attractive forces draw fragments together. Distances between components change continuously, but remain within limits—due to *balance* of attractive and repulsive interactions. *Closure* of relationships enables each chemical entity to retain self-identity through interactions.

Molecules adopt the spatial configuration with lowest potential energy that is consistent with constraints which obtain. For dihydrogen, a minimum of potential energy occurs at *a single* internuclear distance (the H–H bond-length). More-complex chemical entities are described in multidimensional configuration space. Potential-energy minima in configuration space correspond to more or less stable molecular structures: the minimum of lowest energy is the equilibrium structure (Earley 2012b). Vibrations around such structures occur, and are more vigorous at higher temperature. Stability corresponds to *closure of relationships among components* which leads to *a minimum of potential energy* for a specific connectivity of elemental centers. The details of each such "potential-well" determine how each system interacts, and specifies many properties of the coherence. Closure of relationships of components is what accounts for the coherence of diverse properties for each molecular substance. Such closure allows each molecule *to make a characteristic difference* and so have ontological significance (Earley 2008a, Ney 2009).

The factors that determine the variation of potential-energy with internuclear distance for the dihydrogen molecule are well understood. Electrostatic attraction between negatively-charged electrons and positively charged protons, mutual repulsion of like-charged electrons (and protons), the fact that the internal energy of atoms and molecules is restricted to certain specific values (quantized), and the Pauli Exclusion Principle (which limits occupany of each molecular energy-level to two electrons) are all major factors. For the dihydrogen molecule, theory-based (a priori) calculations of molecular properites—dissociation energy, equilibrium internuclear distance, rotational moment of inertia—agree well with experimentally-determined values. Approximate calculations for more-complex molecules—usually involving parameters estimated from experiments—yield approximate predictions of characteristics of chemical molecules, even for some quite complicated ones.

Properties that depend only on the mass and/or volume of molecules derive directly from phenomena that potential-energy–versus–internuclear-distance curves describe. Certain (colligative) properties of solutions depend on the *number* of solute units (molecules or ions) dissolved in a given volume of solvent, rather than on the properties of the individual molecules. Properties such as the melting temperature of a solid, the boiling temperature of a liquid, the critical temperature of a substance all depend on mutual interaction of molecules of a single type. Both colligative and phase-change properties might well be grouped with molecular properties.

2.2 Spectroscopic Properties

Like electronic energies, vibrational and rotational energies are *quantized*. Possible energy-levels are separated more widely for electronic states, less widely for vibrational states, and quite narrowly for rotational states. Transition from one energy level to another involves absorption or emission of energy. If a dihydrogen molecule (H_2) absorbs a photon of appropriate energy, a transition from a lower-energy electronic level to a higher-energy electronic level would occur. Once the system was in the upper level (with a longer equilibrium-distance) vibrations and rotation could occur but sooner or later either a photon would be emitted and the molecule would relax to the lower level, or the molecule would split to produce energetically-excited hydrogen atoms. Transitions between energy levels provide a way by which molecules interact with the rest of the world while retaining integrity (for dihydrogen, by maintaining the H-H bond). Under usual conditions, some chemical coherences persist indefinitely, others have short lifetimes, and many are evanescent. Arguably, persisting long enough to undergo rotation is the lower-limit of molecular existence (Earley 1992).

The geometric structures of molecules and crystals partially determine their spectra. Details of structure can often be inferred from spectral measurements. *Spectroscopy*—study of the energy absorbed or released when chemical systems change from one energy level to another—accounts for much of the effectiveness of modern chemical science.

2.3 Chemical Properties

Chemistry turns less-valuable materials into more-valuable items. When materials are mixed, new substances appear and old ones vanish as component elementary centers rearrange. Every reaction involves a *decrease* in free energy—a lowering of potential for chemical reaction. Reaction continues, however slowly, until a minimum of free energy is reached—at a condition of chemical equilibrium—and no further net change occurs although forward and reverse reactions continue at equal rates.

Any given starting mixture conceivably could produce a variety of final products. Principles governing alternative changes are easy to state: Applying them requires experience and skill. The two main considerations are *thermodynamic stability* and *kinetics* (speed). If all reactions occur rapidly, the final state will be the condition that has the lowest free energy—total energy decrease due to chemical bonding adjusted to take account of the complexity of the structures involved (less-complicated structures have an advantage over more-complex ones). Every chemical change involves decrease in thermodynamic reaction-capacity of the system—reduction of free energy or chemical potential. Chemical reaction corresponds to transition between stable states—movement of a system from one potential-well to another. Chemical reactions correspond to production of new closures—every chemical process is a becoming (Earley 1998).

Transition from reactant to product potential-wells necessarily involves passage through higher-energy arrangements of components which are *intermediate* between the configurations of the starting materials and products. *Slower* rates of chemical reaction involve traversing patterns corresponding to higher potential-energy barriers between reactant and product potential-wells: faster reactions involve lower barriers. For instance, change from the stable *cis*-conformation of 1,2-dichloroethane ($C_2H_6Cl_2$) to the equally-stable *trans*-configuration of the same molecule occurs rapidly. The corresponding reaction of 1,2-dichloroethylene ($C_2H_4Cl_2$) is slower because the stronger double bond between carbon centers resists twisting.

It is a serious error to focus only on *thermodynamic* factors and to ignore *kinetic* (reaction-rate) considerations. Most chemically interesting reactions are *controlled* by kinetic influences, so that history must be taken into account. Quite usually, the product that results from a chemical change is not the thermodynamically most stable product but some higher-free-energy form(s) (Earley 2012b).

Rates of chemical reactions are usually discussed *as if* intermediate "transition" states had real existence. This approach forms the basis of a myriad of chemical explanations of widely varying sophistication. Eugene Wigner suggested that reaction-rate parameters should be computed directly from molecular dynamics, as a preferable alternative to the thermodynamics-related transition-state approach to understanding reaction rates (Jaffé, Kawai, Palacián, Yanguas, and Uzer 2005). Molecular-dynamics calculations can now compute the motions of some tens of individual atoms during reaction. For instance, modeling of the isomerization of HCN to produce HNC (Ezra 2009) shows that quasi-periodic trajectories and multidimensional chaos exert significant influences on chemical reactivity. The transition-state approach does not take account of such effects: To the extent that they are important that approach may lead to wrong conclusions.

Typically, chemical reactions happen through several rather distinct sub-processes, which often involve intermediates of significant stability. Such a series of steps is called a *reaction mechanism*. The reaction rate for each such mechanistic step is subject to many influences, generally different for each step. Sometimes, one step will be so much slower than the others that it will *mainly determine* the overall reaction rate but usually several reaction steps will be significant for determining the overall rate of reaction.

It sometimes happens that reaction-rates for a series of related reactions follow the trend in thermodynamic driving-force (free energy) for the same series—but *there is no necessity* for this to be the case. Reactions with highly favorable free-energy changes may well be slow; reactions with low thermodynamic driving-force might be rapid. *Which product* (of the myriad possible in a particular case) results from a mixture of reactants is usually determined by the relative rates of *many competing reactions*. Rarely does a collection of starting materials generate a single product: usually several products occur in proportions which depend sensitively on environmental conditions. Large networks of reactions (such as occur in biochemistry, astrochemistry, or geochemistry) are governed by the relatively-simple principles just summarized but each type of reaction-network has additional special features.

2.4 Relative Onticity

Early in the twentieth century, a disagreement between physicists attracted attention (Atmanspacher and Primas 2003, 301). Albert Einstein held: "Physics is an attempt conceptually to grasp reality as it is thought independently of its being observed." In contrast, Niels Bohr warned: "It is wrong to think that the task of physics is to find out how nature is. Physics concerns what we can say about nature." It became customary to distinguish scientific statements as either *ontological* or *epistemological*. The former deal with how things *really are*: the latter concern how things *appear to be*. Eventually most scientists came to assume that only the a priori calculations of quantum physics merited the ontological designation. Admixture of a posteriori considerations was considered to relegate a statement to mere epistemological status. The opinion that chemical propositions *could not attain* to ontological status became widespread: this view still has adherents (e.g., McIntyre 2007). But this clear position is subject to Putnam's objection:

> Once we assume that there is, somehow fixed in advance, a single 'real,' a single 'literal' sense of 'exist'—and, by the way, a single 'literal' sense of identity—one which is cast in marble and cannot be either contracted or expanded without defiling the statue of the god, we are already wandering in Cloud Cuckoo Land.
>
> (PUTNAM 2004, 84)

Chemists believe that they clarify the *real order of things* when they provide detailed accounts of the diverse chemical species with which they deal. But they employ a multi-level system of entities (Earley 2003b)—they use a *relative onticity.* Which levels of entities are to be employed in a discourse is decided *during* the formulation of the discourse—not in advance, as many logical systems assume.[3]

> The distinction of epistemic and ontic descriptions can be applied to the entire hierarchy of (perhaps partially overlapping) domains leading from fundamental particles in basic physics to chemistry and even to living system in biology and psychology. Ontic and epistemic descriptions are then considered as relative to two (successive) domains in the hierarchy.... While atoms and molecules are epistemically described within the domains of basic physics they acquire ontic significance within the domain of chemistry...the central point of the concept of relative onticity is that states and properties of a system which belong to an epistemic description in a particular domain can be considered as belonging to an ontologic description from the perspective of another domain.
>
> (ATMANSPACHER AND PRIMAS 2003, 311)

3 Coherence

3.1 Wholes and Parts

According to the dictionary definition given in note 2 any theory of wholes and parts is "a mereology." However, the specific part-whole logical system that Peter Simons (1987, 1) calls Classical Extensional Mereology (CEM) has influenced English-speaking philosophers so deeply that that 'mereology' often refers exclusively to CEM. Three basic axioms of CEM (Lewis 1991, 74) are

> *Transitivity*: If x is part of some part of y, then x is part of y.
>
> *Unrestricted Composition*: Whenever there are some things, there exists a fusion of those things.
>
> *Uniqueness of Composition*: It never happens that the same things have two different fusions.

The first axiom is relatively non-controversial: The knob of a door in a house is a part of the house. The second axiom seems troublesome, since it gives rise to statements such as that of Willard V. Quine: "There is a physical object part of which is a silver dollar now in my pocket and the rest of which is a temporal segment of the Eiffel Tower through its third decade" (Quine 1976, 859). However Quine also states:

> Identification of an object from moment to moment is indeed on a par with identifying an object from world to [possible] world: both identifications are vacuous, pending further directives.... The notion of knowing who someone is, or what

---

[3] Ruth Barcan Marcus (1963) proposed a *substitutional* interpretation of first-order logic: This interpretation does not require specification of a universe of discourse in advance but rather admits all *entities that figure in true statements.* Her interpretation (also called "truth-value logic") should be more appropriate for chemical inquiry than is the standard ("objectual") logical approach.

something is, makes sense only in the light of the situation. It all depends on what more specific question one may have had in mind.

One can understand this as giving the axiom of unrestricted composition a *pragmatic* interpretation: what sorts of items should be in a universe of discourse depends on the *purpose* of the discussion—if it should be useful in some particular inquiry, any grouping of items might properly be taken to comprise a single unit. On that interpretation, the principle of unrestricted composition would be acceptable, but with an odd notion of existence.

The third axiom of CEM (uniqueness of composition) clearly is incompatible with chemical understanding. *Normal butane*, $CH_3CH_2CH_2CH_3$, and *isobutane* $(CH_3)_3CH$, have quite distinct properties but identical atomic-level constituents (C_4H_{10}). This is precisely what the third principle of CEM asserts *does not occur*. CEM is clearly *not appropriate* for *chemical* wholes and parts.

William Wimsatt identified "four conditions [that] seem separately necessary and jointly sufficient for *aggregativity* or non-emergence"—the situation in which the axioms of CEM might apply. One is *absence* of cooperative or anti-cooperative interactions: any such nonlinearity would make CEM inapplicable. The other three conditions require *invariance of all system-properties* under certain operations: (1) rearranging the system parts, (2) decomposing the system into its parts and then recombining the parts to reconstitute the system, and (3) subtracting parts of the system or adding more similar parts. If operations (1) or (2) result in any change—or there is a *qualitative* change under operation (3)—then the system is not a mere aggregation and CEM *cannot* apply. Wimsatt concludes: "*It is rare indeed that all of these conditions are met*" (Wimsatt 2006, 675, emphasis in original).

Wimsatt developed a trichotomy in which *aggregates* (to which CEM may apply) are clearly distinguished from *composed systems*, in which details of connectivity (in what ways components relate to each other) strongly influence the characteristics of the system, and a third class is *evolved systems*—those that have developed through historical selection processes so that no way of identifying component parts has priority. In such systems, there is no unambiguous way to determine what the *parts* of the composite system are: investigations carried out for diverse purposes will identify different items as parts of the system. Investigators who prefer to deal with decomposable coherences avoid such systems—but that does not show that they are either unimportant or uninteresting. Chemical systems are composed systems rather than aggregates: Biological and ecological systems generally fall in the evolved-systems category. Since atoms of most chemical elements were generated by contingent processes in exploding stars, chemical entities might also be properly regarded (for some purposes) as evolved systems.

Philosophical discussions generally assume (e.g., Robb 2005, Vander Laan 2010), often implicitly (Armstrong 2010), that items do not change when they become parts of wholes—so that there is no ambiguity in identifying which individuals are the parts of a composite whole. Chemical and biological experience clearly indicates that entities included as parts of compound individuals

HOW PROPERTIES HOLD TOGETHER IN SUBSTANCES | 209

are greatly changed by their inclusion in such wholes. Items included in coherences often (but not always) retain their individuality—but their properties in the whole are generally not the same as the characteristics of similar items outside such coherences.

Simons (1987) pointed out that the Polish logicians from whose work CEM derives had strong prior commitments to nominalism—the doctrine that *only individuals* actually exist—and that therefore they eschewed recognition of universals (properties and relations). Contemporary nominalists (e.g., Lewis 1991) also regard universals as no more than linguistic conveniences. For them, only particulars *really* exist. In contrast, *ante rem* (or Platonic) realists—perhaps including David Armstrong (2010)—hold that properties and relations, although they are not located in space or time, *really exist* independently of (and in some sense prior to) the individuals that constitute instances of those universals. Several positions (jointly designated moderate or *in rem* realism) are intermediate between nominalism and *ante rem* realism: they typically hold that properties and relations are important—but they only exist as *features of individuals*. Kit Fine (2006) and Kathrin Koslicki (2008) endorse the notion that arrangements of components—structures—are essential to the coherence of objects, and therefore structures should be considered *proper parts* of objects. There is an urgent need for formal development of alternative mereological systems[4] that will have wider applicability than CEM does (Llored 2010, Harré and Llored 2011).

David Vander Laan (2010, 135) reports that the question "'Under what conditions do some objects compose another?' has increasingly been recognized as a central question for the ontology of material objects." This query (known as "the composition question") consists of two sub-questions:

1. How can several items *function as a single unit* in causal interactions?
2. What characteristics must individuals have in order to constitute such a causally-effective unit?

The first is the external (or *epistemological*) sub-question: the second, the *ontological* sub-question.

3.2 Symmetries and Groups

Three of Wimsatt's four conditions for aggregativity involve *invariance of system properties under a specified operation*. This characteristic of a system is called

---

[4]A. N. Whitehead sketched out an alternative mereology even before CEM was developed (Simons 1987, 81–86). Whitehead based his system on an insight that he later summarized as: "However we fix a determinate entity, there is always a narrower determination of something which is presupposed in our first choice, also there is always a wider determination into which our first choice fades by transition beyond itself" (Whitehead 1967, 93). This can be stated more formally as: *for every particular x there exists an entity y that extends over x, and there also exists an individual z that x extends over*. Whitehead did not complete development of his mereological system, but his basic approach seems more consistent with chemical practice and its results than is CEM. Both Whitehead's multi-level mereology and Ruth Barcan Marcus' substitutional logic can accommodate the importance of *purposes* in scientific representation.

symmetry. Wimsatt's conclusion can be rephrased as: CEM applies if the system of interest involves *neither* cooperative or inhibitory interactions, and *also* is *symmetric* with respect to *all three* of the operations, (1) rearranging constituent parts, (2) disassembling and reassembling those fragments, and (3) adding components similar to those already present, or subtracting some of the original parts. Systems to which *all four* of these conditions apply are *extremely rare*.

Symmetry conditions such as Wimsatt invokes are at the center of currently-used tests for the identity of entities. Galileo Galilei's seventeenth-century research led to general and explicit acceptance of the uniformitarian doctrines that physical laws are the same in all parts of the universe, that all times are equivalent, and that no spatial direction is preferred. On this basis, physical objects can properly be described at any specific time (past, present, or future) and using any convenient coordinate system (oriented in any direction). Since Galileo also established that acceleration (rather than velocity) is what is important, objects can be represented as moving with respect to an external reference point at any constant velocity. (This is the Galilean principle of relativity.) On this basis, the validity of a description would not change on alteration of the single time specification, the three spatial coordinates, the three orientation angles, or the three velocity components (boosts). Any entity that is not changed by any of these ten transformations (alone or in combination) is considered to be a Galilean particle (Castellani 1998)—that is, such a coherence is properly regarded as a *single individual existent*.

Each of the entities considered in current particle-physics displays *symmetries* (invariance under appropriate operations) that correspond to (represent) one or another of certain mathematical objects called *groups*. A group is a special kind of set (a collection of elements) for which applying a stated procedure (the group operation) to any two members of the group generates a *member of the group*—and not something else. This closure requirement is a severe one: Groups are rare among sets (Joseph E. Earley 2013).

The ten operations under which Galilean particles are symmetric (single time-specification, three spatial coordinates, three orientation angles, and three velocity components) constitute the defining operations of the Galilei Group, designated G. Each of the many types of Galilean particles corresponds to a representation of G. Since all the component operations of G can be carried out to greater or lesser degrees without restriction (like the rotations of a circle about its center), G is a continuous group rather than a discontinuous or discrete group. (Rotation of a regular polygon around its center corresponds to a discrete group.)

3.3 Extensions of the Symmetry Concept

In 1929 Hermann Weyl revived a proposal he had made in 1918 that descriptions of electromagnetic systems would be unchanged by a gauge transformation—an operation by which λ, the phase[5] of the wave function ψ, was changed

---

[5] The fraction (say, 45°) of the complete (360°) oscillation completed at a given instant.

by an amount that differed in various locations ($\lambda = \lambda(x)$)—and a corresponding term was added to the electromagnetic potential f_p.[6] With respect to this transformation, Cao observed:

> In this way the apparent 'arbitrariness' formerly ascribed to the potential is now understood as freedom to choose any value for the phase of a wave function without affecting the equation. This is exactly what gauge invariance means.
>
> (CAO 1988, 120)

By 1927 Wolfgang Pauli had rationalized Pieter Zeeman's 1896 discovery that magnetic fields *split* spectral lines, on the basis of the proposal that electrons possessed intrinsic angular momentum (spin): he also explained the 1922 demonstration by Otto Stern and Walther Gerlach that magnetic fields split beams of silver atoms, by postulating that silver ions also had spin. In 1928 Paul Dirac rationalized the spins of the electron and of the silver ion by combining gauge transformations with Einstein's Special Theory of Relativity. The Poincaré Group—the product of the ten operations that define the Galilei group G and the four Lorentz transformations[7]—defined each particle, as the Galilei group had done in ordinary mechanics.

After the neutron (discovered in 1932) turned out to be nearly identical to the proton (except for electrical charge), Eugene Wigner (1937) proposed that the proton and neutron should be regarded as *alternative energy states* of a single particle-type—resembling the two (+½ and –½) spin-states of the electron. He suggested the name isospin for whatever characteristic of these entities might be analogous to electron spin. Isospin, so conceived, does not have units of angular momentum as electron spin does—the name is *clearly metaphorical*. In 1954, after the detection of other subatomic particles, Chen Yang and Robert Mills proposed that the proton, the neutron, and other hadrons should be considered to be interrelated by gauge symmetries involving isotopic spin (also called isospin or, preferably, isobaric spin) which could assume values including ½, 1, and 3/2. This approach predicted the existence of a number of previously-unknown particles that subsequently were experimentally observed.

It turns out that the symmetries corresponding to the electronic spin and isobaric spin are *not* continuous, as are direct coordinate transformations. For instance, there are *only two* values of electronic spin (+½ and –½). The more-complicated groups that replaced the Yang-Mills group also have only a rather small number of representations, corresponding to the proton and neutron and to the several excited states (other hadrons) that derive from them. Since only certain specific system-states fulfil the group requirements, the groups pertaining to those symmetries are designated as *discrete or non-continuous* groups.

---

[6] That is, $\psi \rightarrow \psi'' = e^{i\lambda}\,\psi$ and $f_p \rightarrow f_p'' = f_p - \partial\lambda/\partial x_p$.

[7] The Lorentz transformations for a body moving with constant velocity (v) in the x direction are: $z'' = z$: $y'' = y$: $x'' = k\,(x - v\,t)$: $t'' = k\,(t - v\,x\,/\,c^2)$, where $k = (1 - v^2\,/c^2)^{-1/2}$ and c is the velocity of light.

Extensions and modifications of the Yang-Mills proposal (and also subsequent experimental developments) eventually led to the current Standard Model of Particle Physics (Cottingham and Greenwood 2007) which treats hadrons as composed of spin one-half fermions held together by bosons—vector particles of unit spin that arise from quantization of gauge fields (as *the photon*, the boson that carries electromagnetism, arises from quantization of the electromagnetic gauge field).[8] Spatiotemporal symmetries (such as those of spatial rotation) are called external, global, or geometric: Phase-symmetry in electrodynamics and isospin-symmetries are designated internal, local, or dynamic: this usage corresponds to a *major extension* of the concept of symmetry.[9]

The principle that every individual corresponds to a representation of a mathematical group responds to the external (epistemological) part of the composition question: the requirement that operations which combine members of groups *always produce members of the group* answers the internal (ontological) sub-question. Stability is achieved only when relationships internal to each item demonstrate such *closure* that states of the system repeatedly reoccur—so that the system persists through time. This restriction is severe. Section 4 describes several ways in which that restriction may be satisfied.

4 Varieties of Closure

4.1 Closure Louis de Broglie

After the First World War Louis de Broglie (1892–1987) resumed his university studies and collaborated with his older brother Maurice, an accomplished physicist. In 1922, the brothers de Broglie confirmed the astonishing report that electrons produced by impact of X-rays on metals had velocities *just as large* as the velocities of the electrons used to generate those X-rays. This was as if throwing a log into a lake should cause one similar log (out of many on the other side of the lake) to jump up *with equal energy* (de Broglie and de Broglie 1922).

At that time, theories of the internal structure of atoms that had been devised by Bohr and by Arnold Sommerfeld could adequately rationalize available data on the line spectra of atomic hydrogen and ionized helium—but only by making the assumption that electrons within atoms are restricted to having *only certain energy-values*. Those early versions of quantum mechanics (QM) *could not explain*: "why, among the infinity of motions which an electron ought to be able to have in the atom according to classical concepts, only certain ones were possible" (de Broglie 1965, 246). Louis de Broglie deduced from Einstein's theory of relativity and early versions of quantum theory that each material object (including electrons within

---

[8] The Standard Model does not have a single theoretical basis, but rather is a collection of ad hoc sub-models (MacKinnon 2008).

[9] Eugene Wigner remarked: "The concept of symmetry and invariance has been extended into a new area—an area where its roots are much less close to direct experience and observation than in the classical area of space-time symmetry…(Wigner 1967, 15).

atoms) must have wave-properties. He wrote: "The fundamental idea of the theory of the quantum is the impossibility of depicting an isolated quantity of energy that is not accompanied by a certain frequency" (de Broglie 1972, 1316).

In a 1923 *Comptes rendus* note, Louis de Broglie proposed that each electron had *two* characteristic frequencies, an *intrinsic* frequency associated with its rest-mass, and a second frequency that pertained to a wave that moved in the same direction as the electron. With respect to electrons moving on a closed trajectory, he introduced the following postulate:

> It is *almost necessary* to suppose that the trajectory of the electron will be stable *only if* the hypothetical[10] wave passing O' catches up with the electron in phase with it: the wave of frequency v and speed c $/\beta$ has to be in resonance over the length of the trajectory
>
> (DE BROGLIE 1923A, 509).

That is to say, an atomic system would be stable (that atom would persist as a unified entity) *only if* the extrinsic wave remains in phase with the internal vibration of the electron corresponding to its intrinsic frequency. The extrinsic wave was then called the phase wave.

In fulfilling this postulate, each possible arrangement of an atomic system engenders another subsequent state of the system that also fulfills the criterion. This feature insures that the system, as a coherent unit, persists through time. The *only if* in the postulate (emphasized in the original French) identifies a clear requirement that a specific relationship between multiple quantities (phases of two vibrations in this case) must obtain in order for a composite system to be stable. The essential novelty of Louis de Broglie's contribution was to provide *a clear and specific criterion* that served as a basis for understanding why some energy-states of electrons in atoms were capable of persistence while other apparently-equivalent possibilities could not attain such longevity. On the basis of this postulate, Louis de Broglie was able to derive the Bohr-Arnold Sommerfeld criteria for the stable energy-states of the electron in the hydrogen atom. This was a great triumph. In a second 1923 note, he predicted that (material) electron-beams would be diffracted as light-beams are (de Broglie 1923b).

Early in 1927, two groups of experimentalists independently confirmed de Broglie's prediction of electron diffraction—validating a second major achievement for him. On the basis of his 1924 doctoral thesis, Louis de Broglie received the 1929 Nobel Prize for Physics. Through Einstein's mediation, that thesis also had facilitated the 1925 development of Schrödinger's wave equation—from which the many impressive results of quantum mechanics flow. However, even before Stockholm Nobel festivities celebrated de Broglie's award, his realistic but non-local pilot-wave interpretation of QM had been supplanted (Bonk 1994) by the non-realistic but local Copenhagen Interpretation (CI). The main proponents of CI were strongly influenced by anti-realistic philosophical ideas current in postwar (Weimar) Germany (Cushing 1994).

---

[10] This replaces "fictional" as a translation of the French *"fictive."*

Closure of networks of relationships (*Closure Louis de Broglie*) is necessary for persistence of stable things, including chemical entities. This requirement has not been emphasized in past and current philosophical discussion of how properties cohere, but it does respond to the second (ontological) part of the composition question: How is it that several Xs can constitute a Y? Criteria analogous to the one that de Broglie discovered provide the basis by which each of the possible states of a system produces a subsequent state that has the same characteristics. This suggests that it should be possible to interpret each of those states as a representation of some mathematical group. Similar situations should obtain for every persistent chemical entity.

4.2 Closure Henri Poincaré

Philosophic discussion of composite objects generally assumes that properties of composites depend only ("supervene") on the properties of components (Armstrong 2010, 29–32)—the properties of the least-extensive level are held to determine all properties. Whatever its strengths in other fields may be, this doctrine *does not apply* in chemistry. Chemical entities arise from and are sustained by *interactions* among their constituents —and mainly derive their characteristics from those interactions, rather than from properties of the components.

Characteristics of dilute gases can sometimes be inferred from information regarding component molecules—based on the approximation that each molecule acts independently. However, when gases are cooled correlations of molecular motion develop, first over short ranges then over longer distances. Motions of individual molecules become interrelated—so that the properties of the macroscopic sample *cannot* be inferred by mere addition (or other straightforward combination) of the properties of the component molecules.

Similarly, the simple model that chemical reactions occur by *elementary steps*—events that involve instantaneous collisions of pairs of molecules— seldom is even approximately correct. Generally, in reactions of gaseous species "sticky" collisions produce *resonances* (transient aggregates) and give rise to *correlations* among the properties of molecules. Reactions in condensed media (e.g., fluid solutions) are even more subject to cooperative influences than gas-phase reactions are.

In both cool-gas and chemical-reaction cases, as correlation increases application of fundamental theory rapidly becomes unwieldy and impracticable. At certain parameter-values (singularities) computations become *impossible in principle*—as computed quantities go to infinity. In the late nineteenth century, while dealing the motions of the planets in the solar system, Henri Poincaré (1854–1912) encountered situations in which standard methods of physical mechanics failed due to the presence of singularities. In such cases, when near singularities, he replaced variables with divergent series (*asymptotic expansions*). This technique (Berry 1994) allowed otherwise intractable problems to be handled— but it had unexpected consequences. Often, after the variable-replacement, the equations that best described the situation *changed discontinuously* at the singularity. Poincaré found that beyond the singularity relatively simple expressions—that emphasized contextually important features and suppressed irrelevant detail—often applied. Similar situations occur in electrodynamics. Interaction of the charge of the electron with the vacuum complicates computation of the charge and mass of the electron from experimental data (Teller 1988)—but renormalization (a procedure related to the asymptotic expansion method that Poincaré devised) makes a self-consistent theory of electrodynamics possible. Similar approaches facilitated adequate understanding of solid-liquid-gas phase changes, critical behavior for pure substances, and transitions between ferromagnetic and paramagnetic behavior (Batterman 2011): analogous techniques are also used in molecular-dynamics-based approaches to chemical reaction kinetics (Jaffé, Kawai, Palacián, Yanguas, and Uzer. 2005, 194).

The simpler description that asymptotic expansion yields at and after singularities typically has different semantics (another *topology*) than the fundamental-level description that applied before the singularity. Properties of correlated systems *do not supervene* on properties of components—they require quite new and topologically-incommensurable descriptions (Atmanspacher, Amannn, and Müller-Herold 1999; Batterman 2011; Bishop 2012; Bishop and Atmanspacher 2006; Primas 1998, 2000).

When cooperative interaction of units becomes dominant, situations adequately described by fundamental theories change into situations that require approaches that use quite different sets of entities, and other kinds of relationships among entities than are used in the fundamental theories. Transition between these diverse *topologies of explanation* may properly be regarded as *a second kind* of closure (*Closure Henri Poincaré*). Like Closure Louis de Broglie discussed in the previous section, Closure Henri Poincaré vitiates assumptions that properties of entities can adequately be understood on the basis of descriptions based on properties of component parts. That sort of closure also underlies the persistence and effectiveness of higher-level entities that result from lower-level cooperativity.

This situation is well described by Hans Primas: "The task of higher-level theory is not to approximate the fundamental theory but to represent new patterns of reality" (Primas 1998). Physics Nobel laureate Robert Laughlin argued (2005) that aspects of fundamental physics *result from* cooperative interactions.

4.3 Closure Ilya Prigogine

Investigation of the properties of networks of processes (such as interconnected chemical or biochemical reactions) is an active field of twenty-first-century science. Such studies show that *modes of dynamic closure* determine the coherence of properties of important systems, including flames, hurricanes, biological organisms, ecologies, and human societies. Chemical systems conveniently clarify fundamental principles of such coherences.

Networks of chemical processes operating in far-from-equilibrium conditions often give rise to *oscillation* of concentration of components. If the system is *closed*, such repeated concentration-variation eventually dies out. However if the system is *open*, so that reactants can continuously enter and products exit, then oscillations may continue indefinitely. In stirred systems, chemical oscillations may be either gradual or sharp: In the latter case, immense numbers of chemical ions or molecules undergo coordinated chemical change in a fraction of a second. Such dynamic chemical coherences necessarily degrade chemical energy and increase *entropy* (overall disorder). This type of organization is called dissipative structure. Chemists—particularly Belgian physical chemist Ilya Prigogine (1917–2003, Nobel Prize in Chemistry 1997)—have made major contributions to our understanding of such coherent organization of processes in far-from-equilibrium open systems (Prigogine 1977, Kondepudi and Prigogine 1998, Earley 2012a, Lombardi 2012).

Each chemical dissipative structure involves a reaction (or set of reactions) that *gets faster as it goes on*—an autocatalytic process. The simplest example is A + B → 2 B (where the rate of reaction is proportional to the product of concentrations of A and of B). In this case, B is a *direct autocatalyst*. Many kinds of interaction that facilitate coordinated or cooperative functioning produce *indirect* autocatalysis (e.g., Sugihara and Yao 2009). Systems that involve direct or indirect autocatalysis tend to be *unstable*—they readily explode and disperse. However, if such a system also involves one or more processes that *suitably reduce* the autocatalysis (such as B + C → D) then the system may return to its original condition, and even do so repeatedly, thereby generating *continuing oscillations*—a type of long-term persistence (Earley 2003a).

When oscillations occur in open-system chemical systems, variations of concentrations of several components can be followed often over long time periods. Careful study of such *time-series* (along with relationships among times-series for diverse components, and responses of the system to perturbations) yields information on the details of the chemical reactions underlying the continuous oscillations. When concentrations of system-components vary in a *correlated* way, the variation of one may cause the change in another, or alternatively, both may independently depend on some third factor. Studies of relationships among time-series may discriminate between such possibilities if the system is followed for sufficiently long times (Sugihara, May, Ye, Hsieh, Deyle, Fogarty, and Munch 2012). In many cases, explicit mathematical models of chemical mechanisms reproduce main features of observed time-series (and/or illustrate important behaviors of dynamic coherences).

In each such open-system dynamic coherence, the rate of entropy-generation in the presence of the coherence is greater than it would be in the absence of the coherence-defining closure. Also, the effects of the coherence on other items are *quite different* from effects of the same components but without the closure. Therefore the system *as a whole* makes a difference. That coherence must be counted as one of the items that comprise the world (Ney 2009, Earley 2006)—that is, the dissipative structure *as a whole* has *ontological* significance. The effects of the structure are the sums of the effects of the components, but *which components persist* in the system depends on the details of the closure of the network of reactions. The *closed network* of relationships regulates the *composition* of the system. For example, the existence of biological dissipative structures accounts for the production (and therefore the existence) of high-energy molecular species such as sugars, proteins, and DNAs, but those molecules are themselves components of the dissipative structures that produce them. Components are influenced by the characteristics of the coherences which those same components constitute.

Networks of non-linear chemical processes share important characteristics with other complex dynamic coherences, including biochemical oscillations, cardiac rhythms, ecological developmental changes, and financial-market variations. *Closure of networks of processes* is essential for all such persistent dynamic coherences. We designate this as *Closure Ilya Prigogine*.

4.4 Robustness of Biochemical Dissipative Structures

Surprisingly, some chemical species that are involved in networks of biochemical reactions somehow remain at quite constant concentrations for long periods. This *robustness* of concentration occurs both in vitro and in vivo, and can be modeled in silico. Robustness in a network facilitates incorporation of that coherent set of relationships as a reliable component part of more-inclusive dynamic systems. A remarkable theorem (Shinar and Feinberg 2010) clarifies what network characteristics are needed to generate the approximate concentration-robustness that is experimentally observed. Specific network properties provide necessary and sufficient conditions for insuring long-term stability of some concentrations.

Robustness, considered the ability to continue to function in spite of both internal and external fluctuations, occurs at an amazing variety of nested biological levels. Each of the following illustrations of robustness is well documented (Wagner 2005): The sequence of codons in RNA and DNA is insensitive to replication errors; the function of proteins does not depend on point-mutations in codons; RNA secondary structure is immune to changes in nucleotides; the spatial structure and function of proteins does not depend on the amino acid sequence of the protein; the expression function of a gene is robust to mutations in regulatory regions; the outputs of metabolic pathways are little influenced by changes in regulatory genes; genetic networks function even when interactions among network genes alter; metabolic-network function is not sensitive to elimination of specific chemical reactions from the network; development of phenotypic patterns persists even though component genes vary; body-plans of organism survive modifications in embryonic development. Each of these achievements of robustness involves closures that are themselves intricate combinations of simpler sorts of closure.

4.5 Closure Jacques Cauvin

Propensity to engage in cooperative interactions is arguably the most salient characteristic of *human* individuals. Many of the features of the various worlds we all inhabit are institutional facts—objective realities that are held in being by their widespread acceptance (Searle 2010). Human institutions (e.g., specific languages and ways of living, as well as economic, political, and religious systems) are all based on (usually unconscious) human cooperativity. In all such cases, relationships among constituents must themselves be so interrelated that coherences maintain definition and identity over time. In these cases, system properties do not supervene on properties of constituents. Anthropologist Jacques Cauvin (2000) explored how *prior changes* in concepts and social practices *made possible* initial development of agriculture, and thereby grounded the flourishing of subsequent human cultures. We designate any mode of interrelation of processes which makes *further network-formation* possible as *Closure Jacques Cauvin*.

Marjorie Grene (1978, 17) observed: "We do not just *have* rationality or language or symbol systems as our portable property. We *come to ourselves* within symbol systems. They have us as much as we have them." Cauvin's insight has been developed by recent studies based on evolutionary network theory (Atran and Henrich 2010; Richerson, Boyd, and Henrich 2010) that have clarified how human agents are themselves shaped by relational networks—at the same time are each of those networks has been created by the behaviors of such agents.

Louis de Broglie clarified (for his time) how coherence of electrons and protons *constitutes* atoms. Henri Poincaré demonstrated how *new kinds of relationship* (novel topologies) result from cooperative interactions. Ilya Prigogine clarified how far-from-equilibrium dynamic coherences persist in open systems such as when interactions generate large-scale coherences. Jacques Cauvin showed how *conceptual innovations* (influential institutional facts) made possible the distinctive evolution of *human* societies. In each of these diverse cases, the emergence of novel modes of effective functioning (*properties*) depends on defining *closures of relationships* of components. There surely are other significant modes of closure. *Closure* provides the basis for the origin of novel centers of influence—and thereby grounds the *ontological* changes which are brought about by cooperative interaction.

5 Structures and Properties

5.1 The Structuralist Revival

Considerable attention has recently shifted to *structuralist* philosophical approaches to scientific (especially microphysical) questions (e.g., Bokulich and Bokulich 2011, Landry and Rickles 2012).[11] In 1989 John Worrall pointed out that although entities postulated by scientific theories do change over time, *structural* aspects of theories tend to persist through such *ontological* modifications. He proposed that what we learn in scientific investigation is correct, but that we learn only about structures—not about entities. This approach, called

---

[11] Such approaches were widely discussed in the first third of the twentieth century (e.g., Eddington 1935).

Epistemological Structural Realism (ESR) claims that *relationships* are more important than are the *entities* that are related. Others carried this reasoning further and concluded that structure is not only *all that we can know* but also that structure is *all that exists* to be known. Steven French (in French and Ladyman 2011, 30), for instance, "dispenses with objects entirely." This more-radical view is called Ontological Structural Realism (OSR). However, the notion of structures of relationship which do not involve items that are related ("relations without relata") seems paradoxical.

Earlier suggestions of the philosophic importance of structures had specifically focused on chemical problems. David Armstrong (1978, *Vol. 2*, 68–71) recommended that "being a methane molecule" should be considered to involve a structural universal (a multiply realizable polyadic relationship that necessarily involves "being carbon," "being hydrogen," and "being chemically bonded" as correlative universals).

5.2 Are Any Properties Intrinsic?

Some versions of OSR consider that *all* properties of any entity are *structural* properties—characteristics that depend on relationships with other entities within structures. This amounts to a denial that entities have *intrinsic* properties—characteristics that they would possess if fully isolated so as to be independent of all relationships. Two recent moderate versions of OSR provide for intrinsic properties as well as relational ones—and (their adherents claim) thus resolve the relationship-without-relata conundrum and introduce causal necessity (modality) into structural realism (SR).

In one such proposal, Holger Lyre agrees that all properties are structural, but holds that some structural properties are "structurally derived intrinsic properties (invariants of structure)" (Lyre 2010, sec. 6). He had previously considered the relationship between OSR and the symmetry-group (U(1)) which applies to electromagnetism and concluded that, in permutable theories,[12] description of objects as solely group-theoretically constituted "becomes mandatory. For there we only have access to the objects as members of equivalence-classes, under those symmetry transformations which leave the physical properties invariant" (Lyre 2004, 663). In order to establish that some structural properties are intrinsic, Lyre points out: "An object may have its invariant properties according to the world's structure: the structure comes equipped with such properties" (Lyre 2012, 170). Perhaps to allay suspicion that he might be advocating *ante rem* Platonism, he insists: "The world structure must...be an instantiated structure" (Lyre 2012, 170).

In an alternative proposal designed to introduce causal necessity into SR, Michael Esfeld advocates Causal Realism (CR). He initially proposed that all

---

[12] Those which concern particles (such as protons) which cannot be distinguished from other particles of the same sort.

properties are *causal properties* (powers[13]) and rejected the (characteristically Humean) doctrine that some properties are *categorial*—merely qualitative with no relation to causality (Esfeld 2003). Subsequently, he conceded: "we no longer take OSR—at least in the moderate version that we defend, recognizing objects—to be opposed to the acknowledgement of the existence of intrinsic properties, as long as such intrinsic properties do not amount to an intrinsic identity of the objects" (Esfeld and Lam 2011, 155). He claims that structures (as well as properties) must be causally efficacious:

> Structures can be causally efficacious in the same sense as intrinsic properties of events: as events can bring about effects in virtue of having certain intrinsic properties, they can bring about effects in virtue of standing in certain relations with each other so that it is the network of relations—that is, the structure as a whole—that is causally efficacious.
>
> (ESFELD 2012, sec. 3)

He further claims that: "In sum, ontic structural realism is suitable as a form of scientific realism only if it commits itself to causal structure, that is to say, only if the essence of the fundamental physical structures is taken to consist in the power to produce certain effects" (Esfeld 2009, 188).

5.3 Can Structures Be Causes?

There are objections to the entire structural-realist project. F. A. Müller claims that proponents of structural realisms do not adequately specify the meaning of structure—and that the two best-established formal definitions of structure do not have the properties that structural realists require. He claims: "neither set-theoretical nor category-theoretical notions of structure serve the needs of structural realism" (Müller 2010, 399). Stathis Psillos, a persistent critic of SR, argues that OSR incoherently requires that structures be both *abstract* (multiply realizable) and *concrete* (causally efficacious)—but since structures (considered as polyadic properties) are *purely mathematical* entities they cannot have *causal* efficacy—and therefore structures do not have modal force (Psillos 2012).

Rom Harré and E. H. Madden (1975) revived the moribund notion of causal powers, but Harré (2002) urged members of the Critical Realist school of social psychology (an approach based on Harré's prior work) to cease ascribing causal power to impotent fictions such as "the banking system." Subsequently Harré (2009b) identified the tendency toward reification ("substantialism") as lying at the root of the errors he decried—and even went so far as to assert: "Structural models in the human sciences are heuristic models only—there are no structures" (Harré 2009a, 138). This Neo-Thacherian dictum appears to contrast strongly with some of Harré's earlier statements, specifically: "Structured groups, that is, collectives, are ontologically prior to individuals.

---

[13] George Molnar (2003, 60–81) tried to account for "how these properties [powers] have an object toward which they are oriented or directed" and postulated *physical intentionality*, but did not give a convincing account of it.

Human beings are constituted as people by their interpersonal relations" (Harré 1993, 34), and: "Forms of life are the contexts in which personal and social identities are formed" (Harré 2009a, 142).

In his discussion of social causation Harré identified two modes of causal action—*event* causality and *agent* causality. In event causality some happening activates a mechanism that engenders a subsequent result; in agent causality some *continuously existing and active being* brings about consequences without external stimulation (save, perhaps, removal of obstructions). Harré proposed that detection of event causality requires identification of a *mechanism* that connects an initiating occurrence with an ensuing result: Recognition of agent causality demands location of one or more specific *powerful particulars* that act continuously (unless that action is somehow blocked). Harré concluded that only *individual human persons* qualify as causal agents in social psychology—vague constructs such as the banking system certainly do not merit that designation. Both of the modes of causality Harré identifies involve powerful particulars (either events or continuants); neither of them implicates universals (properties, relations, or structures) in causality. If, as Esfeld claims, the validity of OSR depends on structures having causal power, then structuralists need to respond to the objections of Müller, Psillos, and Harré by clarifying how (or in what sense) structures may properly be considered causes.

6 Causes and Determinants

6.1 Concepts of Causation

The word *cause* has several distinct meanings in standard English—it is poly-semic. *The Shorter Oxford English Dictionary* (SOED) identifies *four* connotations of cause corresponding to: (1) agent, (2) reason, (3) lawsuit, and (4) social movement. Chemists use cause to refer to both agents and reasons. In contrast, philosophers generally recognize only the first (agent) meaning of cause given in the SOED, and regard all other uses as suspect—or just wrong.

Ernan McMullin (1999) quoted the section of Aristotle's *Physics* (194b, 18–20) that introduces the four types of cause (material, formal, efficient, and final)—"Knowledge is the object of our inquiry, and men do not think they know a thing until they have grasped the 'why' of it (which is to grasp its primary cause)." This definition appears to identify cause and reason. McMullin concluded that what Aristotle meant by *cause* differs in important ways from how most philosophers now understand that term.

Although Aristotle counted any adequate response to a why-question as a cause (*aitiā*), he also made a clear distinction between *efficient* causes (change-initiating agents) and *formal* causes—arrangements necessary for an event to occur. As Psillos (2012) points out, structures are formal: The objection that structures *cannot be* causes seems related to the distinction between efficient and formal causality. Robert Pasnau (2004) carefully described how, during the rapid development and subsequent slow decline of medieval Scholastic

Philosophy, the understanding of the Aristotelian concept of substantial form (roughly equivalent to the modern notion of structure) gradually changed away from its original (purely formal) Aristotelian meaning, and increasingly acquired overtones of efficient agency. Pasnau concluded that the further modifications in the usual philosophical understanding of cause which subsequently occurred should be interpreted as continuations of that trend.

With the success of Newtonian physics, only interactions similar to events on billiard tables (where precisely determined impacts yielded exactly predictable results) came to be considered as causally significant interactions. *Impact* of impenetrable corpuscles became the paradigmatic causal process; *efficient* causality took over the designation cause. Most philosophers relegated any other factors that might be involved in answers to why-questions to subordinate status—or to oblivion.

6.2 Limits to Agency

Mario Bunge pointed out that:

> Some of the grounds for the Renaissance reduction of causes to the *causa efficiens* were the following: (a) it [the efficient cause] was, of all the four [Aristotelian causes], the sole clearly conceived one; (b) hence it was mathematically expressible; (c) it could be assigned an empirical correlate, namely an event (usually a motion) producing another event (usually another motion) in accordance with fixed rules…; (d) as a consequence, the efficient cause was controllable; moreover its control was regarded as leading to the harnessing of nature.
>
> (BUNGE 1959, 32)

However, although Bunge does "restrict the meaning of the term cause to *efficient cause*, or extrinsic motive agent, or external influence producing change" (33), he also recognizes that "causation…is only one among several types of determination; there are other types of lawful production, other levels of interconnection" (30). He distinguishes between *causes* (effective agents—the *how* of things) and *reasons* (rational explanations—the *why* of things), pointing out that these two notions are often confounded: "The identity of explanation with the disclosing of causes is even rooted in the Greek language, in which *aition* and *logos* are almost interchangeable since both mean cause and reason. The confusion of cause with reason, and that of effect with consequent, are, moreover, common in our everyday speech" (226–227).[14]

Biologist Ernst Mayr—a founder of the modern synthesis in evolutionary theory—urged (1961) that biologists clearly distinguish proximate causes from ultimate causes. The *ultimate causes* of the long-distance migrations of certain birds, for instance, are (in Mayr's view) the historical explanations (largely in terms of natural selection) that account for *why* those birds carry out such journeys: the *proximate causes* are whatever hormonal changes (or other internal

---

[14] Bunge subsequently observed: "From the point of view of cognitive neuroscience, reasons for acting are efficient causes" (Bunge 2010, 224).

mechanisms) account for *how* that behavior-pattern works out in practice. In Bunge's terminology, Mayr's ultimate causes are not causes but reasons; only Mayr's proximate causes would properly be designated causes (as current philosophers use that word).

It is now clear that most interesting biological systems do not fit Mayr's recommendation nearly as well as seasonal bird migrations do. Avian flight patterns have vanishingly small effects on the progression of the seasons—cause and effect are quite distinct—but, in contrast, the vast majority of biological systems involve *reciprocal* (rather than *unidirectional*) causality, so that Mayr's distinction does not apply. In discussion of this point, Kevin Laland and his colleagues report: "When a trait evolves through intersexual selection, the source of selection is itself an evolving character. The peacock's tail evolves through though the mating-preferences in peahens and those preferences coevolve with the male trait" (Laland, Sterelny, Odling-Smee, Hoppitt, and Tuler 2011, 1512). Whenever *reciprocal determination* makes it impossible cleanly to distinguish causes from reasons, restricting causality to *efficient* causes (as philosophers recommend) is not appropriate.

6.3 Peircean Determinants

T. L. Short (2007, 105–107) observed that the narrowness of the contemporary philosophic understanding of causation (a baleful influence, he says, of Hume's ghost) has unfortunate effects—but called attention to an understanding of causality developed by Charles S. Peirce. Stephen Pratten (2009) suggested that Short's interpretation of Peirce's causal theory provided an adequate response to Harré's conclusion that structures cannot be causes.

The interactions that classical mechanics deals with have time-reversal symmetry: Videos of billiard-ball collisions look the same whether running forward or backward. In contrast, many natural processes *proceed in only one direction*. (The spark-induced explosion of a mixture of H_2 and O_2 loudly leads to rapid production of H_2O: the reverse reaction is unobservable.) Peirce calls such unidirectional processes "finious;" Short suggests the designation "anisotropic;" chemists call such changes "irreversible." Pierce held an alternate kind of causal process obtains in irreversible processes—"that mode of bringing facts about according to which a general description of result is made to come about, quite irrespective of any compulsion for it to come about in this or that particular way; although the means may be adapted to the end" (CP 1.211). This corresponds to understanding cause as reason rather than as agent—that is, to using the second denotation of the English word cause given in the SOED, rather than the first.

Peirce considered that Darwin's account of the origin of biological species exemplifies this alternative mode of result-determination. Natural selection gradually (and irreversibly) eliminates those characteristics of organisms that are not suited to the environmental conditions that prevail. Such reduction (culling) of possibilities eventually produces one particular determinate result—which outcome is produced depends on the contingencies of the culling rather than on the action of underlying agents.

There remains little doubt that the Darwinian theory indicates a real cause, which tends to adapt animal and vegetable forms to their environment. A very remarkable feature of it is that it shows how merely fortuitous variations of individuals together with merely fortuitous mishaps to them would, under the action of heredity, result, not in mere irregularity, nor even in a statistical constancy, but in continual and indefinite progress toward a better adaptation of means to ends.

(CP 7.395)

Natural selection works in such a way as to produce adaptation of life-forms to their circumstances (Thompson 2012). This general aim "does not determine in what particular way it is to be brought about, but only that the result shall have a certain general character. The general result may be brought about at one time in one way, and at another time in another way" (CP 1.211).

On this basis, Peirce considers that each effective selection-criterion is a *general* rather than a *particular* (a universal rather than a substance). Each such criterion might be called a *controlling general*—an outcome-determining universal. By this means structures (closures of relationships that have the property of engendering future versions of the same closures) would have result-shaping effects, although they are not agents. In other words, if a certain state-of-affairs results from a *prior selection* on the basis of some criterion, that criterion (a universal) is a determinant (a cause in a general sense) of that state of affairs. To the extent that closure of a network of relationships of components is a prerequisite for the stability of entities, that closure (which corresponds to a structural universal) is also a *necessary determinant* of the states of affairs that it engenders. The key feature is that if an *equivalent to selection* accounts of the existence of a structure, then the influence of that structure (as such) may properly be termed a determinant—a cause, in a sense that is more general is usually recognized by the dialect of the province of the philosophers.

6.4 Non-Agentive Determination in Oseltamivir-Resistant Swine Flu

Several experimental results have recently been reported that show how *structural* features *determine outcomes* as reasons rather than as agents.

Since 1990, millions of people have been sickened by swine flu—infection by the H1N1 influenza virus. From its introduction in 1999 until recently the drug oseltamivir (Tamiflu, Hoffman-LaRoche) has been effective against swine flu. However, during drug-testing, a mutation (named H274Y)[15] made the virus immune to oseltamivir—but also impaired infectivity of the virus and *did not reduce* the drug's effectiveness. In 2007 several oseltamivir-resistant strains of the virus with *no reduction in virulence* appeared. Bloom, Gong, and Baltimore (2010) found that two additional mutations[16] had occurred independently and enabled the virus to tolerate subsequent H274Y mutation without loss of virulence. The first new mutation was ineffective: the second mutation increased

---

[15] Tyrosine (Y) replaced histidine (H) at position 274 in the virus DNA.
[16] L111Q and S106P.

the effectiveness of the otherwise-innocuous first mutation: *combination* of the mutations was result-determining. Such a combination (as a determinant of the outcome) is not an *efficient* cause. It is a *reason* not an *agent*. Determinants of this sort can be identified when some arrangements function while others fail. Such pruning defines a result.

Tenaillon, Rodríguez-Verdugo, Gaut, McDonald, Bennett, Long, and Gaut (2012) and Meyer, Dobias, Weitz, Barrick, Quick, and Lenski (2012) reported other examples of non-agential determination. In each case a result depended on actions of particulars but those actions *did not determine* the result: *Selection constraints* specified the outcome. Rather than actions of components determining the outcome, constraints determined *which* components acted—and thereby specified what actually happened.

6.5 Contextual Determination

Every macroscopic state of affairs involves myriads of microscopic and submicroscopic components. In the absence of external constraints (so all components are independent) each sample would have an immense number of equally-probable possible future states. *But no real system is unconstrained.* Every sample has a history (usually unknown and untold) that specifies its current context and limits the range of available futures (Earley 2012b). Theories that apply at the level of micro-components provide *necessary conditions* for properties of more-inclusive coherences—but they *do not provide sufficient conditions*: Sufficient conditions must be dealt with by less-fundamental approaches. An open vat of nutritious broth quickly changes into a teeming mass of biological organisms, but *which specific type(s)* of organism result depends on the particular species that happen(s) happily to colonize that soup.

"There are properties of the higher-level theories (chemistry and thermodynamics) for which the full arsenal of the fundamental theories (quantum mechanics and statistical mechanics) provide no sufficient conditions for their derivation or definition" (Bishop and Atmanspacher 2006, 1755). By *contextual property emergence*, upper-level properties derive from the *context of constraints* of the system as well as from the underlying level which involves less-extensive components. Upper-level constraints typically remove degeneracies that characterize lower-level situations and thus lead to *stable* states. "Necessary conditions due to the original topology of the basic description are not violated as the new topology is consistent with (though not implied by) the original topology." Such constraints are *contextual determinants*.

The network of relationships that underlies and defines an emergent coherence corresponds to one or more *structural universals*. The problem of the modality of structural realism arises from the concern that mathematical objects (such as structural universals) should not directly have physical effects. In the cases considered here each structural universal corresponds to the closure of a network of relationships—a physical process that has consequences. Closure of networks of interactions among components generally leads to situations that persist and/or recur. This is how behavior-patterns (properties in Peirce's sense) of substances remain coherently related over time.

6.6 Ordinary Things

But what about ordinary entities, such as David Robb's tennis ball Alpha (discussed in section 1.6, above)? How can we understand the coherence of Alpha's properties? That tennis ball coheres and behaves as a unit because the many billions of elemental centers (atoms) that make up that sphere adhere to each other through rather stable chemical bonds—each one a closure of attractive and repulsive forces that is protected from disruption by activation-barriers. But why do those particular bonds exist rather than others? Why is Alpha small and yellow rather than some other size and color?

The International Tennis Federation (ITF) defined the official diameter of a regulation tennis ball as 65.41–68.58 mm (2.575–2.700 inches), and ruled that balls must weigh between 56.0 g and 59.4 g (1.975–2.095 ounces). Yellow and white are the only tennis-ball colors that are approved by the United States Tennis Association (USTA) and the ITF. Since 1972 most tennis balls have been made in a specific fluorescent yellow color—because research showed that balls of that color were more visible on television that those of any other color. The properties of tennis ball Alpha depend on the well-understood physical and chemical factors that determine how chemical bonds relate to each other—but also on contingent historical, economic, and psychological factors that influence decisions of the USTA and ITF committees that specified what properties are necessary for a ball to qualify as a regulation tennis ball. In this sense Alpha is as much an evolved system (in Wimsatt's trichotomy) as is any biological species or individual. Production of tennis balls (what Peirce would have called a *finious* process) is clearly much influenced by determinants of several types, whether or not philosophers would designate such factors as causes.

In Robb's example, and in all other cases, closures of relationships of constituents—determining structural universals—serve as *criteria for selection* among results of activities of powerful particulars. In each case, coherence at any level influences and depends on both wider and narrower (more-inclusive and less-inclusive) integrations. The characteristics of material objects that philosophers discuss depend on closures that involve chemical components—and also on human decisions within more-or-less stable cultural integrations (institutional facts).

Properties stay together in each chemical entity because networks of dynamic relationships among components generate a *closed* set of subsequent states. Each such closure depends on continuance of closures in its components, and also may participate in (and be determined by) closures of more-inclusive relationship-networks. Structures (polyadic relationships) *determine* specific outcomes, even though they are not themselves *agents*.

Throughout human evolution people have modified naturally occurring materials to produce useful items. Major cultural advances occurred when chemical processes (cooking, baking of ceramics, ore-smelting, etc.) were first used to *improve* natural resources. Arguably tin-bronze was the first artificial material *intentionally synthesized* in order to produce *specific properties* important for more-inclusive coherences. In the third millennium BC, techniques for manufacture and use of effective *bronze daggers* spread (remarkably rapidly) throughout lands surrounding the Aegean Sea: "Just as every nation in the Levant need[ed] tanks in the 1960s and 1970s, so every man needed a dagger during the bronze age" (Renfrew 2011, 320).

In our own culture, most of the items with which we deal are made of artificial materials. From one point of view, behavioral characteristics of each such item hang together because of chemical bonding: in a another sense, those properties exist because each item is *useful in some context*—and that *criterion of utility* motivated the designer, manufacturer and purchaser. On that basis, every artifact which contains artificial materials is *an evolved system* in Wimsatt's trichotomy. Diachronic modes of thought must be used to understand all items: *synchronic* methods—attempting to understand coherences in terms of their current composition without attention to how those integrations came to be—yield wrong conclusions.

Persistence of every entity depends on *closure* of relationships among its components (each part must be stable enough) and is also determined by higher-level integrations of which that entity is a component. Creative activity, some involving human action, continually generates "new shapes of value which merge into higher attainments of things beyond themselves" (Whitehead 1967, 94).

References

Armstrong, D. M. (2010). *Sketch for a Systematic Metaphysics*. Oxford: Oxford University Press.

Armstrong, D. M. (1978). *Universals and Scientific Realism*. Cambridge: Cambridge University Press.

Atmanspacher, Harald and Primas, Hans. (2003). "Epistemic and ontic quantum realities." In *Time, Quantum & Information* (Lutz Castell and Otfried Ischenbeck, eds.). Berlin: Springer, 301–321. (http://www.igpp.de/english/tda/pdf/vaexjoe.pdf.)

Atmanspacher, H., Amann, A., and Müller-Herold, U. (1999). *On Quanta, Mind and Matter: Hans Primas in Context*. Dordrecht: Kluwer.

Atran, Scott and Henrich, Joseph. (2010). "The evolution of religion: How cognitive by-products, adaptive learning heuristics, ritual displays, and group competition generate deep commitments to pro-social religions," *Biological Theory*, 5, 18–30.

Bader, R. F. W., and Matta, C. F. (2013). "Atoms in molecules as non-overlapping, bounded, space-filling open quantum systems," *Foundations of Chemistry*, 15, 253–276.

Barnes, Jonathan, ed. (1984). *Complete Works of Aristotle, Volume One*. Princeton, NJ: Princeton University Press.

Batterman, Robert W. (2011). "Emergence, singularities, and symmetry-breaking," *Foundations of Physics*, 41, 1031–1056.

Benovsky, Jiri. (2008). "The bundle theory and the substratum theory: deadly enemies or twin brothers?" *Philosophical Studies*, 141, 175–190.

Berry, Michael. (1994). "Asymptotics, singularities and the reduction of theories." In *Logic, Methodology and Philosophy*, IX (D. Prawitz, B. Skyrms, and D. Westerståhl, eds.). Amsterdam: Elsevier B.V., 597–607.

Bishop, Robert C. (2012). "Fluid convection, constraint and causation," *Interface Focus*, 2, 4–12.

Bloom, J., Gong, L., and Baltimore, D. (2010). "Permissive secondary mutations enable the evolution of influenza oseltamivir resistance," *Science*, 328, 1272–1277.

Bokulich, Peter, and Bokulich, Alisa, eds. (2011). *Scientific Structuralism*. Dordrecht: Springer.

Bonk, Thomas. (1994). "Why has de Broglie's theory been rejected?" *Studies in the History and Philosophy of Science*, 25, 375–396.

Bunge, Mario. (2010). *Matter and Mind*. Dordrecht: Springer.

Bunge, Mario. (1959). *Causality: The Place of the Causal Principle in Modern Science*. Cambridge, MA: Harvard University Press.

Cao, Tian-Yu. (1988). "Gauge Theory and the Geometrization of Fundamental Physics." In *Philosophical Foundations of Quantum Field Theory* (Harvey R Brown and Rom Harré, eds.). Oxford: Oxford University Press.

Castellani, Elena. (1998). *Interpreting Bodies*. Princeton, NJ: Princeton University Press.

Cauvin, Jacques. (2000). *The Birth of the Gods and the Origins of Agriculture* (T. Watkins, trans.). Cambridge: Cambridge University Press.

Cottingham, W. N., and Greenwood, D. A. (2007). *An Introduction to the Standard Model of Particle Physics, Second Edition*. Cambridge: Cambridge University Press.

Cushing, James T. (1994). *Quantum Mechanics: Historical Contingency and the Copenhagen Hegemony*. Chicago: University of Chicago Press.

de Broglie, Louis. (1972). "Phase waves of Louis de Broglie" (J. Haslett, trans.), *American Journal of Physics*, 40, 1315–1320.

de Broglie, Louis. (1965). "The Wave Nature of the Electron." In *Nobel Lectures, Physics 1922–1941*. Amsterdam: Elsevier, 244–256.

de Broglie, Louis. (1923a). "Radiation, Waves and Quanta" (Brigitte Lane and Barton Lane, trans.), *Comptes rendus* 177, 507–510.

de Broglie, Louis. (1923b). "Quanta de lumière, diffraction et interferences," *Comptes rendus*, 177, 548–551.

de Broglie, Maurice, and de Broglie, Louis. (1922). "Remarques sur les spectres corpuscupaires et l'effet photo-électrique." *Comptes rendus*, 175, 1139–1141.

Dürr, S., Fodor, Z., Frison, J., Hoelbling, C., Hoffmann, R., Katz, S. D., et al. (2008). "Ab initio determination of light hadron masses," *Science*, 322, 1224–1227.

Earley, Joseph E. (2013). "Three concepts of chemical closure and their epistemological significance." In *Philosophy of Chemistry: Practices, Methods and Concepts* (Jean-Pierre Llored, ed.). Newcastle: Cambridge Scholars Press, 608–628.

Earley, Joseph E. (2012a). "Ilya Prigogine (1917–2003)." In *Handbook of the Philosophy of Science, Volume 6, Philosophy of Chemistry* (Andrea Woody, Robin Hendry, and Paul Needham, eds.). Amsterdam: Elsevier, 165–175.

Earley, Joseph E. (2012b). "A neglected aspect of the puzzle of chemical structure: How history helps," *Foundations of Chemistry*, 14, 235–244.

Earley, Joseph E. (2008a). "Ontologically significant aggregation: Process structural realism (PSR)." In *Handbook of Whiteheadian Process Thought, Volume 2* (Michel Weber and Will Desmond, eds.). Frankfurt: Ontos Verlag, 179–191.

Earley, Joseph E. (2008b). "Process structural realism, instance ontology, and societal order." In *Researching with Whitehead: System and Adventure* (Franz Riffert and Hans-Joachim Sander, eds.). Berlin: Alber, 190–211.

Earley, Joseph E. (2006). "Chemical 'substances' that are not 'chemical substances'," *Philosophy of Science*, 73, 814–852.

Earley, Joseph E. (2003a). "Constraints on the origin of coherence in far-from-equilibrium chemical systems." In *Physics and Whitehead: Quantum, Process, and Experience* (Timothy E. Eastman and Hank Keeton, eds.). Albany: State University of New York Press, 63–73.

Earley, Joseph E. (2003b). "Varieties of properties: An alternative distinction among qualities." *Annals of the New York Academy of Science*, 988, 80–89.

Earley, Joseph E. (1998). "Modes of *chemical* becoming," *Hyle*, 4, 105–115. (http://www.hyle.org/journal/issues/4/earley.htm)

Earley, Joseph E. (1992). "The nature of chemical existence." In *Metaphysics as Foundation: Essays in Honor of Ivor Leclerc* (Paul Bogaard and Gordon Treash, eds.). Albany: State University of New York Press, 272–284.

Eddington, Arthur. (1935). *The Nature of the Physical World*. London: J. M. Dent & Sons.

Esfeld, Michael. (2012). "Causal realism." In *Probabilities, Laws, and Structures* (D. Dieks, W. J. Gonzalez, S. Hartmann, M. Stöltzner, and M. Weber, eds.). Dordrecht: Springer, 157–169.

Esfeld, Michael. (2009). "The modal nature of structures in ontic structural realism," *International Studies in the Philosophy of Science*, 23, 179–194.

Esfeld, Michael. (2003). "Do relations require underlying intrinsic properties? A physical argument for a metaphysics of relations," *Metaphysica*, 4, 5–25.

Esfeld, Michael, and Lam, Vincent. (2011). "Ontic structural realism as a metaphysics of objects." In Bokulich and Bokulich (2011), 143–160.

Ezra, G. S., Waalkens, H., and Wiggins, S. (2009). "Microcanonical rates, gap times, and phase space dividing surfaces," *Journal of Chemical Physics*, 130, 164118–164133.

Fine, Kit. (2006). "In defense of three-dimensionalism," *Journal of Philosophy*, 103, 699–714.

French, S., and Ladyman, J. (2011). "In Defence of Ontic Structural Realism." In Bokulich and Bokulich (2011), 25–43.

Grene, Marjorie. (1978). "Paradoxes of historicity," *Review of Metaphysics*, 32, 15–36.

Harré, Rom. (2009a). "Saving critical realism," *Journal of the Theory of Social Behavior*, 39, 129–143.

Harré, Rom. (2009b). "The siren song of substantialism," *Journal of the Theory of Social Behavior*, 39, 466–473.

Harré, Rom. (2002). "Social reality and the myth of social structure," *European Journal of Social Theory*, 5, 111–123.

Harré, Rom. (1993). *Social Being*. Oxford: Blackwell.

Harré, Rom, and Llored, Jean-Pierre. (2011). "Mereologies as the grammars of chemical discourses," *Foundations of Chemistry*, 13, 63–76.

Harré, Rom, and Madden, E. H. (1975). *Causal Powers*. Oxford: Blackwell.

Hume, David. (1739). *A Treatise on Human Nature*. London: John Noon.

Humphrys, Paul. (2008). "Synchronic and diachronic emergence," *Mind and Machines*, 18, 431–442.

Humphrys, Paul. (1997). "How properties emerge," *Philosophy of Science*, 64, 1–17.

Jaffé, C., Kawai, S., Palacián, J., Yanguas, P., Uzer, T. (2005). "A new look at the transition state: Wigner's dynamical perspective revisited," *Advances in Chemical Physics A*, 130, 171–216.

Kondepudi, Dilip, and Prigogine, Ilya. (1998). *Modern Thermodynamics: From Heat Engines to Dissipative Structures*. New York: Wiley.

Koslicki, Kathrin. (2008). *The Structure of Objects*. New York: Oxford University Press.

Laland, K. N., Sterelny, K., Odling-Smee, J., Hoppitt, W., and Uler, T. (2011). "Cause and effect in biology revisited: Is Mayr's proximate-ultimate dichotomy still useful?" *Science*, 334, 1512–1516.

Landry, Elaine, and Rickles, D. P., eds. (2012). *Structural Realism: Structure, Object, and Causality*. Dordrecht: Springer.

Laughlin, Robert. (2005). *A Different Universe: Inventing Physics from the Bottom Down*. New York: Basic Books.

Lewis, David. (1991). *Parts of Classes*. Oxford: Blackwell.

Llored, Jean-Pierre. (2010). "Mereology and quantum chemistry: The approximation of molecular orbitals," *Foundations of Chemistry*, 12, 203–221.

Locke, John. (1690). *An Essay Concerning Human Understanding*. London: Thomas Basset.

Lombardi, Olivia. (2012). "Prigogine and the many voices of nature," *Foundations of Chemistry*, 14, 205–220.

Lyre, Holger. (2012). "Structural invariants, structural kinds, structural laws." In *Probabilities, Laws, and Structures* (D. Dieks, W. J. Gonzalez, S. Hartmann, M. Stöltzner, and M. Weber, eds.). Dordrecht: Springer, 169–183.

Lyre, Holger. (2010). "Humean Perspectives on Structural Realism." In *The Present Situation in the Philosophy of Science* (F. Stadler, ed.). Dordrecht: Springer, 381–397.

Lyre, Holger. (2004). "Holism and structuralism in U(1) gauge theory," *Studies in History and Philosophy of Modern Physics*, 35, 643–670.

MacKinnon, Edward. (2008). "The standard model as a philosophical challenge," *Philosophy of Science*, 75, 447–457.

Marcus, Ruth Barcan. (1963). "Modal Logics I: Modalities and Intensional Languages." In *Boston Studies in the Philosophy of Science, Volume 1* (Marx Wartofsky, ed.). Dordrecht: Reidel, 77–116.

Mayr, Ernst. (1961). "Cause and effect in biology," *Science*, 134, 1501–1506.

McIntyre, Lee. (2007). "Emergence and reduction in chemistry: ontological or epistemological concepts?" *Synthese*, 155, 337–343.

McMullin, Ernan. (1999). "Imagination, insight, and inference," *Proceedings of the American Catholic Philosophical Association*, 73, 251–265.

Mellor, D. H. (2006). "Wholes and parts: The limits of composition," *South African Journal of Philosophy*, 25, 138–145.

Mertz, Donald W. (2003). "An instance ontology for structures: Their definition, identity, and indiscernibility," *Metaphysica*, 4, 127–164.

Mertz, Donald W. (1996). *Moderate Realism and Its Logic*. New Haven, CT: Yale University Press.

Meyer, J., Dobias, D., Weitz, J., Barrick, J., Quick, R., and Lenski, R. (2012). "Repeatability and contingency in the evolution of a key innovation in phage lambda," *Science*, 335, 428–432.

Millikan, Ruth Garrett. (2000). *On Clear and Confused Ideas.* New York: Cambridge University Press.

Molnar, George. (2003). *Powers: A Study in Metaphysics.* Oxford: Oxford University Press.

Müller, F. A. (2010). "The Characterization of Structure: Definition versus Axiomatization." In *The Present Situation in the Philosophy of Science* (F. Stadler, ed.). Dordrecht, Springer, 399–417.

Ney, Alyssa. (2009). "Physical causation and difference-making," *British Journal for the Philosophy of Science,* 60, 737–764.

Pasnau, R. (2004). "Form, substance and mechanism," *Philosophical Review,* 13, 31–88.

Pratten, Stephen. (2009). "Critical and realism and causality: Tracing the Aristotelian legacy," *Journal for the Theory of Social Behaviour,* 39, 189–218.

Prigogine, Ilya. (1977). "Time, Structure and Fluctuations." In *Nobel Lectures in Chemistry, 1971–1980* (Sture Forsén, ed.). Singapore: World Scientific. http://www.nobelprize.org/nobel_prizes/chemistry/laureates/1977/prigogine-lecture.html.

Primas, Hans. (2000). "Asymptotically Disjoint Quantum States." In *Decoherence: Theoretical, Experimental and Conceptual Problems* (Philippe Blanchard, Domenico Giulini, Erich Joos, Claus Kiefer, and Ion-Olympia Stamatescu, eds.). Berlin: Springer, 161–178.

Primas, Hans. (1998). "Emergence in the exact natural sciences," *Acta Polytechnica Scandinavica,* 91, 83–98.

Psillos, Stathis. (2012). "Adding Modality to Ontic Structuralism: An Exploration and Critique." In Landry and Rickles (2012), 169–186.

Putnam, H. (2004). *Ethics without Ontology.* Cambridge, MA: Harvard University Press.

Putnam, H. (1969). "On Properties." In *Essays in Honor of Carl G. Hempel* (Nicholas Rescher, ed.). Dordrecht: Reidel, 235–254.

Quine, W. V. (1976). "Worlds away," *Journal of Philosophy,* 73, 859–863.

Renfrew, Colin. (2011). *The Emergence of Civilization: The Cyclades and the Aegean in the Third Millennium bc.* Oxford: Oxbow.

Richerson, Peter J., Boyd, Robert, and Henrich, Joseph. (2010). "Gene-culture coevolution in the age of genomics," *Proceedings of the National Academy of Sciences (USA),* 107, 8985–8992.

Robb, David. (2009). "Substance." In *The Routledge Companion to Metaphysics* (Robin le Poidevin, Peter Simons, Andrew McGonigal, and Ross P. Cameron, eds.). New York: Routledge, 256–265.

Robb, David. (2005). "Qualitative unity and the bundle theory," *The Monist,* 88, 466–493.

Searle, John R. (2010). *Making the Social World.* New York: Oxford University Press.

Shinar, Guy, and Feinberg, Martin. (2010). "Structural sources of robustness in biological networks," *Science,* 327, 1389–1391.

Short, T. L. (2007). *Peirce's Theory of Signs.* Cambridge: Cambridge University Press.

Sider, Theodore. (2006). "Bare particulars," *Philosophical Perspectives,* 20, 387–397.

Simondon, Gilbert. (1964). *L'individu et sa genèse physico-biologique.* Paris: PUF.

Simons, Peter. (1994). "Particulars in particular clothing: Three trope theories of substance," *Philosophy and Phenomenological Research,* 54, 553–575.

Simons, Peter. (1987). *Parts: A Study in Ontology.* Oxford: Oxford University Press.

Sugihara, G., May, R., Ye, H., Hsieh, C., Doyle, E., Fogarty, M., and Munch, S. (2012). "Detecting causality in complex ecosystems," *Science*, 338, 496–500.

Sugihara, G. and Ye, H. (2009). "Cooperative network dynamics," *Nature*, 458, 979–980.

Teller, P. (1988). "Three problems of renormalization." In *Philosophical Foundations of Quantum Field Theory* (Harvey R. Brown and Rom Harré, eds.). Oxford: Clarendon Press, 73–89.

Tenaillon, O., Rodríguez-Verdugo, A., Gaut, R., McDonald, P., Bennett, A., Long, A., and Gaut, B. (2012). "The molecular diversity of adaptive convergence," *Science*, 335, 457–461.

Thompson, John N. (2012). "The role of coevolution," *Science*, 335, 410–411.

Vander Laan, D. (2010). "A relevance constraint on composition," *Australasian Journal of Philosophy*, 88, 135–145.

Wagner, Andreas. (2005). *Robustness and Evolvability in Living Systems*. Princeton, NJ: Princeton University Press.

Weiss, Paul. (1959). *First Considerations*. Carbondale: Southern Illinois University Press.

Whitehead, A. N. (1978). *Process and Reality: An Essay in Cosmology, Corrected Edition* (David Ray Griffin and Donald W. Sherburne, eds.). New York: Free Press.

Whitehead, A. N. (1967). *Science and the Modern World*. New York: Free Press.

Whitehead, A. N., and Russell, B. (1970). *Principia Mathematica* to*56 (2nd ed.). Cambridge: Cambridge University Press.

Wigner, Eugene. (1967). *Symmetries and Reflections*. Bloomington: Indiana University Press.

Wigner, Eugene. (1937). "On the consequences of the symmetry of the nuclear Hamiltonian on the spectroscopy of nuclei," *Physical Review*, 51, 106–130.

Wimsatt, William C. (2006). "Aggregate, composed, and evolved systems: Reductionistic heuristics as means to more holistic theories," *Biology & Philosophy*, 21, 667–702.

CHAPTER II | Scientific Realism and Chemistry

HASOK CHANG

SCIENTIFIC REALISM IS A philosophical issue with relevance to all sciences, but there are some particularly interesting and distinctive ways in which it has manifested itself in chemistry. Paying proper attention to such aspects will deliver two types of benefits: First, it will aid the philosophical understanding of the nature of chemical knowledge; second, it will throw some fresh light on the realism debate in places where it has developed without much attention to chemical practices and chemical concepts. In the following discussion I will attempt to make a reasonably comprehensive survey of relevant literature, while also advancing some original points and viewpoints.

1 *Unobservable Entities*

Recall Bas van Fraassen's now-classic formulation of the realism debate as an argument about whether we can know about *unobservable entities* featuring in scientific theories, and whether we *should try to know* about them (van Fraassen 1980). If this is how we understand realism, and if we take the long view of the history of science, chemistry is the most important science to consider in the realism debate. Until the development of atomic, nuclear, and elementary-particle physics starting in the early twentieth century, chemistry was the science in which debates about the epistemic and ontological status of unobservable theoretical entities took place with most ferocity and most relevance to practice. An interesting contrast is astronomy, in which the Copernican Revolution brought in a long and secure phase of realism about astronomical objects far out of reach of any human senses (including those that do not even register as tiny specks of light to our eyes).[1] In contrast, the achievements of chemistry up

---

[1] This underlying astronomical realism is unwittingly incorporated even in van Fraassen's anti-realist philosophy: that the moons of Jupiter are observable entities because if we went near enough to them we would be able to perceive them with our unaided senses. How do we know that the moons of Jupiter are the sort of things that we will be able to sense if we get close enough to them? How do we even know that the space around Jupiter is the sort of place

to the early nineteenth century only deepened the sense of inaccessibility and unobservability concerning the putative fundamental entities postulated in chemical theories.

Unobservability in relation to chemical theories is not only an issue about atomism, though surely the problem was clearly present with the atomistic particles imagined by a wide range of thinkers from Democritus and Leucippus of ancient times to Descartes and other early-modern mechanical philosophers. More important for early chemistry than atomism were theoretical speculations about elements, conceived without much regard to the question of their atomistic constitution. For example, in the *tria prima* theory of Paracelsus and his followers, the elements "sulphur," "mercury" and "salt" do not refer to the normal substances called by those names, which are perfectly observable (see Brock 1992, 43–48). Mercury, for example, was a theoretical substance embodying perfect fluidity, only imperfectly manifested in liquids such as mercury as we know it in everyday life; alchemists sometimes spoke of "the philosophical mercury" or "our mercury" in order not to be mistaken as meaning ordinary mercury. Hardly anyone imagined that the philosophical mercury could be directly observed.

The situation was similar with other fundamental substances of early chemistry, such as phlogiston; except for a very brief period (roughly speaking, from 1766 to 1784) when Henry Cavendish thought that inflammable air (hydrogen) was pure phlogiston, even chemists working in the phlogistonist tradition did not imagine that phlogiston could be isolated in observable form. The situation was also very similar with heat and electricity, both of which were quite routinely regarded as chemical subjects in those crucial decades around the year 1800. Both were considered material substances capable of chemical combinations—electricity as one kind of fluid or two depending on one's theoretical allegiance, and heat as "caloric" (appearing at the top of Antoine-Laurent Lavoisier's table of chemical elements, along with light).[2] Both were regarded as unobservable in themselves, though their effects were surely observable. Even something as straightforward-sounding as Lavoisier's "oxygen" may fall into the realm of the unobservable. In Lavoisier's own terminology, the substance he was handling daily was *oxygen gas*, which was a chemical compound made up of *oxygen base* and caloric. Only the removal of all caloric would have given the pure form of oxygen base, and such an absolute privation of heat was not thought to be possible. So oxygen (base) was an unobservable entity according to Lavoisier's own theory, in a similar way as free quarks are declared to be impossible by quantum chromodynamics (the theory of quarks) itself.

---

humans can get to? Van Fraassen says, frankly, that we have to rely on what our best scientific theories tell us (and in the post-Voyager era we may have quite strong confidence that humans would be able to go and see similarly as our unmanned spacecraft does). But if we rely on our best scientific theories to tell us about something that is strictly speaking unobservable, isn't that precisely realism by van Fraassen's lights?

[2] This table appeared in Lavoisier (1789), p. 192; see also the discussion of various points concerning this table in Chang (2012), ch. 1.

The question of unobservability only deepened with the advent of chemical atomism, most famously spearheaded by John Dalton and Jöns Jakob Berzelius at the start of the nineteenth century (see Nye 1972, ch. 1; Nye 1976; Gardner 1979). It is not that ideas about atoms were absent in earlier chemistry and physics, but the question about their nature became urgent with Dalton and Berzelius probably because they had finally succeeded to make atomism useful in chemistry.[3] This situation needs a more careful analysis than is usually given. We need to start with a caution regarding the importance of the realism debate: even in the age of atomic chemistry, whether atoms "really existed" was mostly not a pressing concern to most working chemists for about a century following Dalton's first publication (1808) on atoms. Alan Rocke (1984) makes a very useful distinction between chemical atomism and physical atomism. Most knowledgeable historians of chemistry would agree with Rocke that chemical atomism was accepted by most nineteenth-century chemists but physical atomism was not so universally accepted, and moreover not necessary for most of the work that chemists were doing.

On the other hand, it would be too simple to imagine that chemical atomists (those that weren't also physical atomists) were instrumentalists or even positivists. Chemical atomism, like any other theory postulating unobservable entities, could be taken in either a realist or an anti-realist way. Chemists were not always explicit about such issues, but I think many working chemists in the nineteenth century did believe that there was *something* real atomic "out there," entities possessing definite weights that constituted basic units of chemical combination and recombination, though what other properties they had was not known for sure. That is a different attitude from the positivism of saying that there were simply the stoichiometric regularities concerning how much of which substance combined with how much of which other substance (expressed as "equivalents"), or from the instrumentalism or fictionalism of saying that chemical combinations happen *as if* there were chemical atoms. Especially as structural organic chemistry developed in the middle decades of the century, I believe that most working chemists did begin to take chemical atoms as real entities. A crucial moment in this shift was the achievement of consensus on atomic weights and molecular formulas through the efforts of Stanislao Cannizzaro and others (at the emblematic Karlsruhe Congress of 1860, for example)—and slightly earlier, through what Rocke calls the "quiet revolution" of the 1850s in organic chemistry.[4] Confidence in the reality of chemical atoms, whose defining property was weight, understandably went up when various chemists could agree on what their weights were.

What else came to be considered real about atoms? This is a long and complex story. Dalton had originally assigned definite shapes (spherical) and sizes

---

[3] See Chalmers (2009) on the very limited use of atomistic ideas for chemical practice until the nineteenth century.
[4] See Rocke's numerous publications on this subject, including Rocke (1993); see also Chang (2012), ch. 3, and references therein.

to atoms, and attempted to use these properties in various physicochemical explanations of phenomena. Within his own theory he did not achieve convincing determinations of such properties (see Fox 1968), and almost no other chemists took them seriously. Similarly, it was not until the advent of stereochemistry (chemistry in three-dimensional space) in the late nineteenth century that theoretical notions about *molecular* shapes were given realist credence by a large number of chemists, even though there was always a significant realist impulse coming from crystallography. But even with the development and acceptance of stereochemical theories, the shapes, sizes, and other physical properties of individual atoms remained beyond the reach of empirical determination until a whole new era arrived in physics. And when that era arrived and matured, through quantum mechanics, the notion of the shape of an atom became quite nebulous. Similarly, even though structural chemistry gave clear indications of how many bonds each atom could and did form with other atoms, and with what spatial orientation, the specifics of the mechanism of the chemical bond (usually represented with the deceptive simplicity of a line) remained obscure.

What philosophical sense can we make of the practices of structural chemists in the second half of the nineteenth century? Their science was wildly successful in all kinds of ways, including the persuasive elucidation of the constitution of a very wide range of substances and the facilitation of the synthesis of many complex materials. The first thing to recall is that all this was achieved on the basis of a very partial knowledge of the physical properties and structures of atoms. And the nineteenth-century chemists were perhaps wise to be anti-realist in van Fraassen's sense, declining to attempt to figure out such things with the material and conceptual resources at their disposal. When the discovery of electrons and subsequently the development of quantum mechanics shed further light on the structure of atoms and the nature of the chemical bond, the picture that emerged was not like anything that nineteenth-century chemists had imagined, or even could have imagined.

What do these reflections tell us about the realism issue as a debate about our proper epistemic attitude toward theories concerning unobservables? The first point is that realism is not an all-or-nothing matter. One can be committed to the reality of some of the presumed properties of a theoretical entity, without a commitment to other properties. In such situations it is too crude to talk about "unobservable entities" as whole bundles. A related point is that it is possible (and perhaps unavoidable at least in practice) that a theory only comments on some aspects of the entities in its domain and keeps silent about the rest. A theory can be highly successful on such a partial description of reality, and scientists can make successful use of a theory while making a realist commitment only to part of what it says about the unobservable reality. Therefore the success of a science (or some particular system of practice within it) only constitutes an argument for the truth of some part of the theory. This conclusion should be a comfort to those who advocate various positions under the rubric of "partial realism."[5]

<hr />

[5] For a recent overview and assessment of partial realism, see Peters (2012).

Going under the banner of partial realism is only a vague direction. A crucial next step is to discern what is actually responsible for the success of a theory, and that is something very difficult to do convincingly in the abstract. And when we analyze concrete situations, it is not good enough to note continuity or commonality (this point will be discussed further in section 4). Even if the same aspect is present in two theories, and is responsible for the success of one, it does not follow that it is responsible for the success of the other. We need to take a specific successful theory, and see which aspects of it were used successfully—as premises for an inference to a successful conclusion, or as factors motivating a successful practice, and so on. We also need to ask if those successfully used aspects of a theory were used with a realist commitment. Without going into full details here, I note two interesting types of situations. First, chemists have often been realists about an unobservable property that they could engage with in the laboratory, whether it be phlogiston's ability to confer metallic properties to a calx, or the weight of chemical atoms. This is in line with the "operational realism" advanced by Bernadette Bensaude-Vincent and Jonathan Simon (2012, 206–211), and also with the "experimental realism" advocated by Ian Hacking, which will be discussed in detail in section 3. Second, chemists have often successfully employed unobservable properties and entities in their work, even those that are declared to be unreal in the fundamental theories to which they at least pay lip service. Orbitals and curly arrows, both unreal according to the principles of quantum mechanics, are used freely and successfully by modern chemists. The same may be said about the practice of nucleus-clamping in theoretical quantum chemistry itself. All this may be an expression of instrumentalism—or, on the contrary, a low-key operational realism concerning entities about which high theoreticians are anti-realists.

2 The Nature of Models

In philosophy of science recently there has been a recognition that much theoretical work in science happens by means of models, rather than anything billed as "theories" or "laws." This is very much based on how scientists themselves speak about their work these days. In chemistry, compared to the kind of physics that informed traditional philosophy of science, there is a much longer history of conscious uses of models, and chemists have often debated the epistemic function of models and the ontological status of the entities postulated in the models. This is quite interesting from the viewpoint of the realism debate. To call something a "model" already has significant anti-realist connotations—a good "theory" can be imagined to express reality perfectly, but a "model" would seem inherently idealized (used in lieu of a completely accurate theory) or abstract (only featuring some aspects of the full reality). Yet, there is also realist intent in modelling in many cases, a desire to make a partial or approximately true representation of reality, faithful within its acknowledged limits. This is consonant with my view expressed in section 1 that successful scientific work can be, and often is, carried out on the basis of a partial realist commitment to a theory concerning unobservables.

Chemists' attitudes toward their models have been various and changeable, but rarely have they been fully realist or fully anti-realist. Some models, such as the stacking of unit cells in eighteenth- and early nineteenth-century crystallography, were taken at face value, but even there the realist commitment was partial for most of the history, in the sense that the models only concerned the (presumably undeformable) spatial shapes of the cells and not any other properties that the cells might possess. As Christoph Meinel (2004) has shown in illuminating detail, the structural models of organic chemistry began mostly as heuristic and pedagogical tools and only gradually acquired higher realist significance. The most famous type was the "ball-and-stick" models originating in the 1860s with August Hofmann's rigging-up of croquet balls connected by sticks drilled into them, with which he expressed the notion of valency (valence) that he and others were crafting at the time. Hofmann's models are the direct ancestors of the colorful ball-and-stick sets that continue to aid the thinking of chemists, chemistry students, and the public to this day.

Such models are taken quite literally today, as far as the spatial locations of atoms are concerned—despite the quantum-mechanical qualms one would generally have about such a classical representation, and despite the fact that the notion of the ball-shape of atoms has been thoroughly discredited by quantum mechanics. But even concerning the spatial locations of atoms within such models, the realist commitment was not always there, even when the models were quite successful for their purposes. Hofmann himself clearly lacked full realist ambitions for his models, as he kept them 2-dimensional, and he made no attempts to theorize about bond lengths, keeping all the sticks at the same length. The nature of the epistemic moves involved in making and using such models becomes clearer when we consider representations that were definitely not meant to be taken "literally." For example, consider August Kekulé's "sausage" models in which atoms were represented as sausages of different lengths, the length indicating the valency of each atom; it was certainly not Kekulé's intention to argue that atoms really were sausage-shaped (see Chang 2012, 193). There is also the earlier case of Auguste Laurent's "prism" model, which was not intended in a realist way even though it was geometrically a 3-dimensional model.

There have also been important chemical models that were not spatial at all, and in that sense not even trying to be realistic depictions of atoms and molecules. For example, what exactly was the epistemic status of bits of chemical formulas, famously designated as "paper tools" by Ursula Klein (2003)? There were many ways of dissecting the compositional formula of a molecule into its presumed component parts; Kekulé famously noted that even those who all agreed on the empirical formula for acetic acid had given 17 different structural formulas (see Brock 1992, 253). There were various ways of convincing oneself that some presumed parts of molecules were real, but Klein's point is somewhat different: The manipulation of the formulas themselves was a distinct method of chemical reasoning and research, and the formula for each presumed grouping of atoms was a model-element, a conceptual tool with the help of which various theoretical aims were achieved. Did success in such theoretical work confer reality to what those formulas represented? For example, should various Berzelian radicals have been considered real because they facilitated some important explanations and predictions of various chemical reactions?[6] Or should one have reserved reality for only those elements that were manifested in a more direct way in experimental settings, for example as products of electrolysis? There was no consensus among chemists on this issue, and it is difficult to see that there is a clear answer that they should have agreed on.

It is interesting to observe how chemists have sometimes changed their attitudes toward certain models, and to discern what it is that they felt warranted such a transition. For example, the concept of valency began as an abstract notion of combining power expressed with the help of heuristic graphic models (such as Kekulé's sausages and Hofmann's sticks). Even though combining power itself would have been considered real by many chemists, there was little indication that the heuristic pictures should be taken as representations of reality. This all changed with the advent of stereochemistry (see Ramberg 2003). Tetrahedral carbon makes an interesting story. Those who took the quadrivalency of carbon seriously in a realist way could only conclude that the shape of the carbon atom had to be basically tetrahedral, if its bonds were to be distributed symmetrically in 3-dimensional space. The success of tetrahedral carbon marked a clear transition in the ontological status of valency.[7]

Debates about the reality of various models in chemistry did not cease in the twentieth century. When Gilbert Newton Lewis began to make creative use of the newly discovered electron in his theorizing about chemical bonds, starting with his model of the cube that was completed by 8 electrons, it was not clear that this alleged distribution of electrons in the space around the atomic nucleus should be taken seriously in a realist way. The same kind of question arose with other schemes, including J. J. Thomson's and Niels Bohr's, that tried to use the distribution of electrons to explain the periodicity of chemical elements (see Arabatzis 2006, ch. 7). None of these models were based on any credible physics of how an electron moved and interacted with a nucleus and with other electrons, so it was perhaps easy enough to be anti-realist about them.

However, the existence of a worked-out theory behind a model is not necessary for that model to be taken in a realist way; consider the early realist Copernicans, who had no credible theory of kinematics or dynamics about

---

[6] See Ihde (1984), ch. 7, for a convenient summary of relevant developments in early nineteenth-century organic chemistry.

[7] But what *were* these bonds, if they were to be taken in a realist way as physical entities? No good answers were forthcoming within the tradition of nineteenth-century structural chemistry, which did not have any account of the internal structure of atoms. Through the twentieth century the phrase "atomic structure" has become very familiar, but it would have had a clear oxymoronic ring in the nineteenth century, when "atom" still meant an indivisible unit, by definition lacking any meaningful internal structure.

how the earth could be spinning without giving any apparent sign of that motion. Even though the Lewis cube did not survive the test of time and further work, a related idea from him did: electron-pairs. Lewis had no theoretical basis, except a magnetic analogy, for believing that electrons should or even could form stable pairs in the way that his scheme of chemical explanations demanded. It is extraordinary that later developments in theoretical physics provided justification for electron-pairs in a manner that Lewis could not have anticipated, through the concept of spin and the Pauli Exclusion Principle.

The advent of full-fledged quantum mechanics and its application in chemistry did not stop the use of models and modelling techniques of debatable realist credentials (see Gavroglu and Simões 2012). The most obvious case is that of orbitals, which have assumed a curiously strong realist presence in the mind of many chemists and chemistry students. Strictly speaking, orbitals are just mathematical functions.[8] In the simplest form they are the eigenstates of the Hamiltonian (energy) operator in a one-electron atom. In so many textbooks these functions are reified in nice 3-dimensional graphical representations. The explanation of molecular structures and chemical reactions often invoke the shapes and distributions of orbitals. Accounts of the periodic table typically imagine electrons coming in to occupy different orbitals in a specific order, as if the orbitals were so many pigeonholes waiting for different pigeons to occupy them. In the context of the realism debate, it is difficult to know what to make of the great success of orbital-based chemical reasoning.

There would seem to be no good reason why orbital models of multi-electron atoms should work so well. Electrons within a given atom are strongly interacting with each other, so the orbitals derived from the solution of a one-electron Schrödinger equation, or even molecular orbitals, are at best crude approximations for orbitals in a multi-electron atom. Electrons are also truly indistinguishable from each other if we take quantum mechanics seriously, so the talk of different electrons occupying different orbitals within the same atom or molecule does not make sense. And why should we believe that electrons in an atom exist in eigenstates of the Hamiltonian, rather than in a superposition of various eigenstates? So what sense is there in thinking that a given electron is always in one orbital or another? Given all these difficulties with taking orbitals at face value, should we conclude that the success of orbital-based reasoning is just one great lucky coincidence? If we don't want to do that, then we have a challenging task: to spell out exactly in which ways the orbital models get reality correctly in such a way as to underpin extremely effective explanations and predictions. As Eric Scerri puts it (2007, 248), "It is indeed something of a miracle that quantum mechanics [in the orbital-filling mode] explains the periodic table to the extent that it does at present. But we should not let this fact seduce us into believing that it is a deductive explanation."

---

[8] See Scerri (2008) for an interesting critical discussion of a recent claim of a direct observation of orbitals.

3 Interventions and Practical Success

Is there a way to move beyond the typical underdetermined situation, in which we know that a theory or model is successful but cannot pinpoint, in a principled and confident way, which aspects of it deserve realist credence? In that context it is easy to see the appeal of Ian Hacking's doctrine of "experimental realism" that reality can be ascertained by successful experimental intervention. Experimental realism has an initial resonance in chemistry, even though Hacking's own inspiration came from other fields of science such as modern experimental high-energy physics and biological microscopy (see Hacking 1983, esp. chs. 11 and 16). Hacking famously stressed the epistemological importance of practical interventions—"If you can spray them, they're real" (1983, 23). It is commonly acknowledged that chemistry has always been a laboratory science with close ties to technology and industry, a distinctly "impure science" as Bensaude-Vincent and Simon (2012) put it. Chemical knowledge has always been founded on the practical manipulation of substances, and chemists have been proud of their ability to make things and change them.

The history of chemistry would therefore seem to be an exemplary source of object lessons in Hacking's experimental realism, as well as Bensaude-Vincent's operational realism. There are many important examples. From early chemistry, we have the widely accepted norm that the constitution of a chemical substance ought to be confirmed by its decomposition into its presumed components, or by its synthesis from the components. The neatest demonstration of composition was given by a combination of successful decomposition and recomposition. This ideal was exhibited in a whole range of successful chemical research, including Lavoisier's experiments on water, Berzelius's electrochemical theory grounded in results of electrolysis, and right down to the triumphs of industrial chemistry starting in the late nineteenth century.

Slightly different but closely related was the work on substitution: Chemists gained a great deal of constitutional knowledge, starting from the mid nineteenth century, by the technique of substituting one part of a molecule with another atom or radical. This began with the accidental discovery of hydrogen–chlorine substitution in organic chemistry, a reaction deemed rather incredible from the framework of Berzelian dualistic theory, as hydrogen was highly positive and chlorine highly negative. Substitution played a key role in generating realist confidence in molecular structures, and in helping chemists arrive at the concept of valency, which was of enormous benefit to the progress of chemistry and remains essential in much of chemical reasoning (see Chang 2012, ch. 3). Modern synthetic chemistry is perhaps unique among the sciences (at least until synthetic biology develops further) in its ability to create a dizzying array of well-specified entities, many of which are not even known to form spontaneously in nature. Who would dispute the reality of these substances that chemists create by careful design, which they can also use as tools for achieving certain desired effects so well? So it may seem that chemistry, the science of intervention *par excellence*, should give us the best instantiation of Hacking-type experimental realism.

The situation is actually much more complicated than that. Applying Hacking's experimental realism to chemistry in fact generates some strong discomfort. I will argue that there is a good way of defending experimental realism here, but I must first exhibit the discomfort I speak of in some detail. The basic point is that in chemistry Hacking's idea seems especially prone to attack from the pessimistic (meta-)induction from the history of science (Laudan 1981). All sorts of entities such as phlogiston and caloric have been used as bases for successful chemical practices, so it is difficult to deny their reality on experimental-realist grounds. The history of chemistry is full of examples of successful theoretical schemes containing entities that we now take not to exist and assumptions that we now take to be false. For example, take Kant's admiration of the phlogiston theory: When "Stahl changed metals into calx and then changed the latter back into metal by first removing something [phlogiston] and then putting it back again, a light dawned on all those who study nature" (quoted in Chang 2012, 4). Later Joseph Priestley predicted, on the basis of the phlogiston theory, that a metallic calx (oxide, in Lavoisier's terms) combining with inflammable air (hydrogen) would be reduced to pure metal, because calx is de-phlogisticated metal and inflammable air is (rich in) phlogiston. This prediction was brilliantly confirmed in an experiment in which he used a large "burning lens" to focus sunlight on lead calx enclosed in inflammable air (see Chang 2012, ch. 1, esp. 53–54; Musgrave 1976, 199). Isn't this just the kind of intervention that Hacking has in mind as the basis of his experimental realism? Wasn't Priestley "spraying" phlogiston on to the calx, using phlogiston successfully as a tool in order to make a very distinct and practical experimental intervention? If so, there would seem to be no reason not to regard phlogiston as real. Similarly, William Herschel thought he was spraying invisible caloric rays on the thermometer in the experiment in which he discovered infrared rays coming from the sun, and Count Rumford cooled things down, he thought, by spraying cold radiation ("frigorific rays") on them (see Chang 2002).

This is not only about the bygone ages of chemistry. Some of the examples from modern chemistry discussed in section 2 are very instructive. The spectacular practical success of organic structural chemistry in the nineteenth century owed much to models that were not regarded as literally true even at the time. The situation does not change even if we come into the age of quantum chemistry from the 1930s onward (see Gavroglu and Simões 2012). Not only a great deal of theoretical reasoning but numerous experimental interventions rely on the concept of orbitals and on the detailed knowledge of the number and shapes of various types of orbitals, while orbitals have no reality if we take quantum mechanics literally. What these examples illustrate is that experimental success is no guarantee of the reality of the entities that the experimenters in question *presume* to be manipulating when they carry out their experiments. To stay with Hacking's slogan: You may be able to spray something without knowing much about what it is that you're spraying. All you can perhaps be sure about is that there is some "it" that you are spraying, but Hacking's slogan does not quite work as intended if the blank "it" (or "them") is filled in with something specific: "If you can spray phlogiston, then phlogiston is real."

Another way to make the same point is to say that Hacking's separation of two strands of the realism debate—concerning existence and truth, respectively—does not really work. Hacking wants to argue that we can know, at least sometimes, that some unobservable entities postulated in our theories really exist, without needing to know whether the theoretical statements concerning those same unobservables are really true. But even "X exists" can be a serious and substantive theoretical statement concerning X (which may or may not be true). Even in Hacking's own discussion of "entity realism" it is not just the bare existence of something that he thinks we are entitled to claim knowledge of; rather, we are supposed to be secure in our knowledge of some basic causal properties of the entity in question, which is what allows us to interact successfully with them in the first place.

One can conceive of this difficulty by reference to inference to the best explanation (IBE) and its general problem. If Hacking-style arguments are to go beyond "there must be something real here" to "X is real" where X is a particular theoretical entity (such as "positron" or "phlogiston"), then the reasoning that takes one from manipulative success to entity realism is a form of IBE. The reasoning has to be in the following form: "The best way to explain what happens in this manipulation (and why I am successful at it) is that we have X here, X does this-and-that under such-and-such circumstances, and I have designed my experiment on the basis of my knowledge of X's basic causal properties." I have no qualms about IBE as a useful way of thinking; however, it does not work as the basis of Hacking-style experimental realism, because IBE is underdetermined. The history of chemistry amply displays that underdetermination. To return to Priestley and phlogiston: Of course, Lavoisier offered another explanation of the same experiment, which he considered the best; Cavendish came back with a slight modification of Priestley's interpretation, which he and Priestley both considered the best; decades later, Lewis considered that the Cavendish–Priestley explanation was good if one replaced "phlogiston" with "electron"; the best quantum-mechanical explanation is different from Lewis's, too. There are no clear criteria to judge which explanation is *eternally* best here, and the contingency displayed in actual history leads one directly from underdetermination to the pessimistic induction. The best explanation of today won't be the best explanation of tomorrow, and the different entities invoked in the various explanations will come and go, if we follow IBE. Therefore, what Hacking considers "real" will also have to change with the progress of science, and I do not know if he would be troubled by that, but I know that most realists who have found relief in Hacking's arguments will be deeply troubled.

But why do I say that chemistry is particularly prone to the pessimistic induction in relation to experimental realism, rather than just as prone to it as any other science is? My argument on that point concerns the premise of the pessimistic induction argument, namely the fact of the success of past theories.

Paradoxically, chemistry is more troublesome for experimental realism because of its practical nature: In chemistry it is not likely that a theoretical idea would become widely accepted without having proven itself by supporting some measure of practical intervention. This has been the case at least since the advent of affinity-based chemistry in the eighteenth century. Yet, as in all sciences, ideas and theories in chemistry do change, sometimes radically. Then it must be quite easy to find chemical theories that are now rejected (or at least considered false though somehow maintained) but has quite a firm grounding in practical interventions. That practical grounding must still be present unless nature itself has changed in the meantime.

If my argument so far is successful, perhaps it will make people turn away from Hacking's experimental realism; my intention, however, is just the opposite. Rather than despairing about Hacking's experimental realism because it would rule phlogiston in, I want to suggest that we should instead continue to appreciate the cogency of Hacking's position and learn to accept that phlogiston *is* real—at least as much as any other unobservable theoretical entity can be regarded as real. This position obviously needs a careful defense, which is the object of the next section.

4 Preservative Realism versus Conservationist Pluralism

Let us restart with the basic realist intuition that theories providing the underpinnings for successful investigations deserve our realist confidence. As mentioned above, the main challenge in fleshing out that intuition is to discern which parts or aspects of a successful theory are responsible for its success and therefore deserve our confidence. This is a difficult task, especially if we want to understand what enables successful empirical investigations as practices, rather than just trying to see which assumptions lead to which conclusions. If Copernicus attributed his success to his obstinate and purist preference for uniform circular motions that crucially helped him to reject Ptolemaic astronomy (see Kuhn 1957), it would be difficult for anyone to argue with that. In the realm of chemistry, should we say that Lavoisier's success was due to his theory of caloric, which allowed him to believe that he had reached a satisfactory explanation of the heat issuing in combustion, freeing him to discard the phlogiston theory and to focus properly on oxygen and weights? Discerning the real secret of success is bound to be difficult for those mixed up in the creative process while it is in progress. But even with the luxury of 20-20 hindsight it is not easy to tell what was truly responsible for the success of a theory.

The only strategy we can employ in dealing with this difficulty is the same kind of detective work that happens in scientific investigations, and in combatting (scientifically and philosophically) the problem of underdetermination in theory-testing, rather than anything peculiar to the realism debate. In the effort to discern the theoretical elements responsible for success, we should have open minds and be generous to the investigators involved. That means granting operational reality to whatever theoretical conception has been genuinely re-sponsible for our success. This is what we ought to do if success is our only guide in what to be realist about. And since this is not the kind of situation in which we can do systematic large-scale trials, we may be reduced to the kind of homely reasoning exemplified in John Stuart Mill's inductive methods.[9] The method of difference should figure largely in this, where we ask: "Do we (or can we) get the same result without this assumption?" Some of what we need to know here is provided by history, in the form of actual variations among the practices of past scientists; some we can try to reason out through counter-factual history; some we have to learn by doing, by seeing if we can do without certain elements of theory in making calculations or doing experiments.

By such methods we can and should distinguish the operative parts of a theory (the "working posits") from the idle parts (the "presuppositional pos-its"), but without any pretensions to ultramundane confidence.[10] One reason for caution is that what is operative in a given theory will depend on which particular use we are making of it; for example, atomic sizes were operative for certain physical explanations Dalton attempted to make, but they were idle in most reasonings in the atomic chemistry of his time. Besides, in each situation we (both observers and practitioners) may err in identifying what is operative and what is idle; here we can always challenge each other and correct ourselves, the way scientists do. In this context we can, again, appreciate the appeal of Hacking's focus on interventionist success as one clear guideline for pinpoint-ing secrets of success. But how do we deal with the difficulties of experimental realism highlighted in the last section?

Against the difficulty of scientific change (expressed via the pessimistic in-duction), it is tempting to go for what I have previously critiqued as "preserva-tive realism," namely the tendency to argue that what we see preserved through scientific change must reliably correspond to elements of reality (Chang 2003, in response to Psillos 1994). Preservative realism promises some stability, and even holds out the prospect of uniqueness, in case the preservation holds in the long run. But the preservative realist strategy doesn't go so well with suc-cess-based realist strategies, because there is no inherent connection between what is successful and what is preserved. Perhaps there is a "Darwinian" way of correlating the survival of a whole theory or system with its overall fitness, but this does not do the job of pinpointing specific aspects of the system re-sponsible for its success. We need to consider more carefully what kinds of elements or aspects of theory tend to get preserved through major theoretical changes, and why. Only then can we try to discern whether those aspects can be given realist credence. I take it that this is the sort of enterprise that structural realists, for example, are engaged in (see Ladyman 2011; Worrall 1989).

Again, there are some points of caution worth noting. Various aspects of scientific theories may be preserved due to our own habits of thinking, rather

---

[9] For a convenient summary of Mill's methods, see Losee (1993), 153–164.
[10] See Kitcher (1993), esp. 149, for the working/presuppositional distinction.

than anything to do with nature. There are techniques of representation and reasoning (including certain mathematical methods) that we just find convenient or pleasing to use. The employment of linear equations is a very good example. There are also deep-seated metaphysical commitments, including what Holton (1973) has identified as "themata." A good example here is a desire for constancy, which has led physical scientists over the centuries to anchor their most fundamental theories on various conservation principles (concerning mass, motion, *vis viva*, heat, energy). Most of these principles were in the end empirically refuted and rejected, but that has not stopped scientists from simply moving on to some other conservation principle. Even energy conservation does not hold so simply and precisely in the realm of quantum field theory, yet we prefer to say that energy *is* conserved, and then add qualifications. Other popular metaphysical predilections include determinism and atomism. A belief in these principles "can be held true come what may," to borrow Quine's phrasing that was meant to apply more broadly. The fact that certain of these metaphysical beliefs have been preserved by most European scientists for the past few centuries is, in itself, no indication of the way nature is or of what we really know about it.

All in all, the preservationist strategy is deeply problematic. I would like to suggest a different strategy for dealing with the difficulties of scientific change and underdetermination. That strategy is epistemic pluralism, which I have begun to elaborate elsewhere (Chang 2012, chs. 4 and 5). Here I will only present some highlights that have the most relevance to the issues discussed in this chapter.

The pluralism I am promoting here is linked to what I have called the "optimistic rendition of the pessimistic induction" (Chang 2012, 226), which acknowledges the premise of the pessimistic induction, and stops there in appreciation. In the formulation given by Stathis Psillos (1999, 101), the pessimistic induction argument begins: "The history of science is full of theories which at different times and for long periods had been empirically successful..."—let's stop there, and be happy! The history of chemistry (even more than other sciences) is certainly full of theories that were responsible for interventionist and other empirical successes. Why not acknowledge that each and every such theory was true, and remains true in its proper realm? This, of course, requires abandoning the search for an ultimate kind of truth, "Truth with a capital T," which is about the Kantian *Ding an sich*, and which we will never be able to infer from success (or from anything else). If our use of a theory has led to successful outcomes, and if this is not the result of some strange coincidence as far as we are able to ascertain, then we can and should say, modestly and provisionally, that the relevant statements made in this theory are "true" and the entities presupposed in them exist, in the same sense as we say that it is true that rabbits exist, that they have whiskers, and that they live in underground burrows. This "truth" is of the operational, verifiable kind, and is one and the same thing as empirical confirmation taken in a broad sense.[11]

---

[11] This is what I have designated as "truth$_5$" in Chang (2012), 242.

This homely notion of truth is a crucial part of what I call "active realism," which is the commitment to learn about mind-independent reality in all possible ways. That may sound too trivial and obvious even to be worth discussing, but there is at least one controversial aspect to active realism, which is a pluralism inherent in it. This pluralism recommends that *all* theories facilitating successful engagement with nature (including Hacking-style interventions) provide ways of learning from reality, and should be maintained and developed as much as possible, even if they conflict with each other. Success should be appreciated for what it is, and what is responsible for success should be preserved so that it may continue giving us that success. From the active-realist point of view, what success gives is merely a credible promise of more success, a promise accepted with our eyes wide open to the problem of induction. When there is a theory that has time and again supported successful empirical activities, it makes sense to keep that theory for future use. That is a modest and good argument for preserving a successful theory. Such preservation, because it is firmly rooted in practical experience and the kind of basic induction that Hume taught us we can't do without, should be robust in the face of another theory that does something else (or even the same thing) very well. And here we have no need to pinpoint the "secret" of the success of any of the theories, so we do not need to be so troubled by the difficulty of that task.

All these considerations lead to a policy that I have labeled "conservationist pluralism" (Chang 2012, 224): Retain previously successful theories or paradigms for what they were (and are still) good at, and *add* new theories or paradigms that will help us make new and fresh contacts with reality. With conservationist pluralism we can, once again, understand the development of science as a history of cumulative progress—not an accumulation of simple unalterable facts from which more and more general theories would be formed, but of various locally effective systems of practice which somehow continue to be successful. In fact, physics shows this pattern of development more starkly than any other science. If we examine what we actually do in various domains of scientific practice and everyday life, it is easy to recognize that we have retained various successful systems that are good in particular domains: geocentrism (for navigation); Newtonian mechanics (for other terrestrial needs and for space travel within the solar system); ordinary quantum mechanics (for much of microphysics and almost all quantum chemistry); as well as special and general relativity, quantum field theory, and more recent theories. For those upset with the suggestion that the applicability of, say, general relativity, is "local," I can only recommend looking at all the situations in which we would not dream of using general relativity (including rocket science, though not GPS).

A careful look at chemistry shows that this pattern of conservationist pluralism is equally obvious, and perhaps less threatening or disturbing to the reductionists than in physics. Chemical practice includes a great number of different systems in place, each doing the job that it has always done well. The advent of *ab initio* quantum computations has not eliminated the important roles played by systems established earlier, such as orbital-based quantum chemistry, Lewis-style reasoning about electron-pairs shifting around within and between molecules, classic structural chemistry that tells us where the nuclei are in a molecule (without which we can't even set up the molecular Schrödinger equation), classical chemical thermodynamics embodied in the Nernst equation, and so on and so forth, right down to eighteenth-century electrostatic reasoning surviving in our conception of clearly ionic bonds.

5 Reduction and Levels of Analysis

I cannot conclude this discussion of the realism debate concerning chemistry without discussing its relation to another important philosophical issue, namely reductionism, especially the variety of it referred to as "microreductionism."[12] The epistemic or methodological project of (micro)reductionism begins with a belief in ontological (micro)reduction, because it is the latter that confers initial plausibility and promise to the former. That is to say, if we believe that chemical substances are made up of molecules, which are in turn made up of atoms, which are in turn made up of elementary particles, it will be a very tempting thought that the best way to really understand chemical substances and reactions is to consult what elementary particle physics tells us about the nature and interaction of their component parts. It is realism about the microlevel entities and their role as building blocks of macrolevel entities that generates the strongest push for reductionism concerning the chemistry–physics relation. It is because of chemists' realist confidence in the reality of elementary particles and the truth of the theories given by physicists about them, that they are often inclined to accept the epistemic microreduction of their science to physics.

That much is simple enough. But as it often happens, putting the general idea into practice reveals many interesting subtleties and difficulties. In particular, we encounter a question about the most effective or illuminating level of analysis, the answer to which remains open even if one accepts the reductionist ontology fully. The most productive level of analysis in quantum chemistry has turned out to be that of electrons and nuclei. But what ontological level is this? It is not one that any single theory of physics would recognize as a coherent domain. When quantum mechanics took shape as a full-fledged theory in the hands of Werner Heisenberg and Erwin Schrödinger, it turned out to apply very well to electrons in an atom but not to the atomic nucleus. Ordinary quantum mechanics can only use the nuclei as the source of a fixed electrostatic potential, rather than providing answers about the dynamics or internal structure of nuclei, although some fine-tuning of nuclear positions within a molecule will come from the quantum-chemical calculations of bond lengths. The early models of atomic nuclei, such as the "liquid-drop" model or the shell model, had no comment at all to make about how the nuclei would

---

[12] For my view on reductionism and the chemistry–physics relation, see Chang (2015).

interact with electrons, and in this respect even quantum chromodynamics is no better. One could take a more principled and generalized approach and start from the framework of the Standard Model, which encompasses all of the elementary particles including electrons, protons and neutrons. However, finding a "grand unified theory" governing all of these particles has been a challenging task it itself to say the least, and no one has made a credible attempt to set up a single equation or model to describe an atom or a molecule in that framework. As far as the current best practice goes, science treats the atom as a heterogeneous composite made up of electrons and nuclei, much like what John Dupré (1993) says about how biology (rightly) treats an organism as a composite made up of components at various ontological levels—organs, cells, molecules, and even ions. And this heterogeneous ontological picture is the basis of what has turned out to be a spectacularly successful system of chemical practice.

It would be fair to say that chemistry has always been committed to a reductionist ontology, with a degree of realism about the layer(s) of existence "underneath" the observable layer. But it is also clear that this general commitment to ontological reductionism and realism has not dictated very much about the exact directions of chemical practice. First of all, there is the axiological dimension of the realism question: Is the fundamental aim of chemistry to seek correct descriptions of chemical phenomena in terms of the ultimate constituents of matter, or rather to seek the most effective description at any plausible level? On the whole, practicing chemists have tended toward the latter, though that tendency has by no means been universal. At least up to the mid-nineteenth century, those who insisted on instrumental effectiveness tended to have the upper hand in chemistry against the mechanical philosophers, metaphysical Newtonians, Boscovichians, Proutians, and others who prized the attainment of true microphysical descriptions above all. The great success of microphysics in the early twentieth century changed this picture significantly, but the debate continues. For example, many chemists would defend the continuing use of orbitals and electron shells on the grounds of their instrumental usefulness, even though fundamentalists declare them to be non-existent. In having this sort of internal disagreement the chemists are, of course, not unique: even within physics there are plenty of practitioners in areas such as solid state physics who decline to look entirely to elementary particle physics as a source of useful theoretical concepts and assumptions.

Summary

Chemistry, like any other science, is a realist enterprise in the sense that it seeks to learn from reality. But its practices diverge greatly from what standard realist philosophers might imagine. A close attention to chemical practice reveals a highly nuanced and selective epistemological attitude, with a tendency to be realist only about some specific aspects of various theoretical schemes and some unobservable entities. Attention to chemistry can help us refine our views on the realism debate, especially on the meaning and implications of empirical success.

Acknowledgments

I thank Grant Fisher and Georgie Statham for helpful and detailed comments on earlier drafts. Similarly I thank members of the AD HOC group in Cambridge. In ways that I can no longer specify, Robin Hendry and Eric Scerri have influenced various arguments contained in this chapter through many discussions over the years.

References

Arabatzis, Theodore. (2006). *Representing Electrons: A Biographical Approach to Theoretical Entities.* Chicago: University of Chicago Press.

Bensaude-Vincent, Bernadette, and Simon, Jonathan. (2012). *Chemistry: The Impure Science* (2nd ed.). London: Imperial College Press.

Brock, William H. (1992). *The Fontana History of Chemistry.* London: Fontana Press.

Chalmers, Alan. (2009). *The Scientist's Atom and the Philosopher's Stone: How Science Succeeded and Philosophy Failed to Gain Knowledge of Atoms.* Dordrecht: Springer.

Chang, Hasok. (2002). "Rumford and the reflection of radiant cold: Historical reflections and metaphysical reflexes," *Physics in Perspective,* 4, 127–169.

Chang, Hasok. (2003). "Preservative realism and its discontents: Revisiting caloric," *Philosophy of Science,* 70, 902–912.

Chang, Hasok. (2012). *Is Water H_2O? Evidence, Realism and Pluralism.* Dordrecht: Springer.

Chang, Hasok. (2015). "Reductionism and the relation between chemistry and physics." In *Relocating the History of Science: Essays in Honor of Kostas Gavroglu* (Theodore Arabatzis, Jürgen Renn, and Ana Simões, eds.). Dordrecht: Springer, 193–209.

Dalton, John. (1808). *A New System of Chemical Philosophy,* vol. 1, part 1. Manchester: R. Bickerstaff.

Dupré, John. (1993). *The Disorder of Things: Metaphysical Foundations of the Disunity of Science.* Cambridge, MA: Harvard University Press.

Fox, Robert. (1968). "Dalton's Caloric Theory." In *John Dalton and the Progress of Science* (D.S.L. Cardwell, ed.). Manchester: Manchester University Press, 187–202.

Gardner, Michael. (1979). "Realism and instrumentalism in 19th century atomism," *Philosophy of Science* 46, 1–34.

Gavroglu, Kostas, and Simões, Ana. (2012). *Neither Chemistry nor Physics: A History of Quantum Chemistry.* Cambridge, MA: MIT Press.

Hacking, Ian. (1983). *Representing and Intervening.* Cambridge: Cambridge University Press.

Holton, Gerald. (1973). *Thematic Origins of Scientific Thought: Kepler to Einstein.* Cambridge, MA: Harvard University Press.

Ihde, Aaron J. (1984). *The Development of Modern Chemistry.* New York: Dover.

Kitcher, Philip. (1993). *The Advancement of Science*. Oxford: Oxford University Press.

Klein, Ursula. (2003). *Experiments, Models, Paper Tools: Cultures of Organic Chemistry in the Nineteenth Century*. Stanford: Stanford University Press.

Kuhn, Thomas S. (1957). *The Copernican Revolution*. Cambridge, MA: Harvard University Press.

Ladyman, James. (2011). "Structural realism versus standard scientific realism: The case of phlogiston and dephlogisticated air," *Synthese*, 180, 87–101.

Laudan, Larry. (1981). "A confutation of convergent realism," *Philosophy of Science*, 48, 19–49.

Lavoisier, Antoine-Laurent. (1789). *Traité élémentaire de chimie*. Paris: Cuchet.

Losee, John. (1993). *A Historical Introduction to the Philosophy of Science* (3rd ed.) Oxford: Oxford University Press.

Meinel, Christoph. (2004). "Molecules and croquet balls." In *Models: The Third Dimension of Science* (Soraya de Chadarevian and Nick Hopwood, eds.). Stanford: Stanford University Press, 247–275.

Musgrave, Alan. (1976). "Why did oxygen supplant phlogiston? Research programmes in the Chemical Revolution." In *Method and Appraisal in the Physical Sciences* (C. Howson, ed.). Cambridge, England: Cambridge University Press, 181–209.

Nye, Mary Jo. (1972). *Molecular Reality: A Perspective on the Scientific Work of Jean Perrin*. London: Macdonald.

Nye, Mary Jo. (1976). "The nineteenth-century atomic debates and the dilemma of an 'indifferent hypothesis'," *Studies in History and Philosophy of Science*, 7, 245–268.

Peters, Dean. (2012). *How to Be a Scientific Realist (If at All): A Study of Partial Realism*. PhD dissertation, London School of Economics.

Psillos, Stathis. (1994). "A philosophical study of the transition from the caloric theory of heat to thermodynamics: Resisting the pessimistic meta-induction," *Studies in the History and Philosophy of Science*, 25, 159–190.

Psillos, Stathis. (1999). *Scientific Realism: How Science Tracks Truth*. London: Routledge.

Ramberg, Peter J. (2003). *Chemical Structure, Spatial Arrangement: The Early History of Stereochemistry, 1874–1914*. Aldershot, UK: Ashgate.

Rocke, Alan J. (1984). *Chemical Atomism in the Nineteenth Century: From Dalton to Cannizzaro*. Columbus: Ohio State University Press.

Rocke, Alan J. (1993). *The Quiet Revolution: Hermann Kolbe and the Science of Organic Chemistry*. Berkeley and Los Angeles: University of California Press.

Scerri, Eric. (2007). *The Periodic Table: Its Story and Its Significance*. New York: Oxford University Press.

Scerri, Eric. (2008). "The recently claimed observation of atomic orbitals and some related philosophical issues." In *Collected Papers on Philosophy of Chemistry*. London: Imperial College Press, 201–213. (Originally published in 2001 in *Philosophy of Science*, 68, S76–S88.)

van Fraassen, Bas. (1980). *The Scientific Image*. Oxford: Clarendon Press.

Worrall, John. (1989). "Structural realism: The best of both worlds?," *Dialectica*, 43/1–2, 99–124. Reprinted in *The Philosophy of Science* (D. Papineau, ed.). Oxford: Oxford University Press, 1996, 139–165.

CHAPTER 12 | Natural Kinds in Chemistry

ROBIN FINDLAY HENDRY

1 Chemical Kinds

Chemical substances such as gold and water provide paradigm examples of natural kinds: They are so central to philosophical discussions on the topic that they often provide the grounds for quite general philosophical claims—in particular that natural kinds must be hierarchical, discrete, and independent of interests. In this chapter I will argue that chemistry in fact undermines such claims. In what follows I will (i) introduce the main kinds of chemical kinds, namely chemical substances and microstructural species; (ii) critically examine some general criteria for being a natural kind in the light of how they apply to chemical kinds; and finally (iii) present two broad theories of how chemical substances are individuated. The primary purpose of this article is to bring scientific detail and sophistication to a topic—natural kinds—which has a long but not always honorable history in philosophy, but chemists can also learn something from these discussions. Chemistry is in the business of making general claims about substances, a fact which is embodied in the periodic table, as well as in the systems of nomenclature and classification published by the International Union of Pure and Applied Chemistry (IUPAC). At several points in the history of their subject, chemists appear to have faced *choices* about which general categories should appear in these systems. Understanding why these choices were made, and the alternatives rejected, gives us an insight into whether chemistry might have developed differently. This is central to understanding why chemistry looks the way it does today.

So, what are the chemical kinds? Chemists study the structure and behavior of substances such as gold, water and benzene, and also of microscopic species such as gold atoms, and water and benzene molecules. They group together higher kinds of substances: groups of elements such as the halogens and alkali metals, broader groups of elements such as the metals, and classes of compounds that share either an elemental component (e.g., chlorides), a microstructural feature (e.g., carboxylic acids), or merely a pattern of chemical reactivity (e.g., acids). Chemical formulae are often ambiguous, naming both substances and microscopic species. In one sense, "H_2O" names a molecular species: an oxygen atom bonded to two hydrogen atoms. In another sense it names a substance composed of hydrogen and oxygen in the molar ratio 2:1. Not every microscopic species has a corresponding substance though: Some, such as H_3O^+ or NH_4^+, correspond only to (possibly notional) parts of substances. Others, such as He_2, are too short-lived to characterize a stable substance. Yet some unstable species are explanatorily important: Carbonium ions, for instance, are positively charged organic ions formed as intermediates in organic reactions, whose structures and relative stabilities are important in explaining the mechanisms and structures of the products of these reactions. Conversely, not every chemical substance corresponds to a single microscopic species: Common salt contains sodium (Na^+) and chloride (Cl^-) ions arranged in a lattice, but no single microscopic species characterizes the substance. Water, as we shall see, is a case in point.

In this chapter I will concentrate on substances as studied by chemistry. Yet some substances are associated with other scientific disciplines, or indeed craft practices such as gardening or cookery, which have quite different classificatory interests. For instance, even supposing that some sample of artificial silk were chemically identical to a piece of real silk, there is a clear sense in which only one counts as silk, because it came from a silkworm. In this case, it is the causal history of the substance, and its relationship to a particular biological species, that seems more important than its chemical composition. Conversely, what counts as silk because it came from a silkworm might well be heterogeneous from a chemical point of view, depending on the ancestry and environment of the silkworms. The same goes for wool, wood, and many foodstuffs. Jade provides a different kind of example: It is well known to philosophers that "jade" applies to two quite different chemical substances, jadeite and nephrite, application of the term being constrained by appearance and economics as well as constitution (for details see LaPorte 2004, 94–100). The salience of noncompositional factors like causal history or appearance, however, comes from classificatory interests that are foreign to chemistry. In the following discussion I will concentrate on substances from the chemical point of view.

2 Chemical Kinds as Natural Kinds

Are the various chemical kinds *natural* kinds? Clearly so, if the concept of a natural kind is shaped by prior philosophical discussion and paradigm examples: Gold and water were key examples in the essentialist arguments of Saul Kripke and Hilary Putnam (Kripke 1980, Putnam 1975). More recently, chemical elements and compounds crop up frequently in Brian Ellis' arguments for his version of scientific essentialism (see Ellis 2001, 2002, 2009). But the situation is more complicated for, as we shall see, chemical kinds in fact do *not* in general meet the criteria for being a natural kind that Ellis and other metaphysicians apply to the categories of other sciences such as biology. Determining whether chemical kinds are natural kinds involves answering two interrelated questions. (i) The general question: What makes a scientific category a natural kind? (ii) the special question: Do chemical kinds meet the criteria identified in (i)? I will address these questions in turn.

2.1 What the General Issue Is Not: Making Stuff, Naturalness, and Pure Substances

What is a natural kind, and what is not? As it arises in the literature in metaphysics and the philosophy of science, this question is not concerned with natural occurrence. It rather involves the naturalness of *groupings* of objects whose existence may be actual or merely possible. I will explore that issue further in a moment. It is a straightforward category mistake to conflate one issue for the other, and philosophers, historians, and sociologists sometimes do not keep them keep them far enough apart. Thus, for instance, Nalini Bhushan (2006, 328) correctly distinguishes the issue of "natural versus arbitrary groups of objects" from the question of whether something is "'naturally occurring" (found in nature), but she then argues that chemistry's involvement in synthesis, sometimes of exotic new substances, undermines (simple versions of) the idea that chemical kinds can be said to have been discovered. I think the two questions should be kept separate, especially if one wants to address the discovery question. I think they can be, and hope that the following discussion will establish this. The distinction between natural and artificial substances is not without interest, but the most important things to say about it are surely skeptical: that it is a parochial distinction, and of no global significance. Why? Because it must depend on the particular chemical environments and processes that govern the production and survival of substances in the local region surrounding the person who utters the word "natural." From a global point of view, any substance that the laws of nature allow to be made is, in a sense, natural.

Now Ursula Klein and Wolfgang Lefèvre have also recently argued that

> The material productivity of the laboratory sciences challenges the demarcation between objects given by nature versus objects constituted by us.
>
> (KLEIN AND LEFÈVRE 2007, 74)

Bearing in mind our earlier distinction between the naturalness of a category and the naturalness of its members, the material productivity of the laboratory sciences does not thereby challenge the notion of a natural kind. Klein and Lefèvre more generally argue that, to understand the historical emergence of chemistry as a discipline, we need to see it as more than just a branch of natural philosophy that constructed models of how the world might be in order to account for the observed behavior of substances. It was continuous with more practically oriented traditions of manipulating materials:

> The material substances studied by eighteenth-century chemists were for the most part commodities procured, sold, or tested in apothecary's shops, foundries, assaying laboratories, arsenals, dye manufactories, distilleries, coffee shops and so on.
>
> (KLEIN AND LEFÈVRE 2007, 2)

They draw a

> larger picture of chemistry from the late seventeenth century until the early nine-
> teenth century ... extending from its mundane artisanal practices to experimental
> and natural histories and all the way to conceptual or philosophical enquiry.
>
> (KLEIN AND LEFÈVRE 2007, 2)

In the light of this larger picture, one may worry that the causal theory of reference, which is often understood to involve the naming of kinds of natural
object which are just found lying around, seems inapplicable to a science
whose business centrally involves synthesis. One way to address this is to think
of chemistry as a laboratory science that generates new substances, or separates them out of naturally occurring stuffs which are extracted from plants or
dug from the earth. Central to chemistry, on this view, is the project of purification, and the notion of a "pure substance" as a kind of extrapolation from the
impure laboratory substances that surround chemists (van Brakel 2000, ch. 3,
2008; Klein and Lefèvre 2007, Part II). There is certainly something right
about this picture of chemistry as an activity, but I would challenge the claim
that the notion of a pure substance, as a macroscopic body of pure stuff, must
play a substantial role in understanding how chemistry works, whether from a
philosophical or historiographical standpoint. It has no explanatory role in
chemical theory itself. It is sometimes claimed to have a foundational role in
classical thermodynamics, but the notion of substance to be found there is
quite different, and does not map at all well onto chemical classification. Its
significance is therefore unclear (see section 3.2). Is the notion of a pure substance genuinely an actors' category for historical chemists? I have no doubt
that they did indeed use phrases like "pure lead," "pure oxygen," and so on, or
that they mean something in chemistry, but little follows from that. The purity
of lead is a different thing from the purity of water. Perhaps the different puri-
ties have nothing in common except the word "purity." What chemists have
always had, instead, is a painful awareness that the stuffs with which they are
engaged are all, to some extent, impure. To refer to that part of a jar of impure
gas which is oxygen, Lavoisier needed no abstract and generalized concept of
pure substance, but rather the notion of a component of a heterogeneous body
of stuff that is homogeneous in some important way that persists through various chemical changes (and for different substances, the homogeneity may
consist in quite different kinds of similarity). Klein herself has argued persuasively that the notion of a chemical substance emerged when chemists began
to use affinity tables, conceiving of substances as composed of "building
blocks" (1994, 170) that persist through specific kinds of chemical change in
which they are material parts of a succession of different substances.[1] Still,

---

[1] One might think that atomism makes just this assumption, but Klein points out that before
Dalton, atoms were not "substance specific entities" (1994, 169), i.e., theoretically tied to particular chemical substances. Boyle, for instance, gave no account of how many different kinds of
atoms there are, or how these kinds relate to particular chemical substances.

even though "naturalness" for categories does not mean natural occurrence, these discussions of synthesis and purification should convince us that, if the causal theory of reference is to work, it should be understood as applying to parts of impure substances, to elements like sodium even though they are highly reactive and never occur free in the wild, and to exotic or unstable compounds that may never have existed before their artificial synthesis. What's required for that is only to recognize that a *causal* theory of reference claims only that users of a term can acquire an ability to refer via a causal connection to other users of that term. There need be no assumption that the origin of the chain (i.e., the "dubbing") bore some causal relation to a pure macroscopic sample of the substance.

2.2 The General Issue, and How It Applies to Chemistry

We have seen that the naturalness of a *category* is different from the naturalness of the *objects* that fall within it. Let us concentrate on the former, but there are more distinctions to make. Philosophers sometimes distinguish natural kinds from artifactual kinds, giving tables and chairs as examples of the latter. This again invites confusion. What makes chairs and tables "non-natural" is not just the fact that they were made by human hands. "Table" and "chair" also name functional kinds that bear no simple relationship to any non-functional kinds. Substances can be described in functional terms too: Many different chemical substances can act as analgesics, for instance. Philosophers also sometimes distinguish natural kinds from random collections: "Natural kinds are standardly distinguished from arbitrary groups of objects, such as what you had for breakfast" (Daly 2011). This characterization is criticized by Ian Hacking, because when eighteenth-century naturalists worried about the naturalness of the Linnean system, "classes were artificial rather than natural, when they had been invented by botanists, but did not accurately represent the order of living things" (Hacking 2007a, 211). Arbitrariness is not the same as artificiality, but both have been used to make the contrast with natural kinds, and to deny it. Some conventionalists elide the difference between natural and non-natural categories by showing that even the natural ones are interest-relative. All such categories, they argue, are invented or imposed on nature. Others argue that nature does not to favor this category over that, so the choice is arbitrary. Either way, characterizing realism begins with the task of marking off the natural from the non-natural in some way that grounds the distinction in the order of nature.

How might one do that? One approach is to identify, by a priori argument, necessary conditions for being a natural kind, or a system of natural kinds. Because scientific categories (and systems of them) may or may not satisfy these conditions, it is an open question which, if any, turn out to be natural. This, as I understand it, is the approach of Ellis' "scientific essentialism" (2001, 2002, 2009). As we shall see, if the requirements are chosen poorly, then virtually no chemical kinds will count as natural. In what follows I will consider and reject three such a priori requirements: (i) hierarchy (there can be no overlap between two kinds unless one contains the other); (ii) discreteness, and (iii) independence from interests.

(i) The idea that no two natural kinds may overlap unless one includes the other, or both are included within a third, has been widely discussed (Hacking 1991, 2007a; Khalidi 1998, 2013; Ellis 2001, 2002, 2009; Tobin 2010). The broad motivation for the requirement is supposed to be that natural kinds should form a single system. But why should they? There is nothing in the notion of a natural kind that implies hierarchy a priori. The motivation would also seem to be different for realists and anti-realists. Anti-realists about classification hold that classification should broadly be seen as science imposing on the world categories that fit its pragmatic or theoretical interests. If unity is just one more interest, the only question is how it trades off against others. For this very reason, however, it seems wrong for it to be a hard-and-fast requirement. For realists I am even less sure about the motivation. It might be thought that the requirement for unity in a system of kinds reflects the broad realist view that science investigates one world. But one world could contain many systems of kinds, and why must a unified system of kinds be a hierarchical one? A brief look at chemistry undermines the stronger version of the hierarchy requirement, while the weaker one borders on vacuity, as we shall see. Emma Tobin points out (2010, 189) that there is an overlap between tin and the metals, but neither encompasses the other. Tin comes in two common forms, or allotropes: white (metallic), and gray (non-metallic). The transformation from white to gray begins to happen spontaneously as the temperature falls below about 13°C. The objects from which tin is formed—such as buttons or fuel containers—disintegrate, so the process is known as "tin pest" or "tin blight." The defender of the hierarchy requirement could challenge either of these categories as natural kinds, but neither challenge seems convincing. Tin is an element, and what makes something tin is its having a particular nuclear charge (50 atomic units), a property which explains tin's particular profile of chemical and physical behavior (for the argument, see section 3.1). What of the metals? What makes something a metal is its having a particular structure: an array of metal ions surfed by electrons which are relatively free because they are only loosely bound to the atoms. This structure is what explains the common properties of metals, such as their electrical and thermal conductivity, and their lustrous appearance. It is a prima facie objection to any approach to natural kinds that it forces us to reject either of these categories. One might attempt a retreat to the weaker hierarchy requirement (in fact there are many; see Tobin 2010 for the distinctions) by arguing that "tin" and "metal" are both subclasses of the elements, but this fails because some metals (alloys such as steel and bronze) are not elements. Looking yet higher for an encompassing category, we might note that "tin" and "metal" are subclasses of the chemical substances. This, however, relies on two dubious assumptions: that chemical substances themselves form a natural kind, presumably to be contrasted with mixtures; and that no metals are mixtures. It is also such a cheap way of saving the hierarchy requirement that one can seriously doubt its value in providing a philosophical understanding of scientific classification. Far better to drop a requirement whose motivation is anyway unclear.

(ii) The idea that kinds must be discrete has an even longer history, and fewer philosophers seem willing to question it. Plato used the metaphor of "cutting nature at its joints" to argue that some divisions are more natural than others. But it can also be used in precisely the opposite way: to see classification as butchery suggests that nature comes undivided; that it takes cognitive work to make kinds; that there is more than one way to do it; and that the particular one we employ is a practical choice (see Slater and Borghini 2011).

John Locke (*Essay* III.VI.12; 1975, 446–447) thought that the divisions between kinds of things must be imposed by the understanding. First, in nature (and in our imagination) we can see a great variety of things that differ in only very minor ways: "In all the visible corporeal World, we see no Chasms, or Gaps" (*Essay* III.VI.12; 1975, 446). Second, the sheer profusion with which the natural world presents us demonstrates that nature (or the Creator) combines all the different qualities in different ways:

> There are Fishes that have Wings, and are not strangers to the airy Region: and there are some Birds, that are Inhabitants of the Water; whose Blood is as cold as Fishes, and their Flesh so like in taste, that the scrupulous are allow'd them on Fish-days.
>
> (*Essay* III.VI.12; 1975, 447)

Although they agree with Locke that natural kinds must involve natural divisions, some realists see the situation differently. With Locke, Ellis assumes that continuity would entail that there are no real kinds:

> Natural kinds exist if and only if there are objective mind-independent kinds of things in nature. Hence, to believe in natural kinds one must believe that things are divided naturally into categorically distinct classes.
>
> (ELLIS 2009, 57)

But Ellis contrasts the animal species with chemical kinds which, he maintains, are not continuous with each other:

> The elements and their various compounds are all *categorically distinct* from each other. They are distinct in the sense that there is never a gradual transition from any one chemical kind to any other chemical kind. Consequently, it is never an irresolvable issue to which chemical kind a given chemical substance belongs. Where there are such transitions in nature, as there are between the colours, for example, we have to draw a line somewhere if we wish to make a distinction.
>
> (ELLIS 2002, 26)

Elsewhere he argues that "There is no continuous spectrum of chemical variety that we had somehow to categorize," although "what is true of the chemical kinds is not true of biological species. The existing species of animals and plants are clusters of morphologically similar organisms" (2009, 59). Against both Locke and Ellis, I will argue that neither the mere possibility nor the actual occurrence of continuous transformations between two chemical kinds entails that they fail to be objectively distinct (that is, independently of our imposing such boundaries). With Locke and against Ellis, I think that it is entirely clear that chemical variety, with the exception of the elements, is continuous.

First I argue in section 3 that chemical kinds are individuated by their microstructures. The microstructures of compound substances themselves seem to be defined by continuously varying quantities such as distances between atoms.[2] This indicates that continuous transition between distinct chemical substances is possible at least in thought. Second, such continuous transition between distinct chemical species is exactly how, according to chemistry, chemical change takes place. William Goodwin (2012) distinguishes two notions of reaction mechanism: in what he calls the "thick" sense, a mechanism traces the motions of the constituent atomic nuclei and electrons in a continuous path from the reagents' molecular structure to that of the products; a mechanism in the "thin" sense neglects some parts of the process in order to concentrate on a few well-understood and explanatorily important stages. How the two conceptions fit together is a good question (Goodwin 2012, 310–315), but in either case the transitions are continuous. In thick mechanisms this is explicitly the case, but it is also true of the steps that make up a thin mechanism. Consider for instance a textbook example: the reaction of an alkyl halide RX (for instance, bromoethane) with a nucleophilic ion Nuc^- (for instance, the hydroxyl ion OH^-):

$$RX + Nuc^- \rightarrow RNuc + X^-$$

Depending on the nature of the alkyl group R, the "leaving group" X^-, and the conditions under which the reaction takes place (e.g., the nature of the solvent), the reaction proceeds via one of two mechanisms, called S_N1 and S_N2 (which stand for unimolecular and bimolecular nucleophilic substitution, respectively). For kinetic purposes the S_N1 reaction is regarded as taking place via two steps: a (slow) unimolecular dissociation of the alkyl halide:

$$RX \rightarrow R^+ + X^-$$

followed by a rapid attack by the nucleophile:

$$R^+ + Nuc^- \rightarrow RNuc$$

The slow first step involves the breaking of a bond between the halogen atom and the neighboring carbon atom in the alkyl group. Isn't the breaking of a bond a discontinuous process? Not if it involves a gradual lengthening and weakening. It all depends on what a bond is. On one influential account (Bader 1990), bonds are features of a molecule's electron-density distribution, defined in terms of its local maxima and minima. If so, then the making and breaking of bonds consists in gradual rearrangement of the electron density. The discontinuous bond-breaking arises from underlying continuity, much as

---

[2] In fact the notion of a structure is a little more complicated, because one must distinguish geometrical structure from bond structure: see Hendry 2013.

the peak in a mathematical function can be made to disappear discontinuously by continuous transformation of the function of which it is a peak.

The third argument draws on potential energy surfaces (PESs). These are contour maps in which each point is a distinct geometrical configuration of a molecule, and the height of the surface at that point is the energy of that particular configuration. Stable (i.e., low-energy) configurations correspond to minima on the PES (one might think of a valley). A typical transition between one stable substance and another is a journey over a mountain (or through a mountain pass) to a neighboring valley. So once again, chemical change is presented as gradual transition from one chemical kind to another, and chemical kinds cannot be regarded as discrete. As with the hierarchy requirement, one might quibble about the status of natural kinds that violate the discreteness condition. But in this case such a defense is surely disastrous: Apparently, no compound substances are to count as natural kinds, and I take it that we should turn away from any account of natural kinds which has this consequence. Different chemical compounds are readily separable by natural means. The differences between them are recognized by nature's laws, and so are part of nature itself.

(iii) That natural kinds should be independent of the interests of classifiers is an equally venerable requirement. Once again it is applied both in both defense and criticism of the idea that chemical kinds are natural kinds. Thus Ellis 2002 (26–27) argues that "questions of chemical identity can never be dependent on our interests, perceptual apparatus, psychologies, languages, practices or choices" (2002, 26). In contrast Donnellan (1983, 98–104) worries that chemists' interest in atomic number over atomic weight results from their "psychological quirks" (1983, 103) and "historical accident" (1983, 104). Achille Varzi argues that "Even in physics, our microscopic categories seem to suffer from a variety of human contingencies" (2011, 141), although his example is the same as Donnellan's: isotopy and the chemical elements. While the *existence* of classes of atoms of like nuclear charge (or of like atomic weight) is clearly not interest-dependent, chemists' focus on nuclear charge rather than atomic weight clearly is. I agree with Donnellan and Varzi that chemical kinds are interest-dependent in important ways, but do not think that this in any way undermines their status as natural kinds. Elsewhere (2006, 2008, 2010a), I have argued that historical scientists such as Lavoisier should be understood as using the names of elements as referring to classes of atoms of like nuclear charge even though they were agnostic about the existence of atoms and were entirely innocent of the notion of nuclear charge, because this is the main determinant of the kinds of chemical behavior they were seeking to understand. Atomic weight is a much less important factor, and many substances such as mercury, silver and tin that were known as elements in Lavoisier's time are isotopically diverse (that is, heterogeneous in respect of their atomic weight), as were the samples they had of these elements. What of compound substances? I argue below that chemistry individuates them by their microstructures, but that other ways are possible. While I think that it would be incorrect to regard this as a *choice* that the discipline has made, alternatives are available. At the outset I noted that other disciplines and activities have quite different interests: Wool and silk are individuated by the biological species in which they originate, and LaPorte (2004) notes that "jade" refers to two distinct chemical substances. Furthermore, in section 3.2 I examine one further way of individuating substances, macroscopic thermodynamics: My view of this and the other alternatives is that they are all feasible ways of individuating substances, but they are not *chemistry's* ways. Structure infects the discipline's entire approach to substances and the important similarities and differences between them. But the interest-dependence does not end there (see Hendry 2013): "Structure" itself means a number of different things, depending on context. First distinguish the bond structure of a substance (the order of bonded connections between its atoms) from its geometrical structure (the relative positions of atoms and ions in a substance). Neither, I argue, is more fundamental than the other, and geometrical structure itself is different at different (energy, length, or time) scales. Any substance has different structures, and which one is relevant depends on what one is trying to explain. In short, structure, and therefore chemical classification, is interest-dependent.

So our three a priori requirements of natural kinds seem to fail when confronted with chemistry's classificatory practice. Taken together with chemistry's centrality to the very notion of a natural kind, these failures suggest that that notion should not be understood in terms of a priori requirements. A quite different approach starts with the practice of making generalizations and predictions: natural kinds are based on genuine similarities that underwrite scientific understanding. This is central to the broad tradition of natural kinds in the philosophy of science, as Hacking himself traces it (2007a). A scientific realist adds the requirement that there should be a non-trivial explanation of systematically successful prediction in science (Boyd 1991), even though such explanations may proceed quite differently for different natural kinds. (In short, kinds may be natural in quite different ways.) Hence the weak realist view is that a kind is natural insofar as its members share some property which is causally relevant to maintaining the regularities in behavior whose existence underlies the very usefulness of the notion of a natural kind. This approach is naturalistic, in that it begins with scientific practice (cf. Khalidi 2013). General claims about kinds ought to have some empirical basis. Consequently, it may well be inclusive, and hence pluralistic like John Dupré's promiscuous realism (1993): There could be many natural kinds, and many different systems of them, corresponding to all the different ways in which objective differences are causally relevant to understanding the similarities and differences in the behaviors of things. Moreover there are no particular reasons to think that natural kinds should bear any particular relationships to each other, or form a single system. It is possible that they do, but that is a matter of how the world is, and therefore for empirical research. The formation of classificatory *systems* is undeniably an important part of science (see Hendry 2012), but it is something that arises as a feature of highly developed explanatory systems. It cannot be imposed a priori. Human kinds, artifactual kinds, and even functional kinds can all be natural, on this view, although there may be differences among them depending on the depth (or non-triviality) of the scientific explanations they underwrite, and the explanation of the systematically successful predictions they are involved in. None of this implies that all the different possible kinds, or classificatory systems, are on a par: There are limits to pluralism (Hendry 2012).

This approach does give us a quick answer to the question of whether or not chemical kinds are natural: They are, because they are good bases for successful explanation and prediction in a particular empirical science—chemistry—and we understand why they are. In some cases, as in the elements, I believe a full-blown essentialist explanation is available. In others, such as the acids, the explanations are much thinner: There is probably nothing more to being an acid than displaying enough of the typically acidic kinds of behavior. The explanation is thinner because while there are, of course, detailed causal explanations of why each acid behaves the way it does, the explanations diverge, because no single component or structural property gives rise to all the different cases of acidity.

3 Two Views of Chemical Substances

Since chemistry studies both substances, which can be macroscopic bodies of stuff, *and* the molecular species that are found within them, there is one obvious question: How are substances and molecular species related to one another? One attractive view, apparently supported by both chemical theory and classification, is that from the point of view of chemistry substances are individuated in terms of their characteristic molecular constituents. I have characterized and defended this thesis, which I will call "microstructuralism," elsewhere (2006, 2008). It is important to distinguish microstructuralism from what is sometimes called "microessentialism," a view commonly attributed to Kripke (1980) and Putnam (1975) according to which microstructural properties are what make chemical substances what they are. While I am sympathetic to microessentialism, it makes things clearer to focus on microstructuralism first, and then assess the prospects for the stronger essentialist claim. There is an alternative to microstructuralism: that substances are individuated by their macroscopic behavior and properties rather than their microstructures. In this section I will introduce these two positions, and the arguments for and against them.

3.1 Microscopic Conceptions of Substance

Microstructuralism is a claim about a specific scientific discipline, and I will defend it via three arguments which are grounded in the theories, practices, and interests of chemistry itself, rather than a priori metaphysical constraints, or any other transcendental perspective. Hence it might appear that microstructuralism is a purely hermeneutic exercise: an exploration only of how chemistry sees the world. It is true that anything we establish through immanent arguments based on chemistry is discipline-specific, tending to embody only how *chemistry* views its subject matter. But the discipline-specificity of the conclusions doesn't mean that they can just be ignored in favor of other perspectives, for it is within chemistry that one can find the expertise on its particular subject matter: the nature and transformations of chemical substances. Even if there are other perspectives on substances and their transformations, it can't be assumed that they are any good for any purpose, or on a par with chemistry's, or even that they meet minimal standards of coherence and nontriviality. These are things for which positive arguments are required.[3]

The first argument for microstructuralism is that microstructure is the basis of chemistry's own classification of, and nomenclature for, chemical substances. Take for instance 2 4 6-trinitrotoluene, better known as TNT:

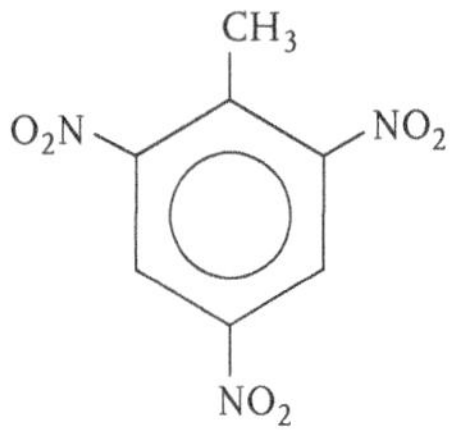

FIGURE 1 The structure of TNT.

The name "2 4 6-trinitrotoluene" comes from its being regarded as a substituted version of toluene (or methylbenzene), which consists of six carbon atoms bound together in a six-membered structure called a benzene ring, with a methyl (-CH$_3$) group attached. It is *trinitro*toluene because it contains three substituent nitro- groups (-NO$_2$), and it is 2 4 6-trinitrotoluene because these three groups are placed at the second, fourth, and sixth places, counting clockwise around the benzene ring, starting from the methyl group as 1. (Positions 3 and 5 are occupied by hydrogen atoms, which are usually left out for clarity.) IUPAC has developed systematic ways of naming chemical substances, referring exclusively to microstructural properties (see Thurlow 1998). The name "2 4 6-trinitrotoluene" predates this nomenclature, but I mention it because it is the origin of the widely used abbreviation "TNT," and it too is based entirely on microstructural features. The systematic nomenclatures for *in*organic chemistry are similarly based on microstructural properties and relations.

The second argument for microstructuralism is that microstructural properties and relations are involved indispensably in explanations of the physical properties, chemical reactivity, and spectroscopic behavior of chemical substances. Starting with the physical properties, the boiling point of a substance depends on the strength of the forces that need to be overcome in liberating its molecules from the liquid: The stronger the forces, the greater the energy required to overcome them. These intermolecular interactions depend on the size and structure of the molecules, and how charge is distributed within them. Moving on to chemical reactivity, every aspect of a chemical reaction, including the products it yields, how much heat it generates or absorbs, and

---

[3] For further discussion of immanence in the philosophy of a particular science, see Hendry forthcoming, ch. 1. For further discussion of pluralism about chemical substances, and its limits, see Hendry 2012.

how fast it happens, is understood in terms of the structures of the reactants and products, and the mechanism by which one transforms into the other. What is a reaction mechanism? As we saw earlier, Goodwin (2012) identifies two conceptions at work in chemical explanation, but under either conception, a mechanism begins and ends with structures, and consists of structural changes. Turning at last to spectroscopy, a substance's absorption or emission of light, from the ultraviolet down (in energetic terms) through visible light to the infrared, is understood via the interaction of specific parts of its structure with light from the relevant parts of the electromagnetic spectrum. Thus the presence of an -OH group (for instance in alcohols or phenols) can be detected by the broad absorption of light in a specific part of the infrared region, with wavelength from about 2700nm to 3100nm (see Silverstein, Bassler, and Morrill 1981). This arises from the stretching of the O-H bond, giving rise to vibrating charges that interact with the electromagnetic field. Since structure plays a central role in understanding the physical properties, chemical reactions and spectroscopic behavior of substances, it is no surprise that chemists think about substances in terms of their microstructures. Why would they do anything else?

The third argument for microstructuralism is that there is no alternative. There is no conception of the sameness and difference of chemical substances that is both independent of microstructure and consistent with the ways in which chemistry *in fact* classifies substances, and how it explains their behavior. For instance, one promising alternative, canvassed by Paul Needham (2010, 2011), is that substances can be individuated by macroscopic relations of sameness and difference of substance provided by classical thermodynamics, on the principle that a body of matter behaves differently, from a thermodynamic point of view, if it is a mixture of different substances rather than a pure substance. But these thermodynamic relations are either unable in themselves to distinguish among substances, or they distinguish among substances more finely than does chemistry, to the extent that they track differences between physical states of substances, rather than differences between substances themselves (for some relevant arguments see Hendry 2010b). So while there are other ways to differentiate substances than their microstructure, these non-microstructural ways are not chemistry's ways. This argument depends for its detail on the conclusion to the next section, where I critically examine macroscopic properties and behavior as a basis for individuating chemical substances.

I conclude that, from the point of view of chemistry, it is the molecular structure of a substance that pretty much determines what it is and what it does. I say "pretty much determines" because some pairs of substances are pretty much the same microstructurally, yet can be distinguished macroscopically. Take for instance the red and yellow forms of mercury (II) chloride: both consist of the same repeating units (-O-Hg-O-), but they differ in the size of the lumps into which these units are aggregated, which is affected by the specific process by which the substance was formed. Thus, for instance, the red oxide is made by direct oxidation of mercury, while the yellow oxide is precipitated from a solution of a soluble Hg^{2+} salt. One might be tempted to see the difference between these two forms as a failure of the supervenience of substance identity on microstructure, but the differences in lump size are themselves microstructural differences (how many of the repeating units there are per lump), and the red and yellow forms need not be regarded as distinct substances: They can be regarded as the same stuff aggregated into differently sized lumps, like a lump of iron versus iron filings.

I turn now to certain objections to microstructuralism. One longstanding criticism is that elements and compounds can be heterogeneous at the microstructural level. Historically, atomism has often been assumed to see pure substances as collections of qualitatively identical atoms and molecules. Pierre Duhem (2002, 86) and Jean Timmermans (1941, ch. 8) criticized atomism on just these grounds. But elements as defined by IUPAC consist of atoms of like nuclear charge, which allows that they may differ in other respects, such as their mass. Turning to compounds, Putnam seemed to imply a requirement of homogeneity in saying that the extension of "water" is "the set of all wholes consisting of H_2O molecules" (1975, 224). But liquid water is far from homogeneous at the molecular level (see Needham 2000, 2002; van Brakel 2000, ch. 3): a small but significant proportion of H_2O molecules dissociate, forming H_3O^+ and OH^- ions:

$$2H_2O \rightleftharpoons H_3O^+ + OH^-$$

Furthermore, H_2O molecules are polar and form hydrogen-bonded chains which are similar in structure to ice. One might regard the ionic dissociation products as impurities, but the presence of these charged species is central to understanding water's electrical conductivity. Chemists would regard the electrical conductivity they measure as a property of water, and it seems gratuitous to interpret it instead as a property of an aqueous solution of water's ionic dissociation products. Paul Needham and Jaap van Brakel's response to molecular heterogeneity is to reject the microstructuralist contention that water is identified by its composition by H_2O molecules, which raises the question of just what makes something water, if not its molecular composition (I will address some macroscopic alternatives shortly). But molecular heterogeneity is consistent with microstructuralism if one sees water not as a mere assemblage of H_2O molecules, but as a heterogeneous molecular population that is generated by bringing H_2O molecules together (for details see Hendry 2006). On this view, H_2O is the characteristic molecule for water not because pure water is composed just of H_2O molecules, but because samples of water consist of molecular populations that arise when H_2O molecules come together and interact. Microstructuralism need not require that substances are homogeneous populations of molecular or atomic species, or even that every member shares some property, although elements are in fact homogeneous in that way. The requirement is only that the populations as a whole have some significant microstructural property. It is perhaps ironic that Putnam chose water as an example of the microstructuralist view. Other molecular compounds are less problematic, simply because they are more homogeneous at the molecular level.

Another important kind of criticism of microstructuralism is directed at the *arguments for* the thesis, rather than the thesis itself. Both van Brakel and Needham (2011) are skeptical of the Kripke-Putnam view of the reference of kind-terms, which likens them to proper names, and resist what philosophers usually take to be the key metaphysical consequences of that view: that kind-terms have associated essences, and that the essential properties are instantiated at the microphysical level. What is normally understood as "Kripke-Putnam essentialism" has a number of elements, although Kripke and Putnam advanced slightly different bodies of claims (see Hacking 2007b): that the reference of kind-terms is direct (unmediated by "stereotypical" knowledge or belief associated with a kind term) and rigid (applying across all possible worlds). Applied to chemical kind terms, and *given* some claims about the aims and history of chemical theorizing and classification, it is widely taken to support, although it does not entail, microphysical essentialism, a view about what kind of property makes substances what they are. It is important to distinguish the microstructuralism, the view about reference, and the essentialism, as van Brakel and Needham are both aware.

Kripke and Putnam take the necessary truth of (for instance) "Krypton has atomic number 36" to entail that having atomic number 36 is what makes krypton the substance it is. However, van Brakel argues (2000, 109) that there are many necessary truths about krypton: those describing its ground-state electronic structure, and also its chemical, thermodynamic and spectroscopic behavior. Why shouldn't the necessity of these truths qualify them as indicating essences? Which is "the" essence? However, the fact that atomic number, ground-state electronic structure and spectroscopic behavior are all necessary doesn't show that they are all on a par as candidate essences. For a start, "Krypton has atomic number 36" is, on Kripke and Putnam's view, metaphysically necessary, while krypton's ground-state electronic structure and spectroscopic behavior are only nomically necessary. For those who see metaphysical necessity as akin to (or even a species of) logical necessity, since it is semantic in origin, the differences are important. Now if, as nomic necessitarians argue, laws of nature are themselves metaphysically necessary, these distinctions might be thought to collapse (although see Hendry and Rowbottom 2009 for a response), but even nomic necessitarians should be able to distinguish essence from mere necessity. The original Kripke-Putnam view did indeed seem to move illegitimately from necessity to essence (Salmon 1982), and Kit Fine has argued convincingly that necessity of any kind is a poor grounding for essence (Fine 1994). However, necessity is not really the issue: Nuclear charge is privileged by its asymmetric determination of the other properties. Given relevant laws of nature—quantum mechanics, the exclusion principle—nuclear charge determines and explains electronic structure and spectroscopic behavior, but not vice versa. If an argument for essentialism is to be made out on the basis of this priority, it should be done on metaphysical grounds. In the case of the elements, it might go as follows: what is it that is present across all the different complex chemical environments in which an element can be said to be present? The answer is the nucleus, and its charge, which mediates and (to a large extent) determines its interactions with other microstructural entities such as electrons and other nuclei in the formation of the complex things we call chemical substances. Since the eighteenth century, chemistry has built up an explanatory framework in which chemical composition, including microstructure, has become the basis for understanding the properties and behavior of chemical substances. The empirical basis on which this has been established is the analysis of substances into their components, and the explanatory assumption is that more complex situations (compound substances) are understood as composed from the simpler situations into which chemical artifice allows us to analyze them. There is a clear analogy with Nancy Cartwright's defense of Aristotelian natures as part of what she calls "*the analytic method* in physics":

> To understand what happens in the world, we take things apart into their fundamental pieces; to control a situation we reassemble the pieces, we reorder them so they will work together to make things happen as we will. You carry the pieces from place to place, assembling them together in new ways and new contexts. But you always assume that they will try to behave in new arrangements as they have tried to behave in others. They will, in each case, act in accordance with their nature.
>
> (1992, 49)

The analytical method reasons as if there are natures (or essences), and the very success of the analytical project is therefore an argument for the reality of such natures (or essences). Nuclear charge, on this argument, is the ground of an element's tendencies in the many and varied situations in which we find it. It is the nature or essence of an element. Although the situation is more complex, the same argument can be applied to the structures of compound substances.

Paul Needham has more recently argued (2011) that Kripke and Putnam's arguments for microessentialism fail in another way: They depend on the assumption that such stereotypical properties as a substance's appearance, or its chemical, thermodynamic or spectroscopic behavior are not sufficient to individuate chemical substances. Needham points out that the notion of a stereotypical property is far too vague to do any important work in an argument of this sort (2011, Section 2). Moreover, he argues that macroscopic properties (such as chemical, thermodynamic, or spectroscopic behavior) *are* sufficient to individuate substances, pointing out that this, in fact, is how chemists identify them (van Brakel agrees on this point). But the microstructuralist will respond that Needham is not comparing like with like. Kripke and Putnam were seeking to individuate substances across *all possible worlds*; this is the point of their modal arguments. They did not deny that, within the actual world, there are many ways to individuate chemical substances. "The element best known by philosophers and historians of science for having been named by Antoine Lavoisier on the basis of a false theory" uniquely describes oxygen, but that description fails for Kripke and Putnam's purposes because the same description might not apply to oxygen in other possible worlds. We might now call oxygen a name that some other historical scientist called it (e.g., "Henry," or even "dephlogisticated air"). Similarly, if one thinks that the laws of nature that govern the behavior of chemical substances are metaphysically contingent, then Needham's proposed individuation by macroscopic properties will not hold across all possible worlds. One might at this point wax skeptical about metaphysical necessity and possible worlds, but as I have argued elsewhere, chemists do provide explanations that, on the face of it, presuppose the metaphysical contingency of laws (see Hendry and Rowbottom 2009). Moreover, in other contexts Needham himself reasons about hydrogen bonding in ways that seem to indicate that the boiling point of water is metaphysically contingent:

> Whereas purely on the strength of its molecular weight water would be expected to boil around −70°C, it doesn't even melt until 0°C, which gives some indication of the strength of the intermolecular force.

> (2008b, 98)

If it is coherently supposable that something can be water even though it does not have the boiling point that water actually has, the microstructuralist will ask what makes the stuff referred to in the example water, if not its composition by H_2O molecules. This brings us on to alternatives to microstructuralism, because Needham and van Brakel's criticisms are not sufficient in themselves for a rejection of microstructuralism, even if one accepts them: that would require some alternative account of how substances are individuated, independently of microstructural properties. To that project I now turn.

3.2 Macroscopic Conceptions of Substance

What, if not nuclear charge and molecular structure, determines the identity of substances? One might consider the aims and interests of chemistry as a discipline: according to van Brakel (2000, Chapter 3), chemistry is a "science of stuffs" that manipulates and transforms macroscopic quantities of substances, investigating their location in a network that "contains all possible substances" (2000, 72). Individual substances are the nodes, while the relations of production and mutual reaction form the connecting relations. This is an interesting suggestion, but further discussion would require that it be developed well beyond this metaphor. While it is true that synthesis is central to chemistry (see Schummer 1997), the microstructuralist will object that no such structure as van Brakel's network of substances appears explicitly in any work of chemical classification and nomenclature. Operational definitions of substances are also conspicuous by their absence. As we have already seen, the IUPAC rules of nomenclature concern molecular structures. Moreover, as scientific understanding has advanced during the nineteenth and twentieth centuries, chemists have increasingly described substances, and understood the transformations between them, in terms of the way their constituent atoms are bonded together. This suggests that the aims and interests of chemistry point to microstructural properties.

An alternative macroscopic approach is to employ classical thermodynamics, which provides the general theoretical framework for understanding the processes by which chemical substances are separated, and so is the most likely source for a macroscopic conception of substance. The thermodynamic behavior of a system is dependent on the number of distinct chemical substances present in it, which provides a criterion for when a substance is pure rather than mixed. This, in turn, implies a criterion for sameness and difference for substances: Two bodies of stuff are the same chemical substance if they act as a substance when mixed; they are different chemical substances if they act as a mixture. This will be a macroscopic criterion insofar as classical thermodynamics describes the behavior of matter independently of its microscopic structure (Zemansky 1957, 1–3; Denbigh 1966, 3; Vemulapalli 2010).

In fact the thermodynamic behavior of a system depends in a number of ways on the number of substances or components present in it. In what follows I will consider three attempts to ground the identity of substances in thermodynamic relationships.

(i) One might regard some portion of matter as a mixture just in case it can be separated physically into different components. Wilhelm Ostwald (1904) and Jean Timmermans (1941) offered just such conceptions of substance, as more recently has Paul Needham (2008a, 2010). When a mixture of substances is separated by distillation, it is boiled and the vapor is collected and cooled, providing a liquid that is enriched in one component. The distillate can successively be redistilled, purifying the relevant component to whatever degree is required. This practice relies on the fact that a mixture and its vapor typically differ in composition, while a pure substance has the same constitution as its vapor. Constancy of composition over phase change (solid to liquid, or liquid to gas) might seem to provide a thermodynamic criterion for distinguishing pure substances from mixtures (Ostwald 1904; Timmermans 1941, ch. 2), but an immediate problem concerns azeotropes. Azeotropes are mixtures of different compounds (e.g., water and ethanol, or water and hydrogen chloride) which, under certain conditions, can be distilled without affecting their composition. Ostwald distinguished between azeotropes and compounds proper on the grounds that azeotrope composition varies with pressure (Ostwald 1904, 514–516; Timmermans 1941, ch. 2). In fact altering the pressure at which distillation occurs is one way to separate the components of an azeotrope. In this respect azeotropes are more like solutions and other mixtures: The composition of a saturated solution of salt in water, for instance, depends on temperature. In contrast, the composition of most compounds is constant within the range of physical conditions in which they exist. But why does variable composition disqualify an azeotrope from being a compound? There was an extended debate in the first half of the nineteenth century over whether compounds must have a composition that is fixed and independent of physical conditions, because a range of substances (Berthollides) display variable composition and not all can be regarded as solutions or mixtures (see Needham 2008a, 67–68). Constancy of composition over phase transitions therefore seems to fail as a criterion for being a pure substance since it requires appeal to an independent, and indeed contentious criterion for purity of substance.

(ii) A phase is a homogeneous body of matter with a definite boundary. Gibbs' elegant and abstract phase rule relates the number of components (C) in a multiphase system to the number of phases (P) present in it, and its variance (F), which is the number of independent physical variables required to characterize the thermodynamic state of the system:

$$F = C - P + 2$$

Does the phase rule provide a criterion of sameness and difference for substances? The difficulty is that "components" cannot simply be equated with chemical substances, and in some systems it is a matter of some delicacy how to map one onto the other so as to preserve the truth of the phase rule. Needham (2010) discusses the three-component system of calcium carbonate ($CaCO_3$), carbon dioxide (CO_2), and calcium oxide (CaO), generated by heating calcium carbonate in a closed container:

$$CaCO_3 \rightleftharpoons CO_2 + CaO$$

Given its thermodynamic behavior, the phase rule accords this system only two components. The truth of the phase rule can be preserved if we deny that one of the substances is an independent component. Thus CaO and CO_2 might be the components, with $CaCO_3$ a combination of them; or $CaCO_3$ and CO_2 might be the components, CaO now regarded as $CaCO_3$ less CO_2. This suggests that the phase rule doesn't itself say how many substances are present in a system, but only *given* a case-by-case matching of phase-rule components to what we are willing, on independent grounds, to regard as chemical substances. If so, then the phase rule doesn't really offer a criterion for distinguishing substances and mixtures, but only a series of relationships upon which we can impose a carefully constructed interpretation which does so. The question is whether or not such an interpretation amounts to a book-keeping exercise rather than an autonomous macroscopic conception of substance.

(iii) Whenever samples of two distinct substances in thermal equilibrium are allowed to mix, there is an increase in entropy. In contrast, samples of the very same substance generate no entropy change on isothermal mixing. Entropy of mixing generates a criterion for sameness and difference of substance in an obvious way, one which Needham (2008a) applies to water, arguing that it is a mixture because it contains traces of isotopic variants such as $^1H_2^{18}O$ and $^2H_2^{16}O$ as well as the more common $^1H_2^{16}O$: different isotopic variants display non-zero entropy of isothermal mixing. However, entropy change on mixing is displayed by many pairs of species that chemists do not normally regard as different substances, but rather as the same substance in different physical states: Examples include spin isomers such as ortho- and para-hydrogen and populations of otherwise identical atoms in orthogonal quantum states, such as the two beams of silver atoms emerging from a Stern-Gerlach apparatus. If one accepts that all these are different substances, it seems that, on the entropic criterion, any physical difference must count as a difference of substance. This may be logically consistent but is hardly consonant with the aims and classificatory interests of chemistry, according to which chemical substances are continuants that may vary in their physical properties. This is an important consideration when the question is how to understand sameness and difference as judged by the discipline of chemistry (for more detailed discussion of the entropic criterion see Hendry 2010b).

What classical thermodynamics has to say about the sameness and difference of substances is a matter of some interest and significance in its own right. But the resultant criteria are not extensionally equivalent to those that chemistry has developed for itself, and which are reflected in its systems of naming and classification, and its theoretical explanations. The thermodynamic view of chemical sameness and difference is not the same as chemistry's.

Conclusion

In this chapter I have defended the usefulness of the notion of a natural kind to a philosophical understanding of chemistry. Although it was born in the philosophy of science, that notion has been appropriated and adapted by various projects in the metaphysics of properties, causation, and laws of nature. For this reason Hacking would be happy if the term were eliminated from the philosophy of science (2007a). I think this would be a shame, for it would mean losing an important connection with historical tradition for no good reason. It would be far better to resist the appropriation by metaphysics, and challenge the a priori presuppositions about natural kinds that come with it. In no sense do I resist connections between philosophy of science and metaphysics. It is just that neither should impose its own presuppositions on terms which are shared between them.

In the second half of the chapter I addressed a question that is far more specific to chemistry: Whether chemical substances—broadly, stuffs conceived from the point of view of chemistry—are individuated by their microstructural properties, or in some other way. My conclusion was in favor of microstructuralism, simply because no macroscopic approach is consistent with the classificatory theory and practice of chemists themselves. On that hermeneutic basis, I also sketched a causal argument for microstructural essentialism, which is a stronger, and metaphysically more substantial thesis: microstructure is what makes a chemical substance what it is.

References

Bader, Richard. (1990). *Atoms in Molecules: A Quantum Theory*. Oxford: Oxford University Press.

Bhushan, Nalini. (2006). "Are Chemical Kinds Natural Kinds?" In *Philosophy of Chemistry: Synthesis of a New Discipline* (Davis Baird, Eric Scerri, and Lee McIntyre, eds.). Dordrecht: Springer, 327–336.

Boyd, Richard. (1991). "Realism, anti-foundationalism and the enthusiasm for natural kinds," *Philosophical Studies*, 61, 127–148.

Cartwright, Nancy. (1992). "Aristotelian Natures and the Modern Experimental Method" In *Inference, Explanation and other Frustrations: Essays in the Philosophy of Science* (John Earman, ed.). Berkeley: University of California Press.

Daly, Chris. (2011). "Natural Kinds" In *Routledge Encyclopedia of Philosophy* (Edward Craig, ed.). London: Routledge.

Denbigh, K.G. (1966). *The Principles of Chemical Equilibrium* (2nd ed.). Cambridge: Cambridge University Press.

Donnellan, Keith S. (1983). "Kripke and Putnam on Natural Kind Terms" In *Knowledge and Mind: Philosophical Essays* (C. Ginet and S. Shoemaker, eds.). Oxford: Oxford University Press, 84–104.

Duhem, Pierre. (2002). *Mixture and Chemical Combination*. Dordrecht: Kluwer. Translation by Paul Needham of *Le Mixte et la Combinaison Chimique: Essai sur l'Évolution d'une Idée* (Paris: C. Naud, 1902).

Dupré, John. (1993). *The Disorder of Things*. Cambridge, MA: Harvard University Press.

Ellis, Brian. (2001). *Scientific Essentialism*. Cambridge: Cambridge University Press.

Ellis, Brian. (2002). *The Philosophy of Nature: A Guide to the New Essentialism*. Chesham, UK: Acumen.

Ellis, Brian. (2009). *The Metaphysics of Scientific Realism*. Durham, UK: Acumen.

Fine, Kit. (1994). "Essence and modality." *Philosophical Perspectives*, 8, 1–16.

Goodwin, William. (2012). "Mechanisms and Chemical Reaction." In *Handbook of the Philosophy of Science 6: Philosophy of Chemistry* (Robin Findlay Hendry, Paul Needham, and Andrea I. Woody, eds.). Amsterdam: Elsevier, 309–327.

Hacking, Ian. (1991). "A tradition of natural kinds." *Philosophical Studies*, 61, 109–26.

Hacking, Ian. (2007a). "Natural kinds: rosy dawn, scholastic twilight." *Royal Institute of Philosophy Supplement*, 61, 203–239.

Hacking, Ian. (2007b). "Putnam's theory of natural kinds and their names is not the same as Kripke's," *Principia*, 11, 1–24.

Hendry, Robin Findlay. (2006). "Elements, compounds and other chemical kinds," *Philosophy of Science*, 73, 864–875.

Hendry, Robin Findlay. (2008). "Microstructuralism: Problems And Prospects." In *Stuff: The Nature of Chemical Substances* (Klaus Ruthenberg and Jaap van Brakel, eds.). Würzburg: Königshausen und von Neumann), 107–120.

Hendry, Robin Findlay. (2010a). "The Elements and Conceptual Change." In *The Semantics and Metaphysics of Natural Kinds* (Helen Beebee and Nigel Sabbarton-Leary, eds.). London: Routledge, 137–158.

Hendry, Robin Findlay. (2010b). "Entropy and chemical substance," *Philosophy of Science*, 77, 921–932.

Hendry, Robin Findlay. (2012). "Chemical substances and the limits of pluralism," *Foundations of Chemistry*, 14, 55–68.

Hendry, Robin Findlay. (2013). "The Metaphysics of Molecular Structure." In *EPSA11 Perspectives and Foundational Problems* (Vassilios Karakostas and Dennis Dieks, eds.). Berlin: Springer, 331–342.

Hendry, Robin Findlay. (forthcoming). *The Metaphysics of Chemistry*. Oxford: Oxford University Press.

Hendry, Robin Findlay and Rowbottom. Darrell. (2009). "Dispositional essentialism and the necessity of laws," *Analysis,* 69, 668–677.

Khalidi, Muhammad Ali. (1998). "Natural kinds and crosscutting categories," *Journal of Philosophy,* 95, 33–50.

Khalidi, Muhammad Ali. (2013). *Natural Categories and Human Kinds: Classification in the Natural and Social Sciences.* Cambridge: Cambridge University Press.

Klein, Ursula. (1994). "Origin of the concept of chemical compound," *Science in Context,* 7, 163–204.

Klein, Ursula and Lefèvre, Wolfgang. (2007). *Materials in Eighteenth-Century Science: A Historical Ontology.* Cambridge, MA: MIT Press.

Kripke, Saul. (1980). *Naming and Necessity.* Cambridge, MA: Harvard University Press.

LaPorte, Joseph. (2004). *Natural Kinds and Conceptual Change.* Cambridge: Cambridge University Press.

Locke, John. (1975). *An Essay Concerning Human Understanding.* (P.H. Nidditch, ed.) Oxford: Clarendon Press.

Needham, Paul. (2000). "What is water?," *Analysis,* 60, 13–21.

Needham, Paul. (2002). "The discovery that water is H_2O," *International Studies in the Philosophy of Science,* 16, 205–226.

Needham, Paul. (2008a). "Is water a mixture? Bridging the distinction between physical and chemical properties," *Studies in History and Philosophy of Science,* 39, 66–77.

Needham, Paul. (2008b). "A Critique of the Kripke/Putnam Conception of Water." In *Stuff: The Nature of Chemical Substances* (Klaus Ruthenberg and Jaap van Brakel, eds.). Würzburg: Königshausen und von Neumann, 93–105.

Needham, Paul. (2010). "Substance and time," *British Journal for the Philosophy of Science,* 61, 485–512.

Needham, Paul. (2011). "Microessentialism: What is the argument?," *Noûs,* 45, 1–21.

Ostwald, Wilhelm. (1904). "Elements and compounds," *Journal of the Chemical Society, Transactions,* 85, 506–522.

Putnam, Hilary. (1975). "The Meaning of 'Meaning'" In *Mind Language and Reality.* Cambridge: Cambridge University Press, 215–271.

Salmon, Nathan. (1982). *Reference and Essence.* Oxford: Blackwell.

Schummer, Joachim. (1997). "Scientometric studies on chemistry I & II," *Scientometrics,* 39, 107–23, 125–40.

Silverstein, R.M., Bassler, G.C., and Morrill, T.C. (1981). *Spectrometric Identification of Organic Compounds* (4th ed.). New York: Wiley.

Slater, Matthew and Borghini, Andrea. (2011). "Lessons from the Scientific Butchery" In *Carving Nature at its Joints: Natural Kinds in Metaphysics and Science* (J.K. Campbell, M. O'Rourke, and M.H. Slater, eds.). Cambridge, MA: MIT Press, 1–31.

Thurlow, K.J. (1998). "IUPAC Nomenclature Part 1, Organic." In *Chemical Nomenclature* (K.J. Thurlow, ed.). Dordrecht: Kluwer, 103–126.

Timmermans, Jean. (1941). *Chemical Species.* London: Macmillan. Translation by Ralph E. Oesper of a revised version of *La Notion d'Espèce en Chimie* (Paris: Gauthier-Villars et Cie, 1928).

Tobin, Emma. (2010). "Crosscutting Natural Kinds and the Hierarchy Thesis." In *The Semantics and Metaphysics of Natural Kinds* (Helen Beebee and Nigel Sabbarton-Leary, eds.). London: Routledge, 179–191.

van Brakel, Jaap. (2000). *Philosophy of Chemistry*. Leuven: Leuven University Press.

van Brakel, Jaap. (2008). "Pure Chemical Substances." In *Stuff: The Nature of Chemical Substances* (Klaus Ruthenberg and Jaap van Brakel, eds.). Würzburg: Königshausen und von Neumann, 145–161.

Varzi, A.C. (2011). "Boundaries, Conventions and Realism." In *Carving Nature at its Joints: Natural Kinds in Metaphysics and Science* (J.K. Campbell, M. O'Rourke, and M.H. Slater, eds.). Cambridge, MA: MIT Press, 129–153.

Vemulapalli, G.K. (2010). "Thermodynamics and chemistry: How does a theory formulated without reference to matter explain the properties of matter?," *Philosophy of Science*, 77, 911–920.

Zemansky, Mark. (1957). *Heat and Thermodynamics*. New York: McGraw-Hill.

| Theory and Practice

CHAPTER 13 | Chemistry and "The Theoretician's Dilemma"

HINNE HETTEMA

"It is a familiar fact that theories in the sciences (especially though not exclusively in mathematical physics) are generally formulated with painstaking care and that the relations of theoretical notions to each other (whether the notions are primitives in the system or are defined in terms of the primitives) are stated with great precision. Such care and precision are essential if the deductive consequences of theoretical assumptions are to be rigorously explored."

(NAGEL 1961, 99)

This contribution addresses Hempel's well-known "The Theoretician's Dilemma" from the viewpoint of philosophy of chemistry. While from the viewpoint of mainstream philosophy of science it might appear that the issues raised by this paper, published in 1958, are well settled, philosophy of chemistry has the potential to reopen the debate on theoretical terms in an interesting way. In this contribution I will reopen the debate and approach the problem of theoretical terms in a fashion which may be instructive to the wider philosophy of science.

In "The Theoretician's Dilemma" the argument hinges on the *purpose* of theoretical terms. Theoretical terms either serve their purpose (that is, they form part of a deductive chain that establishes definite connections between observables), or they don't. Hempel then mounts an argument to show that if theoretical terms serve their purpose, they *can* be dispensed with. On the other hand, of course, if the theoretical terms don't serve their purpose, they *should* be dispensed with. Hence the dilemma shows that theoretical terms are unnecessary. Hempel's way out of the dilemma is to attack its premise. Hempel argues that theoretical terms do *more* than just establish a convenient shorthand to describe observations. Theoretical terms, argues Hempel, serve an *ontological* function in addition to theoretical systematization. Theoretical terms pick out some essential feature of nature such that they allow theories to "track truth" (in the words of Psillos 1999).

From the viewpoint of philosophy of chemistry, the issue is this. Chemical theories frequently refer to entities, such as "atoms," "chemical elements," "electrons," and "orbitals" that have some counterpart of the same name in theories of physics. Such chemical theories, as per the quote from Nagel above, are generally formulated with great care, as are their counterparts in physics. Yet is also the case that the use of such terms in the theories of chemistry is in many cases inconsistent with how these same terms are conceived in physics. Even worse, it not clear at the outset that the manner in which these terms are used both in the theories of chemistry and physics is even *compatible*. Hence for the theoretical terms in use in chemistry, it is not that easy to show that they have some sort of realist import, and hence the theoretician's dilemma might just be in force.

A good illustration of the issue is found in the notion of "orbital." In quantum theory, an orbital is a part of a many electron wave function. As such, the notion of an orbital is a key component in the understanding of the structure of many-electron wave functions and energetic properties of molecules. But the way in which the concept of orbital is used in many theories of chemistry can take matters much further than that. For instance, in the textbooks of organic chemists, an orbital can easily reappear as a structural component of a molecule, somewhat comparable to the struts that hold the ball-and-stick models of molecules together; hence, there is significant potential for a somewhat unsettling identification of an orbital and a bond. Such an identification is not supported by quantum theory.

The ontological discontinuity resulting from this situation is one of the central problems in current philosophy of chemistry. Chemistry and physics are closely intertwined, even to the degree where, as Needham (2010) pointed out, it is hard to imagine chemistry with the physics removed. Yet if the use and meaning of terms in physics and chemistry are inconsistent between the two theories, then one might reasonably ask what the theoretical entities of physics and chemistry actually *do*. It thus remains an open question whether Hempel's resolution of the dilemma is available to sciences, like chemistry, that are ontologically promiscuous.[1] Chemistry's ontological promiscuity leaves Hempel's realist resolution of the dilemma in considerable difficulty.

Precisely because theories of chemistry, and their theoretical entities, do not seem to typically "fit" with standard (philosophical) accounts of scientific theories, the theoretical entities and admissible inter-theory relationships for theories of chemistry have important additional insights to offer to the philosophy of science in general. This chapter aims to establish part of that program.

I will argue for a number of things. First, I hope to make clear that the problem of ontological discontinuity is one of the central problems in the philosophy of chemistry, around which many other problems, such as that of

---

[1] One could take this one step further, and argue, contrary to Hoffmann (2007), that chemists do not typically engage in theory formation and deductive testing at all. I will not actively pursue this line of enquiry in this contribution.

reduction, revolve. While these latter problems have a bearing on the sort of solutions to the problem of ontological discontinuity that might be acceptable, it is my contention that the problem of ontological discontinuity requires to be resolved prior to these other problems. Second, I will argue that we need a robust method to derive an *ontology* from a scientific theory to have a realistic chance of settling such matters. Third, I will present an outline for a specific ontology of chemical *objects* which can overcome these difficulties.

My proposal is to start this discussion with a return to the central problem posed in "The Theoretician's Dilemma." This chapter therefore addresses a number of key questions that need to be considered afresh in the context of the philosophy of chemistry. These questions concern the role of scientific theories, the ontological status of theoretical terms, matters of realism and instrumentalism, and the use and value of axiomatization in the philosophy of chemistry.

This chapter is structured as follows. In section 2 I revisit "The Theoretician's Dilemma." I argue that Hempel's somewhat hasty way out of the dilemma, which is essentially a hard form of scientific realism, is problematic for the philosophy of chemistry. The quest is then to find a way of resolving the dilemma that can deal with the ontological promiscuity of the theories of chemistry.

I argue that the philosophical ground tilled since Hempel suggests a number of ways out of the dilemma, even while it cannot complete resolve the issue. Specifically, I discuss the notion of ontological commitment and the truth-making alternatives in section 3 to argue that an ontology for science engages with objects in two separate but complementary ways: it provides existence claims as well as grounding claims.

I then (in section 4) discuss two specific examples from chemistry which highlight the interplay between existence and grounding, and expand the issue of ontological promiscuity in more detail. The examples are Paneth's classification of an element, and Fukui's account of frontier orbitals and stereo-selection. Both of these, I will argue, are particularly problematic for standard accounts of theoretical terms. In the first case it can be argued that Paneth's classification of elements in simple substances and basic substances spans the observational/theoretical divide in an interesting way. Simple substances are more or less observables while basic substances are theoretical terms in the sense of being non-observable. The question then is what type of substance chemical laws are about. The second example is particularly interesting because it relies on a view of molecular orbitals that is inconsistent with quantum theory proper, yet provides a convincing mechanistic account of chemical reaction pathways in organic chemistry.

In section 5 I propose my specific resolution of Hempel's dilemma in the light of ontological promiscuity. The specific resolution hinges on my specific proposal for the ontology of chemical objects, which is based on both existence and grounding claims. This section sketches the outline of a program, or approach, which in my opinion will be a fruitful confrontation of philosophy of chemistry with standard philosophy of science, as well as sketch a way out of various niches in which philosophy of chemistry has managed to entrap itself: My approach is based on a number of suggestive terms from object-oriented programming, but fits with a significant recent body of work in the philosophy of science, especially the monographs of Ladyman et al. (2007), Strevens (2008), Bokulich (2008), and Wimsatt (2007), as well as recent work by Colyvan (2008) and Floridi (2011).

In the conclusion I outline what the consequences of this resolution might be in the wider context of philosophy of chemistry and philosophy of science.

1 "The Theoretician's Dilemma"

There are a number of ways in which "The Theoretician's Dilemma" can be seen as both a template and a problem for the philosophy of chemistry.

The paper can be characterized as a template in the sense that it forms one of the core papers of modern philosophy of science, and states, in the form of a dilemma, some key questions on the nature of theories and theoretical terms. Moreover, Hempel relates these issues back to a more fundamental question on the role of scientific theories as either inductive generalizations or explanatory frameworks. On the other hand, the solution Hempel proposes to resolve the dilemma—the idea that theoretical terms do more than just act as connective tissue between empirical observations—poses its own problems for the philosophy of chemistry. The problems arise because of ontological discontinuity: Whereas in the theories of physics it could be maintained that terms once deemed "theoretical"—such as the term electron—could over time migrate into an "observational" category by further scientific advances, this notion is much harder to sustain for the theories of chemistry.

In this section I will briefly discuss some of the key aspects of Hempel's paper. I will then propose a way in which Hempel's notion of realism can be reconstructed so that its potential applicability to the theories of chemistry can be understood.

1.1 The Theoretician's Dilemma

Before publishing "The Theoretician's Dilemma," Hempel had worked on the problem of theoretical terms for some years. In 1952 he published, as part of the *International Encyclopedia* project, the small book *Fundamentals of Concept Formation in the Physical Sciences* (Hempel 1952). In 1958 he published "The Theoretician's Dilemma" around the same time that he wrote the contribution on Carnap's philosophy for the Schilpp volume on Rudolf Carnap (Hempel 1963). As was pointed out by Psillos (2000b) Carnap's later work on the Ramsey sentence was influenced significantly by Hempel's work on theoretical terms. Thus, "The Theoretician's Dilemma" forms a pivotal point in the conception of theoretical terms in logical positivist philosophy of science. Its views on theoretical terms straddle what Craig (1953); Craig (1956) has called the "replacement program." which argues that theoretical terms are somehow definable, and the Ramsey (1960) view on theoretical terms, which maintains that observational languages are meaningfully extendable. It also (arguably) led Carnap to the Ramsey sentence.

"The Theoretician's Dilemma" is subtitled "A Study in the Logic of Theory Construction" and focuses on the role and function of theoretical terms. As a starting point, Hempel views theories largely as bodies of systematized knowledge, which connect observations to each other through theoretical machinery. Specifically, Hempel states that

> The assumption of non-observable entities serves the purposes of systematization: it provides connections among observables in the form of laws containing theoretical terms, and this detour via the domain of hypothetical entities offers certain advantages.
>
> (HEMPEL 1958, 45)

Thus, the role and function of theoretical terms in scientific theories allow (Hempel quotes Feigl here), "a nomologically coherent formulation on the level of hypothetical construction" (68).

Hempel distinguishes between empirical generalization (which generally proceeds without the use of theoretical terms) and theory formation, which does use theoretical terms. His statement of the theoretician's dilemma is a direct consequence of how he conceives the role and function of a scientific theory. Thus (on page 43), we find the (somewhat rhetorical) question

> Why should science resort to the assumption of hypothetical entities when it is interested in establishing predictive and explanatory connections among observables? Would it not be sufficient for the purpose, and much less extravagant at that, to search for a system of general laws mentioning only observables, and thus expressed in term of observational vocabulary alone?

This leads Hempel to formulate the theoretician's dilemma as follows:

> If the terms and principles of a theory serve their purpose they are unnecessary [...] and if they don't serve their purpose they are surely unnecessary. But given any theory, its terms and principles either serve their purpose or they don't. Hence, the terms and principles of any theory are unnecessary.
>
> (HEMPEL 1958, pp. 49–50)

Hempel then goes on to investigate the ways in which the theoretical vocabulary can be avoided (or is *eliminable*) in more formal terms. Specifically, Hempel discusses the suggestions by Craig (1953) and Ramsey (1960) for the treatment of theoretical terms in a formalized system as two approaches to this problem. In the first case, it can be maintained that theoretical terms are *definable* in terms of observational concepts (even if this definability is only partial). This approach relates to the notion of reduction sentences as introduced by Carnap (1936, 1937), and what Craig (1953, 1956) has called the "replacement program."

Second, theoretical terms can be seen as useful extensions to our observational language. This approach depends on the concept of a "Ramsey sentence." The Ramsey view on theoretical terms is based on F. P. Ramsey's notion of theoretical terms as it was introduced in Ramsey's well-known paper "Theories" (reprinted in Ramsey 1960).[2] Ramsey's proposal relies on a two-step approach to the introduction of theoretical terms. In the first step, we consider a sentence of the form $T(o_1,...,o_m;t_1,...,t_n)$, which contains both observational terms o and theoretical terms t, and we replace all theoretical entities t_i by variables x_i. Then, in the second step, we introduce existential quantifiers for the variables x_i, so that we obtain a Ramsey sentence of the form

$$\exists x_1 ... \exists x_n T (o_1,...,o_m;x_1,...,x_n) \tag{1}$$

The Ramsey sentence thus introduces an element of ontology, or, more precisely, "ontological commitment"[3] into its construction: it makes specific existence claims for entities x_i that satisfy the theoretical constructs of the theory, but at the same time is also careful to extensionally restrict the scope of that existence claim.

Under Hempel's reconstruction, the theoretician's dilemma asks whether it is not possible to avoid these ontological commitments altogether. Thus, "The Theoretician's Dilemma" ultimately poses a question about the *purpose* of scientific theories and theoretical terms. If theoretical terms actually fulfil their function as the connective tissue between observations, they *can* be dispensed with, and if they do *not* fulfil this function, they *should* be dispensed with. In response to this, Hempel argues that theoretical terms do *more* than fulfil a role as a connective glue between observational terms: following Ramsey's ontological suggestion, they actually describe, or refer to, "essential constituents or aspects of the world we live in" (p. 87), a contention which is supported by the fact that theories with such theoretical terms lend themselves to inductive systematization.

1.2 From the Theoretician's Dilemma to Realism

The road from "The Theoretician's Dilemma" to scientific realism (SR) is not a direct one. Hempel lays out the theoretician's dilemma in great detail, but resolves it rather hastily. Hempel's principled realist way out of the dilemma is to argue that is premise is misconstrued: theoretical terms do *more* than serve as mere instrumental devices in derivational chains linking observables: Following the Ramsey view of theoretical terms they have an *ontological* import and are thus "truth tracking" (in the terminology of Psillos 1999). As Hempel sees it, asserting that partially interpreted theoretical terms actually have factual

---

[2] Psillos (1999, 2000a) gives an interesting account of how the Ramsey view on theories came to be introduced in the philosophy of science. Ramsey sentences are also discussed by Carnap (1966) and Lewis (1967).
[3] Hempel uses this term on p. 86.

reference, is tantamount to saying that the theoretical statements concerning them are true:

> To assert that the terms of a given theory have factual reference, that the entities they purport to refer to actually exist, is tantamount to asserting that what the theory tells us is true, and this in turn is tantamount to asserting the theory.
>
> (HEMPEL 1958, 84)

Thus, in the words of Suppe (1989), Hempel's theoretician's dilemma could evolve into the "classic positivistic statement of scientific realism" (p. 21).[4]

Hempel's purported realist resolution of the dilemma leads to a number of questions. First, the statement is somewhat difficult to unpack. Hempel does not specify in detail the sort of "factual reference" and "truth-making" that he is after. Furthermore, it is not clear entirely that Hempel is committed to an overly hard realism for the resolution of the dilemma. Hempel's argument is more subtle, and relies on ontological commitment of the Ramsey type, as well as on the feature of inductive generalizations allowed by the theoretical entities. In this sense, Hempel's rejection of the theoretician's dilemma is not an unconstrained acceptance of scientific realism, but rather something more modest.

To make this point more clear, let us specify what SR is. Michael Devitt (2007) has characterized SR as the realization that science is committed to the existence of a variety of unobservable entities, in terms of the following thesis:

> Most of the essential unobservables of well-established current scientific theories exist mind-independently.

SR is a thesis about entities, but not about the properties of these entities. Hence, a stronger version of SR is possible, which also includes a "fact" realism claiming that not only do the unobservable entities exist, they also have the properties attributed to them by our scientific theories. The "fact" realism then runs as follows:

> Most of the essential unobservables of well-established current scientific theories exist mind-independently and mostly have the properties attributed to them by science.

Devitt argues that these theses can in turn come in epistemic and semantic varieties. The epistemic variety claims that a belief in the unobservables of science is justified, the semantic variety claims that (i) the theoretical terms of theories refer to the sort of mind-independent entities featuring in SR and

---

[4] As an aside it might be noted that in his contribution to the Carnap volume, Hempel (1963) mounts a slightly different argument against the avoidability of theoretical terms, arguing that statements in which theoretical terms are avoided are not capable of featuring in statements of an inductive character. As Hempel argues there, these inductive steps are key to the use of scientific theories, and hence scientific theories are *more* than just deductive inferences between sets of observations. However, in the remainder of this chapter I will focus on Hempel's realist resolution of the dilemma.

(ii) the theories' statements about its theoretical entities are approximately true. It is however not entirely clear how Hempel's resolution of the dilemma can be mapped onto Devitt's classification.

I conclude that a simple solution to "The Theoretician's Dilemma" along these realist lines is not available to philosophers of chemistry. There are several reasons for this. The first one is that many chemists take the entities of their theories with varying degrees of commitment, and it is unclear whether the sort of hard realism that Hempel seems to require to avoid the consequences of "The Theoretician's Dilemma" can be put to work in chemical theories. Second, the hard-realist way out of the dilemma drives us headlong into the ontological discontinuity of many chemical concepts and physical concepts: It should be clear that a hard-realist line of the sort needed by Hempel is not possible in cases of such strong discontinuities unless the theories themselves are competitors. In the case of physics and chemistry the two sciences are clearly not competitors, but rather cooperators: As Needham (2010) has remarked, the two sciences cooperate to the degree that it is hard to imagine chemistry with the physics removed. Third, for sciences like chemistry that are theoretical foragers (and have a somewhat promiscuous attitude to theoretical concepts as a result), a hard realism of the sort ascribed to Hempel seems out of the question.

1.3 A Proposal for Resolution

There seem to be three ways out of this conundrum. In the first place one could find a type of reductive argument that shows that the discontinuity is apparent, and that the two theoretical entities are in fact the same. This approach has given rise to the notion of "ontological reduction." Second, one could retain a realist resolution of the dilemma but argue for a form of *internal realism* coupled with theoretical disunity.[5]

The last approach, which I will further develop here, is to take a step back in Hempel's somewhat hasty realist resolution of the dilemma. This proposal recognizes that theoretical terms *do more* than just connect observations to each other, but leaves it open, for the moment, what it precisely *is* that they do. This step therefore involves taking "The Theoretician's Dilemma" as posing an interesting question on the role and function of the theoretical terms featuring in our scientific theories, and then to apply a carefully constructed ontological response. I have already indicated my reasons for believing that Hempel's resolution of the dilemma invites such a response.

To this purpose, I propose that we frame a minimal resolution of the theoretician's dilemma as follows:

> The theoretical terms featuring in our scientific theories do *more* than just connect observations to each other.

---

[5] Internal realism was introduced by Putnam (1981).

This restatement, by leaving the content of the "do more" deliberately vague, invites us to reconsider how we might make the move from theory to ontology, a move which I will consider in the next section.

2 Quantification and Truth-maker Theory: Two Varieties of Ontology

Hempel does not specifically consider ontology in "The Theoretician's Dilemma," apart from mentioning, on page 86, that the Ramsey construction of theoretical terms involves an "ontological commitment." In this section my main line of argument will be that the exact nature of this ontological commitment needs to be clarified if we are to be successful in applying the theoretician's dilemma to the theories of chemistry. Specifically, two ontological theories are of interest here. The first is Quine's extensional view of ontological commitments of the theory, and the second is the notion of truth-making. In a nutshell, ontological commitment is what a theory maintains exists. Ontological commitment thus builds an ontology on top of epistemology. In opposition, truth-making focuses on the sort of ontologies a theory needs in order to be true.

As a philosophical way out of the theoretician's dilemma, Hempel's solution raises various issues. Among those are fundamental questions on the nature of truth, scientific realism in its many varieties, truth-making and the ontological nature of theoretical entities. The aim of this section will be to review Hempel's way out of the theoretician's dilemma in the light of a robust theory of ontology. As it stands, Hempel's ontological resolution reads as a variety of truth-making, in which truth depends on being. In this section, it will be my aim to unpack Hempel's statement with the assistance of clearer definitions of ontology, and then proceed to discuss the sort of realism that is defensible on Hempel's view.

2.1 Quantifiers

The commonly utilized conception of "ontological commitment" was first introduced by Quine (1949) (though the standard reference is usually Quine 1961[6]). The idea is that a theory can be regimented in extensionalist terms, where the ontological commitments of a statement are specified by the existence claims of some suitable paraphrase of that statement. Hence, for instance, a theory that speaks of "orbitals" (choosing a more apt example for the philosophy of chemistry than the ubiquitous unicorn), is ontologically committed to the existence of an orbital to the degree that for a suitable domain of objects there is such an object as an "orbital".

The logical step from statement to ontological commitments of a statement require a *regimentation* which was first introduced in Quine (1949). This regimen-

---

[6] Quine (1949) is essentially a chapter in this collection.

tation is based on a first-order logic in which, for instance, a statement Pa is regimented by its paraphrase $(\exists x)Px$, using an existential quantifier. We furthermore assume all this talk is relative to a domain D, it follows that one of the ontological commitments of Pa is the existence of an $x \in D$ such that Px. Moreover, if we take Pa to be true, then a has to exist.

The ontological commitment is thus what a theory maintains exists on a suitable domain of objects. Ontological commitment thus builds an ontology on top of an epistemology by determining whether the objects of which our theories speak are variables in the paraphrase.

This formulation of ontological commitment is formally similar to the Ramsey construction for theoretical terms. Ramsey's formulation of theoretical terms in the form of a specific ontological commitment of scientific theories formed the inspiration for the realist way out of the theoretician's dilemma. On the other hand, it seems unlikely that Quinean quantification is by itself strong enough to entirely avoid its consequences: The resolution of the theoretician's dilemma requires us to specify what theoretical terms actually *do*, whereas quantification specifies what we are *committed to* in our theoretical sentences, and that only on a suitable domain. While such commitment is a part of what theoretical terms *do*, it cannot be the whole story.

2.2 Truth-making

More recently, varieties of a rival theory of *truth-making* have been proposed as alternatives to quantification.[7] In the truth-maker view a theory is only ontologically committed to those entities which are required to make the theory true, whereas in the quantifier view a theory is committed to the entities it says exist.

Truth-making has been discussed by a number of authors, and it is not my aim to discuss the varieties of this theory in detail. Instead, I focus on the critical evaluation of the various options in an article by Jonathan Schaffer (2008b). Schaffer's starting point in this article is a critical evaluation of the ontological commitments that are made (or, as the case may be, avoided) by various truth-maker theories. The truth-maker view as considered by Schaffer comes in three varieties: the necessitation view, the supervenience view and the grounding view. Their basic statements are summarized in Table 1.

The "necessitation view" (TNec) is, in the words of Merricks (2007), "truth-maker orthodoxy." The necessitation view states that "for all x and all p, x is a truth-maker for p only if x's mere existence is metaphysically sufficient for p's truth" (Merricks 2007, 5). Schaffer provides detailed arguments against the necessitation view. Specifically, he argues that TNec is not in general a correct capture of our intuition that "truth depends on being," and is not an apt theory to measure ontological commitment. The reason for this, according to Schaffer,

---

[7] On the face of it, it seems that Hempel's realist resolution of the theoretician's dilemma could well be a variety of truth-maker theory, though this would require more detailed analysis.

TABLE 13.1 Overview of ontological commitment theories as presented in Schaffer (2008b).

VIEW	STATEMENT OF ONTOLOGICAL COMMITMENT
Quantification	Pa implies $\exists x \in D{:}Px$.
TNec	$(\forall p)\,(\forall w_1)$ (if p is true at w_1 then $(\exists x)$ (x exists at w_1 & $(\forall w_2)$ (if x exists at w_2 then p is true at w_2)))
TSup	$(\forall w_1)\,(\forall w_2)\,(\forall p)$ (if p is true at w_1 and false at w_2, then w_1 and w_2 differ in being [either in what there is, or how it is])
TGro	$(\forall p)\,(\forall w)$ (if p is true at w, then p's truth at w is grounded in the fundamental features of w)

is that TNec fails to make the right ontological commitments on four specific counts[8].

In contrast, a "supervenience view" (TSup) makes global supervenience claims about worlds without making ontological commitments, and is thus not a theory that can get ontological commitment right.

Schaffer is most positive about a "grounding view" (TGro) on truth-making. He sees some redeeming features in TGro, but argues that TGro only argues what is fundamental (as opposed to making specific commitments on what there is). Thus Schaffer concludes that the truth-making view, in its grounding variety, is complementary to ontological commitment: The truth-maker view can furnish us useful additional insight into what a theory says is fundamental via the grounding view TGro, but there is no convincing truth-maker view of ontological commitment.

In the general context of ontology, as Schaffer (2008a) has argued, the intuition of the "grounding" account TGro is that truth depends on being. Specifically, TGro introduces an ordering on what there is in the world, in Schaffer's words, a great tree of being. The grounding relation is characterized as follows:

> By way of gloss, where x grounds y, we can equally say that x is ontologically prior to y, or say that y depends on, or derives from, x. Likewise we can say that y exists and has its nature in virtue of, on the basis of, or because of, x. Where x is funda-

---

[8] The details of Schaffer's account matter little here, but for completeness, they are the following: The first issue is that TNec falls into what Schaffer calls "trivial truth-making" if it accepts facts of the form "p is true." For in that case, if p is any proposition and x is the fact that p is true, then x is a truth-maker for p, but of course a trivial one. The second issue is that TNec handles necessary truths wrongly. For the statement $2 + 3 = 5$ any fact in the world will do as a truth-maker. Third, truth-maker does not handle contingent statements well. If the cat is on the mat, then the existence of a cat would seem to be a truth-necessitator for the statement "the cat is on the mat." Until the cat jumps up and walks away, that is. Finally, TNec does not handle negative existentials. The statement "there is no phlogiston" cannot be made true by any common entities.

mental, we can equally say that x is basic, independent, and primary (as opposed to derivative, dependent, and secondary). Anyone who understands any of these common locutions will understand what is meant by fundamentality and grounding.

(SCHAFFER 2008a, 311)

Schaffer thus concludes that the quantifier view and the grounding view are complementary. Put in another way, a robust ontological theory should contain existence commitments and grounding commitments.

2.3 Resolving the Theoretician's Dilemma

We now return to the main line of argument and consider what use such on-tological considerations have for the philosophy of chemistry and how they might be used in a resolution of the theoretician's dilemma that is sympathetic to the theories of chemistry. Let us start by considering them as a proposed resolution of the theoretician's dilemma *per se*.

A combination of the notion of existence with that of grounding seems to be capable of at least providing us with a number of hints as to what theoretical terms actually *do* in our theories: theoretical terms, on this view, may be said to provide grounds for a critical sort of linkage between the apparatus of one theory to another. This resolution can also be mapped onto Hempel's proposed resolution of the dilemma, which seems to include aspects of both existence claims and truth-maker claims.

While this notion is suggestive, it does not, however, resolve the dilemma be-yond doubt: The "grounding" account does not provide sufficient reason to con-clude that theoretical terms cannot be eliminated. As an account of what theoretical terms actually do, its main obstacle is imprecision; hence the notion of a robust ontological theory gives us some clues, but does not provide us with the conclu-sive evidence needed to resolve the theoretician's dilemma.

For the philosophy of chemistry, their use seems even more limited. One issue with the current ontological discussion in the philosophy of chemistry is that the statements they purport to relate to are not reflective of realistic theo-ries of chemistry or physics. The toy statements of the ontological theories are things such as "you exist," "there is beer in the fridge," or "I am sitting" with an occasional excursion into "electrons." In terms of Schaffer's contention that Quinean quantification and the grounding view of truth-making are compat-ible and complementary, this means that the toy theories of the ontologists give us little further information about how such dependency relations may be cashed out for actual scientific theories.

Thus the notion of grounding by itself is not capable of further analysis in a general ontological framework. However, in the context of the philosophy of chemistry (and also in the more general context of the philosophy of science), "grounding" *is* analyzable roughly in the *epistemic* terms of scientific explana-tion and theory formation. In this sense, ontologically commitment states *that* a certain entity exists, while grounding further explicates *how* an entity exists.

Grounding equates to something like "there being an explanation or theory that of how an x is a fundamental feature of y, or of how y's are made up out of x's, to specify how x may 'ground' y." Two examples from chemical theories may help to lend the notion more precision.

3 Two Chemical Examples

In this section I discuss two examples that illustrate the notion of chemistry's "promiscuous ontology" in greater detail, and will, through these examples, provide a lead-in to the notion of chemical objects that I will develop in the next section. The two examples are the well-known examples of chemical elements and the equally well-known example of orbitals. It is now well-known in the philosophy of chemistry that the notion of chemical element was disambiguated by the chemist Paneth in the 1930s (the two papers were translated in the 1960s). The disambiguation is important in the recognition how the term element has been used in chemical theories.

In opposition, the notion of "orbital" has been the subject of extended discussions, based on the claim by Zuo et al. (1999) that orbitals had been observed. While there has been significant discussion on the ontological implications of this claim, philosophers of chemistry have thereby missed the, in my opinion, more interesting question on how orbitals have been used in the theories of chemistry. I will take the frontier orbital theory of the Japanese chemist Kenichi Fukui as my main guidance.

3.1 Chemical Elements

There is a well-known and influential historical disambiguation of the notion of chemical element found in Paneth (1962a,b). The core of the disambiguation is formed by a distinction between the notion of chemical element into that of a *simple substance* and that of a *basic substance*. For Paneth, the following describes the difference:

> I have referred throughout to *'basic substance'* whenever *the indestructible substance present in compounds and simple substances was to be denoted*, and to *'simple substance'* whenever *that form of occurrence was meant in which an isolated basic substance uncombined with any other appears to our senses*.
>
> (PANETH 1962b, 150)

Paneth moreover distinguishes the practical definition of a chemical element stemming from Lavoisier, in which a chemical element is considered as a non-decomposable substance, from the more (if one must) "metaphysical" considerations of the philosophers (which for instance sometimes limited the number of available elements to some predetermined number). Thus, as a simple substance, gold is a yellow metal, while as a basic substance it denotes the element "gold."

Although much has been made of Paneth's distinction in these two notions of a chemical element, my contention is that the key to the story lies rather in Paneth's notion that chemistry was born of the *identification* of these two notions of element; the realization that "simple substances" manifest themselves as "basic substances" in nature. Thus, Paneth writes that

> The great *significance* of elements for the whole body of chemical theory lies in the assumption that the same substances which produce the phenomenon of 'simple substances' serve in the quality-less, objectively real sphere of nature as 'basic substances' for the construction not only of simple but also of composite substances. *The fundamental principle of chemistry that elements persist in their compounds refers to the quality-less basic substances only.*
>
> (PANETH 1962b, 151)

Paneth argues that chemists employ a certain "naive realism" with respect to simple substances, but that this naive realism cannot be carried over into the realm of basic substances, which are abstract entities in themselves devoid of properties. Thus, the notion that "basic substances" somehow persist in chemical compounds requires a leap into a "transcendental world devoid of qualities" (p. 150). The notion of a chemical element refers primarily to the "basic substance," whereas the "simple substance" is a leftover from a more naive realist period in chemistry, in which, for instance,

> The properties of sulphur are just those properties which it exhibits to us in the special case when it is not combined with any other basic substance.
>
> (PANETH 1962b, 151)

The issue is that for the notion of "chemical element" as a theoretical term the realist way out of the dilemma is not available, since it is not entirely clear, in the light of Paneth's disambiguation, what a chemical element exactly *is*. It could be argued that simple substances refer to observables while basic substances are "theoretical" in the sense that they do provide the sort of organizational characteristics that Hempel requires from theoretical terms. This reading of Paneth's disambiguation suggests that the notion of a chemical element as a "simple" substance has gradually disappeared from our vocabulary, in a manner similar to how Sellars (1963) has suggested that a "manifest" image of nature is over time gradually replaced by a "scientific" image.

This replacement model has been criticized in the philosophy of chemistry, most notably by Jaap van Brakel (2000) and, from a different viewpoint, by Hoffmann (2007).

From the viewpoint of classical philosophy of science, this replacement model also seems to get its direction wrong: It replaces, over time, an "observational" term (i.e., the chemical element as a simple substance) with a "theoretical" term (i.e., the chemical element as a basic substance). The standard route for the realist view on science, as for instance argued in Nola (2007) runs however in the other direction: science usually *postulates* the existence of particular objects which over time can be observed.

Paneth's disambiguation rather suggests that both views of the chemical element have a role to play, and, as Ruthenberg (2009) argues, Paneth's realism develops a "multiperspective" view in which "we can only properly interpret the dualistic view by thinking of *both* worlds as unity" (Ruthenberg 2009, 88). Scerri (2000b) argues that the tension between these two views of the elements[9] can be resolved by adopting an "intermediate position," which is

> a form of realism, [...] tempered by an understanding of the viewpoint of the reducing science but which does not adopt every conclusion from that science.
>
> (SCERRI 2000b, 52)

Viewed in the light of the issues posed by Hempel's "The Theoretician's Dilemma" these issues take on exceptional poignancy: While the terms of theories of chemistry indeed "do more" than connect observations to each other, the sort of realism that flows from both the "intermediate position" and the dual view contain both existence and grounding components. Their main difference seems to lie in a disagreement about how these components should be weighed. However, the sort of hard realism suggested by Devitt (2007) is not in this sense available.

In the practice of chemistry, chemists shift pragmatically between the two ontological concepts of element, dependent on context. This is not in itself a conceptual problem for the science of chemistry, but it poses interesting questions for the philosophical conception of chemical objects, to which I return in the final section.

3.2 Orbitals and Chemical Explanation

One particular problematic aspect in the relationship of chemistry to physics is the considerable success of simple approaches such as the simplified orbital models that liberally sprinkle chemistry textbooks. For physicists, orbitals are (mathematical) components of a many-electron wave function, which have to obey specific symmetry relationships. In the chemical textbooks, however, the orbitals are viewed as direct explanatory elements. In more extreme cases, molecular orbitals are sometimes even decomposed down into their atomic orbitals, with explanations depending specifically on the atomic orbitals on different molecular centers.

The dilemma is thus that the concept of "orbitals" has explanatory traction in the theories of chemistry on the one hand, while on the other hand the proper "grounding" of this concept in the theories of physics is lacking. The lack of this grounding may well explain some of the ongoing discussions in the philosophy of chemistry that followed on the claim by Zuo et al. (1999) that orbitals had been observed. Philosophers of chemistry have discussed this claim at length, for instance in the contributions by Scerri (2000a, 2001),

---

[9] Scerri discusses orbitals as well, but his "intermediate position" is developed primarily on the basis of the chemical elements and the periodic table.

Lombardi and Labarca (2005), and Labarca and Lombardi (2010), as well as Mulder (2011). The concept of an orbital, then, is a case where a "grounding" argument seems to fail.

To make the case for the explanatory efficacy of the orbital concept as strong as possible, I will focus on the concept of a "frontier orbital" and its explanation of the phenomenon of stereo-selection. The interesting aspect of this particular explanation of chemical reactions is that the explanation is truly based on quantum mechanics, is not really intelligible without it, and the fact that the *signs* of the atomic orbital coefficients in the orbital that makes up the frontier orbital play a key part in the explanation. The concept of the frontier orbital was originally introduced through the concept of a "frontier electron" in Fukui et al. (1952). A frontier electron is the molecular analogue of the concept of a "valence electron" in atoms.[10] The frontier orbital concept is a further development of the concept of frontier *electron*. The latter is introduced as follows:

> The two electrons occupying the highest π orbital in the ground state, distinct from the other π-electrons, are essential to the reactivity of the system. We shall refer hereafter to this sort of π-electrons as *frontier electrons*.
>
> (FUKUI ET AL. 1952, 723)

The concept of a "frontier orbital" is introduced by considering the density of the specific orbitals of the frontier electrons as well as their specific symmetry properties. The frontier electrons generally occupy the "highest occupied molecular orbital" (commonly abbreviated as HOMO) and interact with the "lowest unoccupied molecular orbital" (commonly abbreviated as LUMO). The structure and shape of both the HOMOs and LUMOs is then used to determine the propensity for a chemical reaction.

The frontier orbitals provide a reaction mechanism for a large number of pathways in organic chemistry, such as an assessment of the carcinogenic nature of various aromatic compounds, organometallic complexes, stereo-selection, and more. For the purposes of the present chapter I will discuss the role of frontier orbitals in the context of stereo-selection, which is discussed in more detail in Fukui (1971).

The core idea is that the mechanisms of chemical reactions can be described in terms of the interaction between the HOMO and the LUMO of each of the reactants. The reaction is then the result of an "electron transfer" between these two.

The use of this system to describe stereo-selection rules was introduced by Woodward and Hoffmann (1969), and the scheme of Fukui is both a simplification and generalization of this scheme. The frontier orbital approach to stereo-selection is based on the approximate theory of Hückel. In 1931 Erich

---

[10] See the collection of papers in Fukui and Fujimoto (1997) for a historical overview of the concept and its further development. This volume contains reprints of most of the important papers in the development of Fukui's theory of frontier orbitals, as well as a small number of reflections on the relative importance of these papers.

Hückel formulated an approximate theory for the description of the electronic structure of aromatic molecules. The development of such methods allowed for the application of quantum theoretical methods to large systems long before the computational machinery that we currently use was developed.[11]

In what follows, I will discuss the explanation of stereo-selection from Fukui (1966), which is based on simple Hückel Molecular Orbitals. In the Hückel approximation the orbitals that feature in the description are the π (or carbon P_z) orbitals, and the matrix elements over these orbitals are approximated as follows. The overlap integral $\langle \pi_i | \pi_j \rangle = \delta_{ij}$; and the atomic levels are given by $\alpha_i = \langle \pi_i | H | \pi_i \rangle$ while the interaction for neighboring atoms is approximated by $\beta_{ij} = \langle \pi_i | H | \pi_j \rangle$ where i and j label the atoms. The interaction for non-neighboring atoms is neglected and the β is sometimes referred to as a "resonance integral". For a linear conjugated polyene the Hückel molecular orbitals (MOs) are given by

$$\phi_j = \sum_{r=1}^{k} C_r^k \phi_r = \left(\frac{2}{k+1} \right)^{1/2} \phi_r \sin \frac{jr\pi}{k+1} \quad (j=1,2,\ldots,k) \tag{2}$$

For a ring closure, the change in the electronic energy due to the interaction between the first and last AO is

$$\Delta E = 2 \sum_{j=1}^{k} v_j C_1^j c_k^j \gamma = 2 P_{1k} \gamma \tag{3}$$

where v_j is the occupancy of the jth MO and γ is the energetic integral obtained by closing the ring; the P_{1k} correspond to a density. The quantity P_{1k} is related to the bond-order.

As Fukui points out, for linear conjugated polyenes, the P_{1k} can be written as

$$P_{1k} = \frac{(-1)^{k/2+1}}{k+1} \left(1 + \frac{1}{\cos(\pi / (k+1))} \right) \tag{4}$$

Hence, for $k=4m+2$ (where $m=1, 2,\ldots$), $P_{1k} > 0$ and hence ring closing occurs for $\gamma<0$ (since $\Delta E<0$ is the requirement for the closing to occur). Similarly, for $k=4m$, $P_{1k} < 0$ and hence rung closing occurs in the mode $\gamma>0$. The cases $\gamma<0$ and $\gamma>0$ correspond to two different types of ring closures, as in figure Fig. 13.1. Hence the ring closures are different for different length of the chain, with

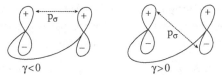

$$\gamma<0 \qquad\qquad \gamma>0$$

FIGURE 13.1 Stereo-selection.

[11] "Machinery" here can of course be partially taken literally, it refers both to computer programs and to the computers they run on.

multiples of 4 closing in a conrotatory fashion and multiples of $4m + 2$ closing in a disrotatory fashion for a thermal reaction.

In the case of a photo-induced reaction, one of the binding electrons is excited to the lowest unoccupied MO and the reactivity changes. The "overlap stabilization" is given by

$$P_{1k} = \frac{(-1)^{k/2}}{k+1} \frac{\cos(\pi/(k+1))+1}{\cos(\pi/(k+1))} \left(2\cos\frac{\pi}{k+1} - 1 \right) \tag{5}$$

which is positive when $k = 4m$ and negative for $k = 4m + 2$ for $k \geq 4$. Hence the situation is reversed for reactions that are photo-induced.

The situation for stereo-selection of ring closures may be summed up as follows:

	THERMAL REACTION	PHOTO-INDUCED REACTION
$k = 4m$	Type II	Type I
$k = 4m + 2$	Type I	Type II

Hence the "frontier orbital" concept is immediately instrumental in the prediction of a chemical reaction path.

As stated before, the issue is that the particular object of a "frontier orbital" cannot be grounded in a more fundamental quantum theory without further ado. This situation has led Paul van der Vet (1987) for instance to postulate a "chemical theory of the atom," which acts as a "proxy defense" in the sense of Hendry (1998). A specific formulation of the concept of a "chemical theory of the atom" (CTA) was introduced in van der Vet (1987) to capture this highly approximate quantum notion of the atom in which, in the words of van der Vet (1987, 170–171):

1. "The chemical properties of an (unbonded) atom are determined by the electron arrangement of the atom in a state which will be called the atom's *basic state*.
2. The chemical properties of molecules are determined by their electron arrangements.
3. The electron arrangement of a molecule depends on the constituent atoms and is determined
   (a) by the basic configurations of the constituent atoms directly
   (b) by the molecule's structure, which is partly determined by the basic configurations of the constituent atoms."

While frontier orbitals cannot be grounded in the concept of a quantum theory of the atom, the "chemical theory of the atom" is something that, in the terminology of Hendry can "stand in" as a proxy for the original concept and supply some of the desirable features that are lacking in the original. With Hendry, I believe this situation to be somewhat unsatisfactory, not in the least because there seems to be no satisfactory resolution of Hempel's "Theoretician's

Dilemma" with the proxy defense as a starting point. In response, I will argue for a concept of objects that is based on chemistry's ontological promiscuity.

4 Theoretical Promiscuity and the Chemical Object

In this final section I will develop an outline for an ontology for chemistry that is capable of dealing with the ontological issues presented by chemical theories. The aim of this section is to propose an answer to Hempel's "The Theoretician's Dilemma" by specifying what, in the theories of chemistry, theoretical terms *do* in addition to serving as the theoretical machinery to tie observations together, and elucidating the interplay between existence claims and grounding. In brief, my answer is that they export and import various concepts to and from other scientific theories, and hence tie chemistry as a science to other sciences.

4.1 Chemical Objects

My proposal is that we extend and build on Paneth's dual notion of a chemical element to construct a theory of chemical objects as "multiperspectival" objects. The basic idea builds on a discussion between Rae Langton (1998) and Henry Allison (2004) (but going back to an analysis by Strawson 1975) on the interpretation of Kantian Transcendental Idealism, which, as Ruthenberg (2009); Ruthenberg (2010) has argued, forms the core of Paneth's dual notion of chemical element. While it would take us too far here to discuss the origins and resolution of that conflict in detail, Allison's proposal is that Kant's distinction between the *phaenomenal* and *noumenal* world should be read as a distinction between how we *consider* things, as opposed to marking an ontological distinction between two non-overlapping sets of properties (i.e., the intrinsic properties marking the thing in itself and the relational properties marking the appearance of the thing). My contention is that Allison's epistemic reading sits well with Paneth's disambiguation of the notion of a chemical element.

An example, illustrating the chemical element as a "simple" and "basic" substance is given in Fig. 13.2. My proposal is to connect the two notions of element with a set of specific *links* between the objects, of which three examples are given in the figure. The links L_1, L_2 and L_3 form part of the grounding relation of the chemical element, and act as "information" and "conceptual" channels that link the two notions of elements together. Also note that these links make specific *epistemological* commitments: they link a concept of "valence" in the "basic" notion of the element for instance to the issue of whether the element as a simple substance is a metal or not.

Epistemologically, therefore, my proposal turns on the notion of *encapsulation*[12]: both chemistry and physics populate their relevant concepts—say, "atom"

---

[12] In object-oriented programming a large complex program is split up into "objects," say, transactions in a banking system, or personal records, which perform certain functions. The object "person" in the computer program may "expose" certain methods, such as "age," "address,"

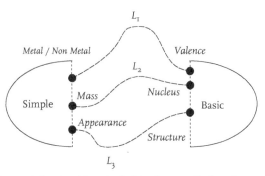

FIGURE 13.2 Proposal for the "chemical object" as applied to the notion of element, disambiguated between a "simple" and "basic" object. The links between the objects are epistemic links that connect, for instance, the internal (atomic) *structure* of a *basic* substance to its *appearance* as a *simple* substance. The links are not derivations.

or "chemical element"—with "methods" such as concepts and data whose precise "implementation" is hidden from view. To develop a quantum theory of atoms and molecules, the early quantum physicists and quantum chemists developed a number of concepts such as "orbitals," "atomic nucleus," "electron," and so on, which were later, in abstracted form and with significant loss of context, *imported* into the science of chemistry. Similarly, chemists developed the concept of "valency," "directed bonding," and so on which were in turn imported into the early theories of quantum chemistry as explanatory questions. Inside the field of either chemistry or physics, these concepts are necessary elements of theorizing, though their relationship with the "exporting" theory may be tenuous and relevant context is lost in the process.

This theory of chemical objects fits the bill both as a resolution of Hempel's "The Theoretician's Dilemma" and as an ontology for chemistry. As a resolution of "The Theoretician's Dilemma" it is capable of outlining how theoretical terms "do more" than just connect observations to each other: It suggests that theoretical terms that are particularly efficacious in an explanatory sense are capable of the sort of epistemic linking commitments that the model requires. The latter are part of the ontological grounding commitments of the theory. Also note that the *logical* structure of these linking commitments is open to multiple interpretations and formulations.[13] Somewhat trivially, the model allows for simple existence claims for the theoretical terms.

"gender," "income" (if the program in question is run by the IRD), and so on. Other parts of the program can "consume" these methods, but do not have to know how they are implemented, the "method" itself is "encapsulated." The internal definition is hidden from view, but the results are accessible to the component that wants to use the method.

[13] I will discuss an interpretation in terms of the "rainforest realism" of Ladyman et al. (2007) later. Other potential logical forms are for instance the "chunk and permeate" approach of Brown and Priest (2004, 2008), a causal approach by Strevens (2008), a structural form by Bokulich (2008), or an "engineering" interpretation in Wimsatt (2007). These models have so far not been applied to concepts and theories of chemistry, but look to hold out some promise for progress in various ways. At the more speculative end of the scale, this reassessment of the

Philosophy of chemistry has much to gain by reexamining the relationship between physics and chemistry in this light. The hope for a reduction relation which consists of "correspondence qua identity" and strict derivation (a particularly austere interpretation, due to Causey (1977), of the reduction conditions given by Nagel (1961) has now vanished, and the prospects of disunity are unattractive. However, the suggestion that the ontological relationship is one of grounding, through a process of importing and exporting concepts in and out of theories, may just create the necessary middle way.

4.2 Relating Chemical Objects to Philosophy of Science

The current proposal fits well with recent work done in both the philosophy of inconsistent objects (Colyvan 2008) and the philosophy of computing (Floridi 2011), although a detailed discussion of the linkages, similarities and differences would be too involved for the present chapter.

To bring out two salient features of my proposal, I will conclude by briefly discussing the consequences of my proposed notion of a chemical object for the notion of ontological reduction, as well as briefly connect this notion of object to (and contrast it with aspects of) "rainforest realism" as developed by Ladyman, Ross, Spurrett, and Collier (2007).

4.2.1 Ontological reduction

I wish to be brief about the consequences of this view for the notion of ontological reduction. Labarca and Lombardi (2010) claim that there is no sense in which it can be claimed that chemistry has been "ontologically" reduced to physics. Specifically, Labarca and Lombardi set up the ontological argument as follows:

> Two theories T_1 and T_2, both containing a term 'C', which is non-referring in T_1, but refers to the entity C in T_2. Moreover, T_2 cannot be epistemologically reduced to T_1. For what reason can we say that the entity C does not exist simpliciter? Since epistemological reduction fails, the entity C described in T_2 does not exist only under the assumption that T_1 is the "true" theory or, at least, the theory most appropriate to describe reality.

> (LABARCA AND LOMBARDI 2010, 155)

Labarca and Lombardi are not entirely clear here on what sense of "refer" and what sense of "epistemologically reduced" they use in their definition of ontological reduction. I will take it that they mean some strict sense of identity *cum* "derivability" by "epistemologically reduced." Taking their notion of "refer" as indicating a dual notion of ontological commitment, as a view on both existence and grounding, the issue dissolves: Both T_1 and T_2 are ontologically committed to C (though in a trivial sense) while a grounding commitment in terms of encapsulation and epistemic links does not require the existence of identity

---

chemical object may also relate to the topic of "object oriented ontology" (see for instance Harman 2011)) which is currently in progress in the area of "continental" philosophy.

and strict derivability.[14] Also note that while the linking commitments are not necessarily formal derivations, they *do* require formal specification. In this sense the dissolution of the issue of ontological reduction does not rely on a blank check to metaphysics.

The grounding notion does include *directionality*. Labarca and Lombardi take "orbitals" as a key example of their term *C*. In my proposal, the concept of orbital is introduced in chemistry through a linking commitment with the theories of physics, and hence has logical ground, even as it is capable of standing on its own as a concept in a theory such as the "frontier orbital" theory. The fundamentality implied by grounding does not commit us to any form of "metaphysical reductionism": rather, it claims that the term "orbital" is transferred as an encapsulated entity from quantum theory to a theory of stereo-selection with a significant loss of context. The transfer is *epistemic* though in this particular case its logical form is not one of derivability.

4.2.2 Ladyman and Ross' rainforest realism

To conclude, I will draw out some parallels and give some contrast of my proposed notion of chemical object with the idea of "rainforest realism" and the resulting notion of unity of science that results from rainforest realism.

Ladyman and Ross' view on the sciences hinges on two specific principles which may be somewhat distasteful to philosophers of chemistry. The first principle is the principle of Principle of Naturalistic Closure (see Ladyman et al. 2007, 37):

> (PNC) Principle of Naturalistic Closure: Any new metaphysical claim that is to be taken seriously at time *t* should be motivated by, and only by, the service it would perform, if true, in showing how two or more specific scientific hypotheses, at least one of which is drawn from fundamental physics, jointly explain more than the sum of what is explained by the two hypotheses taken separately.[15]

The second principle is the Primacy of Physics Constraint, which is stated as follows:

> (PPC) PPC articulates the sense in which evidence acceptable to naturalists confers epistemic priority on physics over other sciences. (p. 38)

Disciplinary boundaries between the sciences, for Ladyman and Ross, are based on what they call the "scale relativity of ontology," picked out by "real patterns." These patterns in turn are based on "information channels" be-

---

[14] One might even wish to take up a suggestion from Brown and Priest (2004, 2008) here, who argue that there is a meaningful way in which both theories can be inconsistent while a meaningful link between them is still possible. What I have referred to as "encapsulation" is somewhat similar to their "chunking" procedure, which is a preservationist variety of para-consistent logic.

[15] Ladyman et al. (2007) also note that this is interpreted by reference to a significant number of terminological stipulations (see Ladyman et al. 2007, 38).

tween structures—channels that carry *data* from one theory to the other in the sense of the Shannon theory of communication—and that at the "receiver" end are algorithmically compressed, to, so to say, "rediscover" the pattern that existed in the initial data. At this point Ladyman and Ross discuss a number of technical notions from the theory of information, namely projectibility and depth.

For the purposes of our discussion, we do not need to delve in the logic of this in detail, but it will suffice to remark that "projectibility" refers to any pattern $x \rightarrow y$ in which y is obtained from a computation with x as input. The notion of "depth" results from the pattern being "compressible" in some useful way. The text by Ladyman and Ross is technical, and here I will only attempt to convey the basic notion.

A real pattern satisfies two requirements. The first one is "projectibility." Ladyman and Ross define projectibility intuitively as "just better-than-chance estimatability by a physically possible computer running a non-trivial program" (Ladyman et al. 2007, 224). The second requirement reads that the pattern

> encodes information about at least one structure of events or entities S where that encoding is more efficient, in information-theoretic terms, than the bit-map [i.e., uncompressed, direct] encoding of S [...].
>
> (LADYMAN ET AL. 2007, 226)

Stripped of its information-theoretic clothing, this statement expresses the requirement that the pattern is generalizable—that is, the pattern has logical depth; it is meaningfully predictive. A theory like Fukui's theory of frontier orbitals satisfies that particular condition. Somewhat similarly, this notion explains how it is that we can infer "basic" substances from "simple" substances, because "basic" substances, over time, form a coherent inference pattern that "simple" substances provide less efficiently, at least from the specific epistemic perspective of how to account for the phenomenon of isotopes.

I take it as a given that in the sense of Ladyman and Ross' "real patterns" the patterns of chemical theories are real. Of course, Ladyman and Ross' rainforest realism is a realism about patterns (or "structures"), not about objects. Yet as it stands, its application to theories of chemistry seems possible (though likely to throw up some interesting issues).

The situation becomes more interesting once we start considering their model of the unity of the sciences. For the "scale relativity of ontology" which forms the basis of the unity of sciences is less in force between chemistry and physics than between, say, biology and fundamental particle physics, or everyday life and fundamental particle physics. Hence, one of the issues with a *direct* application of Ladyman and Ross' notions to science is that chemistry and physics may be too close. The very fact that it is hard to imagine a chemistry with the physics removed limits the applicability of the "scale relativity of ontology" to an interesting extent.

In this sense, their first requirement for a real pattern, that of projectibility, has an interesting twist. Specifically, it would seem that, for instance, the notion of "orbital" is projectible from "fundamental" quantum theory to a "chemical" theory such as the frontier theory of orbitals. For all that the principle of projectibility seems to require is that frontier orbitals, or elements as simple substances are computable in an approximate sense from quantum theory, which they obviously are: even while there exist significant *conceptual* discontinuities between concepts such as frontier orbitals and the basic structure of ab initio quantum chemistry, patterns such as frontier orbitals do result as the endpoint of calculation in a surprising number of cases.

Ontologically speaking, Ladyman and Ross deny the "notional-world idea of cohesive things" (p. 256). Whether such an outright denial is still possible when the "notional world idea" is about the cohesive things of chemistry is still an open question. Chemical theories may be just too close to physical theories for comfort, and there is no sense yet in which Ladyman and Ross have allowed their "real patterns" to be "dual" in the sense in which chemical objects are dual and in the sense in which chemistry may require this.

Conclusion

Theories of chemistry form an interesting challenge for philosophers of science, and hold out the possibility of reopening debates, such as Hempel's "The Theoretician's Dilemma," that philosophers had considered closed for some time. This chapter has argued that robust philosophical engagement with the theories of chemistry might lead to new insights on such old concepts as theoretical terms.

This chapter has argued for an assessment of chemical theories in terms of a philosophy of science based on information: the idea that chemical objects are related to physical objects through specific epistemic links—information channels of limited bandwidth—that in an ontological sense not only specify *that* a chemical object exists, but also *how* it exists. In Hettema (2012) I have outlined how this view of chemical objects is also compatible with (and to a significant extent derives from) a relatively liberal view on Nagelian reduction.

The reassessment of "The Theoretician's Dilemma" proposed in this chapter fits in a wider context of similar work. The recent books by Ladyman et al. (2007), Wimsatt (2007), Strevens (2008), and Bokulich (2008), as well as Floridi (2011) all contain first steps in a similar direction, notwithstanding the significant differences in outlook and their starting positions. In this sense, philosophy of science stands much to gain from a renewed engagement with the theories of chemistry.

References

Allison, H. E. (2004). *Kant's Transcendental Idealism: An Interpretation and Defense* (2nd ed.). New Haven, CT: Yale University Press.

Bokulich, A. (2008). *Reexamining the Quantum-Classical Relation: Beyond Reductionism and Pluralism*. Cambridge: Cambridge University Press.

Brown, B. and Priest, G. (2004). "Chunk and permeate, a paraconsistent inference strategy. Part I: The infinitesimal calculus," *Journal of Philosophical Logic*, 33(4), 379–388.

Brown, B. and Priest, G. (2008). "Chunk and permeate II: Weak aggregation, permeation and old quantum theory." Presented at WCP4, The Fourth World Congress of Paraconsistency, July 13 to 18, 2008, University of Melbourne.

Carnap, R. (1936). "Testability and meaning," *Philosophy of Science*, 3(4), 419.

Carnap, R. (1937). "Testability and meaning–continued," *Philosophy of Science*, 4(1), 1.

Carnap, R. (1966). *Philosophical Foundations of Physics: An Introduction to the Philosophy of Science*. New York: Basic Books.

Causey, R. C. (1977). *Unity of Science*. Dordrecht: Reidel.

Colyvan, M. (2008, October). "The ontological commitments of inconsistent theories," *Philosophical Studies*, 141(1), 115–123.

Craig, W. (1953, March). "On axiomatizability within a system," *The Journal of Symbolic Logic*, 18(1), 30–32.

Craig, W. (1956). "Replacement of auxiliary expressions," *The Philosophical Review*, 65(1), 38–55.

Devitt, M. (2007). "Scientific Realism." In *Truth and Realism* (P. Greenough and M. P. Lynch, eds.). New York: Oxford University Press, 100–124.

Floridi, L. (2011). *The Philosophy of Information*. Oxford: Oxford University Press.

Fukui, K. (1966). "An MO-theoretical illumination for the principle of stereoselection," *Bulletin of the Chemical Society of Japan*, 39(3), 498–503.

Fukui, K. (1971). "Recognition of stereochemical paths by orbital interaction," *Accounts of Chemical Research*, 4(2), 57–64.

Fukui, K. and Fujimoto, H. (eds.) (1997). *Frontier Oribtals and Reaction Paths*. Singapore: World Scientific.

Fukui, K., Yonezawa, T., and Shingu, H. (1952). "A molecular-orbital theory of reactivity in aromatic hydrocarbons," *Journal of Chemical Physics*, 20, 722–725.

Harman, G. (2011). *The Quadruple Object*. Alresford: Zero Books.

Hempel, C. G. (1952). *Fundamentals of Concept Formation in Empirical Science*. Chicago: University of Chicago Press.

Hempel, C. G. (1958). "The Theoretician's Dilemma." In *Concepts, Theories, and the Mind-Body Problem*, Volume 2 of *Minnesota Studies in the Philosophy of Science* (H. Feigl, M. Scriven, and G. Maxwell, eds.). Minneapolis: University of Minnesota Press, 37–98.

Hempel, C. G. (1963). "Implications of Carnap's Work for the Philosophy of Science." In *The Philosophy of Rudolf Carnap*, Volume XI of *The Library of Living Philosophers* (P. A. Schilpp, ed.). La Salle, IL: Open Court, 685–709.

Hendry, R. F. (1998). "Models and Approximations in Quantum Chemistry." In *Idealization IX: Idealization in Contemporary Physics*, Volume 63 of *Poznan Studies in the Philosophy of the Sciences and the Humanities* (N. Shanks, ed.). Amsterdam: Rodopi, 123–142.

Hettema, H. (2012). *Reducing Chemistry to Physics: Limits, Models, Consequences*. North Charleston, SC: Createspace.

Hoffmann, R. (2007). "What might philosophy of science look like if chemists built it?," *Synthese*, 155(3), 321–336.

Labarca, M. and Lombardi, O. (2010). "Why orbitals do not exist?," *Foundations of Chemistry*, 12, 147–157.

Ladyman, J., Ross, D., Spurrett, D., and Collier, J. (2007). *Every Thing Must Go: Metaphysics Naturalized*. New York: Oxford University Press.

Langton, R. (1998). *Kantian Humility: Our Ignorance of Things in Themselves*. Oxford: Clarendon Press.

Lewis, D. (1967). "How to define theoretical terms," *The Journal of Philosophy*, 67, 427–446.

Lombardi, O. and Labarca, M. (2005). "The ontological autonomy of the chemical world," *Foundations of Chemistry*, 7, 125–148.

Merricks, T. (2007). *Truth and Ontology*. Oxford: Oxford University Press.

Mulder, P. (2011). "Are orbitals observable?," *Hyle*, 17(1), 24–35.

Nagel, E. (1961). *The Structure of Science: Problems in the Logic of Scientific Explanation*. London: Routledge and Kegan Paul.

Needham, P. (2010). "Nagel's analysis of reduction: Comments in defense as well as critique," *Studies in History and Philosophy of Science Part B: Studies in History and Philosophy of Modern Physics*, 41(2), 163–170.

Nola, R. (2007). "From Reality to Truth: The Case of Electrons and Electron Theory." In *From Truth to Reality: New Essays in Logic and Metaphysics* (H. Dyke, ed.). New York: Routledge.

Paneth, F. A. (1962b). "The epistemological status of the chemical concept of element (ii)," *The British Journal for the Philosophy of Science*, 13(50), 144–160.

Paneth, F. A. (1962a). "The epistemological status of the chemical concept of element (i)," *The British Journal for the Philosophy of Science*, 13(49), 1–14.

Psillos, S. (1999). *Scientific Realism: How Science Tracks Truth*. New York: Routledge.

Psillos, S. (2000a). "Carnap, the Ramsey-sentence and realistic empiricism," *Erkenntnis*, 52(2), 253–279.

Psillos, S. (2000b). "Rudolf Carnap's 'Theoretical Concepts in Science'," *Studies in History and Philosophy of Science Part A*, 31(1), 151–172.

Putnam, H. (1981). *Reason, Truth and History*. Cambridge: Cambridge University Press.

Quine, W. V. (1961). *From a Logical Point of View*. New York: Harper and Row

Quine, W. V. O. (1948/1949). "On what there is," *Review of Metaphysics*, 2, 21.

Ramsey, F. P. (1960). "Theories." In *The Foundations of Mathematics* (R. B. Braithwaite, ed.). Paterson, NJ: Littlefield, Adams and Co., 212–236.

Ruthenberg, K. (2009). "Paneth, Kant, and the philosophy of chemistry," *Foundations of Chemistry*, 11, 79–91.

Ruthenberg, K. (2010). "Das Kant'sche Echo in Paneths Philosophie der Chemie," *Kant-Studien*, 101(4), 465–479.

Scerri, E. (2000a). "Have orbitals really been observed?," *Journal of Chemical Education*, 77(11), 1492–1494.

Scerri, E. (2000b). "Realism, Reduction and the "Intermediate Position," In *Of Minds and Molecules: New Philosophical Perspectives on Chemistry* (N. Bushan and S. Rosenfeld, eds.). New York: Oxford University Press, 51–72.

Scerri, E. R. (2001). "The recently claimed observation of atomic orbitals and some related philosophical issues," *Philosophy of Science*, 68(3), S76–S88.

Schaffer, J. (2008a). "Truth and fundamentality: On Merricks's truth and ontology," *Philosophical Books*, 49(4), 302–316.

Schaffer, J. (2008b). "Truthmaker commitments," *Philosophical Studies*, 141(1), 7–19.

Sellars, W. (1963). *Science, Perception and Reality*. London: Routledge and Kegan Paul.

Strawson, P. F. (1975). *The Bounds of Sense: An Essay on Kant's Critique of Pure Reason*. London: Methuen.

Strevens, M. (2008). *Depth: An Account of Scientific Explanation*. Cambridge, MA: Harvard University Press.

Suppe, F. (1989). *The Semantic Conception of Theories and Scientific Realism*. Chicago: University of Illinois Press.

van Brakel, J. (2000). *Philosophy of Chemistry*. Leuven: Leuven University Press.

van der Vet, P. (1987). *The Aborted Takeover of Chemistry by Physics: A Study of the Relations between Chemistry and Physics in the Present Century*. Ph.D. thesis, University of Amsterdam.

Wimsatt, W. C. (2007). *Re-engineering Philosophy for Limited Beings: Piecewise Approximations to Reality*. Cambridge, MA: Harvard University Press.

Woodward, R. B. and R. Hoffmann (1969). "The conservation of orbital symmetry," *Angewandte chemie*, 8, 781–932.

Zuo, J. M., Kim, M., O'Keeffe, M., and Spence, J. C. H. (1999). "Direct observation of d-orbital holes and Cu-Cu bonding in Cu_2O," *Nature*, 401, 49–52.

| Divergence, Diagnostics,
and a Dichotomy of Methods

GRANT FISHER

COMPUTATIONAL MODELING IN ORGANIC chemistry employs multiple methods
of approximation and idealization. Coordinating and integrating methods can
be challenging because even if a common theoretical basis is assumed, the
computational result can depend on the choice of method. This can result in
epistemic dissent as practitioners draw incompatible inferences about the
mechanisms of organic reactions. These problems arose in the latter part of
the twentieth century as quantum chemists attempted to extend their models
and methods to the study of pericyclic reactions. The Woodward-Hoffmann
rules were introduced in the mid-1960s to rationalize and predict the energetic
requirements of a number of reactions of considerable synthetic significance.
Soon after, quantitative quantum chemical approaches developed apace. But
alternative methods of approximation yielded divergent quantitative predictions of transition state geometries and energies.

   This chapter explores the difficulties facing quantum chemists in the late
twentieth century as they attempted to construct computational models of
pericyclic reactions. Divergent model predictions resulted in the methods
used to construct computational models becoming the focus of epistemic scrutiny and dissent. The failure to achieve robust quantitative results across quantitative methods prompted practitioners to scrutinize the consequences of
pragmatic tradeoffs between computational manageability and predictive accuracy. I call the strategies employed to probe pragmatic tradeoffs *diagnostics*.
Diagnostics provides the means to probe manageability—accuracy tradeoffs
for sources of predictive divergence and to determine the reliability and applicability of approximation procedures, idealizations, and even techniques of
parametrization. Furthermore, although technological developments in computing power continues to increase, and indeed that there is now a general
consensus on the veracity of high level ab initio and density functional methods applied to pericyclic reactions, diagnostics imposes non-contingent pragmatic constraints on computational modelling. What counts as a "manageable"

model is characterized by two dimensions: *computational tractability* and *cognitive accessibility*. While the former is a contingent feature of technological development the latter is not because cognitive skills are an ineliminable feature of computational modelling in organic chemistry.

In the next section I outline the development of semi-empirical and ab initio approaches to modeling pericyclic reactions and how divergent predictions were regarded as a "dichotomy of methods." Section 3 concerns computational modeling in quantum chemistry and how the problem of divergent predictions arose in spite of significant connections between semi-empirical and ab initio methods. In section 4 I frame the dichotomy of methods in relation to robustness analysis and tradeoffs, arguing that pragmatic tradeoffs between computational tractability and predictive accuracy come to the fore as the source of divergence and dissent in computational modeling of pericyclic reactions. I then outline the methodological and epistemic significance of diagnostic strategies in section 5. I close in section 6 with some brief remarks on how cognitive accessibility and diagnostics might connect with understanding in chemistry.

1 Modeling Pericyclic Reactions

One of the most important developments in physical organic chemistry in the latter half of the twentieth century was the idea of orbital symmetry conservation in organic reactions. Introduced by Robert Burns Woodward and Roald Hoffmann in the mid-1960s, it represented a development of molecular orbital theory in qualitative form, rationalizing and unifying a number of previously unrelated organic reactions of great importance to synthetic chemistry. Pericyclic reactions, as Woodward and Hoffmann called them, included the relatively simple addition of two ethylene molecules, the Diels-Alder reaction, and the Cope rearrangement (an intramolecular rearrangement in which a σ-bond adjacent to one or more π-bonds shifts to a new position) (Hoffmann and Woodward, 1965; Woodward and Hoffmann, 1965a, 1965b, 1969).

By focusing on the relative phase symmetries of the molecular orbital wave functions corresponding to bonds broken and formed during a given reaction, Woodward and Hoffmann proposed that one could predict whether a reaction is symmetry "allowed" or "forbidden." An allowed reaction requires less energy because the symmetry of the molecular orbitals representing the bonds that break and form is conserved in the transition from reactants to products. Symmetry allowed reactions take place preferentially in a single kinetic step via a cyclical arrangement of bonds that break simultaneously or synchronously with the formation of the bonds in the product. Symmetry forbidden reactions require significant input of energy for the photochemical promotion of electrons to higher energy molecular orbitals.

Woodward and Hoffmann's relatively simple, qualitative approach has great utility. It allows theoretically inclined experimental chemists to use some of the conceptual resources of molecular orbital theory to predict the outcome of many important organic reactions. Woodward and Hoffmann's approach is "qualitative" in sense that it does not demand chemists perform complex calculations of transition state geometries and energies plotted as a reaction profile on a potential energy surface.[1] In principle, one merely has to determine whether molecular orbital symmetry is conserved in order to predict the synthesis of the desired product with minimal energetic requirements. The predictive virtues of the approach are underwritten by orbital symmetry and embodied in Woodward and Hoffmann's famous selection rules, known as the Woodward-Hoffmann rules.[2]

By the mid-1970s, Woodward and Hoffmann's idea of orbital symmetry conservation attracted considerable interest from quantum chemists seeking to construct computational quantum chemical models of the reaction energy profile—specifically the fleeting transition state structures representing an energy maxima on a potential energy surface as reactants converted to products. The Schrödinger equation was analytically insoluble for all but the very simplest of molecules. Exact solutions by a brute force approach would mean a solution to the quantum mechanical formulation of the many-body problem, which increases exponentially with the complexity of the system (numbers of electrons). Exact computations of activation energies for the Diels-Alder reaction were beyond even automated computational techniques. So computational quantum chemists had to draw on approximation procedures in order to compute the geometries and therefore the energies of the experimentally inaccessible transition states. Two approximation procedures stand out in this case: semi-empirical and ab initio ("from the beginning") methods.

Defenders of semi-empirical methods for the quantitative study of organic reaction mechanisms—most notably Michael Dewar and allies—argued that one could not solve the Schrödinger equation for complex organic molecules and their reactions from first principles of quantum mechanics due to the computational intractability of the complex target systems, inherent errors in calculations, and a lack of computing power and resources (Dewar 1984). Semi-empirical methods eased calculations and the consumption of resources because one could "take a treatment simple enough for calculations to be carried out at reasonable cost and to try to upgrade its accuracy by introduction of adjustable parameters" (Bingham et al. 1975, 1286). By "introducing parameters whose values are adjusted to experiment" (Dewar and Jie 1992, 537), one could make computations more *manageable* by undercutting the need to compute the many, complex integrals required to solve the Schrödinger equation.

---

[1] Drawing on Hoffmann, Weisberg (2004, 1071) argues that "qualitative" does not mean a lack of numbers—it concerns "degrees of approximation and idealization." Orbital symmetry models correspond to Weisberg's idea of "qualitative" since they depend in part on quantitative molecular orbital methods and because of their highly idealized character. But they are also qualitative in the sense that they generate non-numerical predictions.

[2] To give one example, the Diels-Alder reaction of ethene and butadiene is a thermally allowed $4 + 2$ cycloaddition. The selection rule in this case is: $m + n = 4q + 2$, where m and n are numbers of π-electrons, and q is an integer 0, 1, 2,... (Hoffmann and Woodward 1965).

At least some (sometimes all) of the terms in the expressions for total energy (coulombic interactions, electron-core attractions, core-core repulsions, one-center exchange or resonance terms, and the two-center resonance terms) were set equal to parametric functions and at least some of these (again, perhaps all) were adjusted to fit experimental data (Bingham et al 1975, 1286). Geometric parameters such as bond lengths, bond angles, and dihedral angles were chosen and their precision iteratively upgraded by referring to experimental data from spectroscopy, the known properties of ground state molecules, and antecedent computations. The method was an attempt to balance the manageability of computations with the desire for predictive accuracy. One needed solvable computations while retaining respectably accurate predictions.

By contrast, ab initio methods constitute computational procedures used to construct computational models approximately solving the Schrödinger equation without parameterization—without replacing integrals with parameters adjusted to fit experimental data or antecedent computations. Whatever is learned about the chemical system of interest, it is supposed to be established solely by constructing a computational model in which an approximate solution to the Schrödinger equation can be performed from first principles. Ab initio rigor lies in its methodological and epistemological "purity." One assumes as little as possible about the target system of interest, thereby approximately solving the Schrödinger equation without, or more realistically, by minimizing the use of experimental data.

From the mid-1970s, ab initio calculations of the properties of pericyclic reactions tended to predict transition state geometries, activation energies, and bond lengths consistent with a synchronous reaction taking place in a single kinetic step. These single step or *concerted* reactions were precisely the kind of reactions qualitatively predicted by Woodward and Hoffmann (Townshend et al. 1976). Semi-empirical calculations, on the other hand, predicted the energetic favorability of two alternative mechanisms. The first of these mechanisms is the most extreme relative to ab initio methods: Pericyclic reactions could take place in two distinct kinetic steps via an unstable transition state, followed by a relatively stable diradical or "diradical-like" intermediate structure, represented by a second peak on the potential energy surface. This *non-concerted* mechanism conflicted with the Woodward-Hoffmann rules and ab initio predictions. Less extreme was another possible reaction pathway: A pericyclic reactions might be concerted but nonetheless bond-breaking and formation could still be *asynchronous*. Dewar called this a "two-*stage*" mechanism because while it took place in a *single* kinetic step, the transition state was unsymmetrical. In other words, a pericyclic reaction may be concerted and yet still conflict with ab initio and orbital symmetry predictions since bonds do not break and form synchronously. Hence there were three proposed mechanisms for the Diels-Alder reaction: a concerted synchronous mechanism predicted by Woodward-Hoffmann and ab initio methods, a concerted and *asynchronous* mechanism predicted by semi-empirical methods, and a non-concerted mechanism also predicted by semi-empirical methods. Dewar provocatively argued that pericyclic reactions could not take place preferentially via a synchronous process, proposing a counter-rule opposing Woodward and Hoffmann: "*synchronous multibond mechanisms are normally prohibited*" (Dewar 1984, 211; original emphasis).

The divergence of semi-empirical and ab initio predictions of pericyclic reaction mechanisms was sometimes referred to as a "dichotomy of methods" by defenders of ab initio approaches (Caramella et al. 1977; Houk et al. 1995). But tensions between approximation methods in quantum chemistry were nothing new. According to Buhm Soon Park (2003), a schism had emerged by the late 1950s between different communities of quantum chemists employing semi-empirical and ab initio methods to study the properties of molecular structure and bonding. The development of digital computing had made it feasible to use ab initio approximations to study relatively simple molecules. The pencil-and-paper approach of semi-empirical calculation were in marked contrast to ab initio approaches not only technologically, but also methodologically and even in terms of ontological commitments (Park 2003). As Kostas Gavroglu and Ana Simões (2012) point out, tensions between practitioners employing plural computational techniques must also be considered in light of the positive methodological and epistemological contribution multiple quantitative approaches have made to quantum chemistry. By the late 1970s, productive tensions arose in the quantum chemistry of pericyclic reactions.

Divergent predictions generated puzzles for the development of computational models of pericyclic reactions. For one thing, both methods were deployed within the theoretical framework of molecular orbital theory. Furthermore, Dewar's semi-empirical methods were a simplification of the ab initio Hartree-Fock self-consistent field (SCF) approach (Lewars 2003, 340).[3] The "simplifications" employed, and the role of parameterization in semi-empirical methods, mark the points of departure from ab initio methods. Semi-empirical SCF approaches, unlike ab initio methods, treat only the valence electrons which are conceived as moving within a force field of atomic nuclei. This results in smaller basis sets—Slater functions—considered to be more accurate than ab initio basis functions (Gaussian functions) because they have been parameterized. On the other hand, ab initio methods are not restricted to minimal basis sets, and the Gaussian functions can be computed more quickly for electron-electron repulsion two-electron integrals. Dewar's semi-empirical approach and the ab initio study of organic reactions both make use of the linear combination of atomic orbital approximation, but differ markedly in how they treat core and electron repulsion integrals. Semi-empirical SCF methods tend to ignore many of these integrals (i.e., they are taken as zero) so that they are not calculated from an explicit Hamiltonian or basis functions. The integrals used are evaluated experimentally.

The Dewar group's development of semi-empirical approaches marks the development of different choices concerning what overlap integrals to ignore.

---

[3] The following draws from Lewars (2003).

One important feature of the Dewar group's various approaches as it developed into and beyond the late 1970s concerned the extent to which differential overlap is neglected and how the zero differential overlap approximation was applied to atomic orbitals on different atoms as opposed to orbitals on the same atom. Although this wasn't employed exclusively during the Dewar group's career, by ignoring all of the three- and four-center two-electron repulsion integrals and thereby cutting down on the number of two-electron integrals requiring calculation, the neglect of diatomic differential overlap approximation is an important feature of the Dewar group's methods.

Dewar's provocative claims concerning the veracity of semi-empirical methods in comparison to ab initio methods, and his "multi-bond rule," served as an impetus to the development of ab initio methods in the study of pericyclic reactions like the Diels-Alder reaction and the Cope rearrangement. There are a plethora of approximation procedures even within ab initio methods and these have been joined by the development of density function methods in recent years.[4] Discrepancies in results between semi-empirical and ab initio methods are often matched by discrepancies in the computed activation energies of the concerted process *within* ab initio methods, leading one author to describe "how the concerted activation barrier [of the Diels-Alder reaction] depends on the computational method" (Bachrach 2014, 198). Lower-level ab initio computations of Diels-Alder activation barriers are often overestimated compared with experiment (sometime drastically) because they ignore electron correlations, and this can favor the non-concerted mechanism. This is not to detract from confidence in the veracity of ab initio methods as they have been developed to an increasingly sophisticated, high-level of approximation with larger basis sets as computing power has increased and costs reduced. In ab initio computational studies of the Diels-Alder reaction and the Cope rearrangement, the desire has been to match computations to experimental estimates of activation energies and to experimental and theoretical studies of secondary kinetic isotope effects in order to ascertain the synchroneity or asynchroneity of pericyclic reactions.[5] The development of density functional theories by Kendal Houk's group have perhaps instilled the greatest confidence. The general trend is that ab initio methods, broadly construed, have demonstrated the favorability of the synchronous concerted Diels-Alder reaction:

> All of the ab initio calculations that include electron correlation to some extent clearly favor the concerted pathway for the [reaction of 1,3 butadiene with ethylene]. All of these computations also identified a transition state with CS symmetry, indicating perfectly synchronous bond formation.
>
> (BACHRACH 2014, 209)

There are, however, puzzles remaining from studies of secondary kinetic isotope effects. For example, replacing 1,3-butadiene with isoprene results in

---

[4] See Lenhard (2014b) on the computational turn in quantum chemistry and the central role of density functional theory to the rise of computational quantum chemistry.
[5] The following draws from Bachrach (2014).

transition states with some asymmetry. Other reactions can indicate greater asymmetry in the transition state, for example the cyclization of acrylic acid with 2,4-pentadienoic acid. More radically, it is thought that there are some reactions (the cyclization of 1,3-butadiene with maleic acid) that have the potential to preserve the symmetry of the transition state and yet the asynchronous mechanism is actually preferred. In other cases one cannot tell the difference. For example, experimental and computed secondary kinetic isotope effects can sometimes fail to distinguish asynchronous and synchronous pathways for the reaction of isoprene with dimethyl maleate (Bachrach 2014, 214). Furthermore, some studies appear to challenge the intrinsic bias for synchroneity in some Diels-Alder reactions (Singleton et al. 2001).

One should be careful in responding to these problems and to avoid merely cherry picking anomalies. Caution is especially important given that the confidence in ab initio studies of the Diels-Alder reaction is mirrored in studies of the Cope rearrangement. My aim is not to highlight uncertainties in computational modeling of pericyclic reactions as a prelude to demonstrating their inadequacies but rather to motivate an analysis of what it takes to determine the reliability of approximation methods and the computational models constructed using those methods. Hopefully it is clear from the above just how important this issue is. Computational modelling of the Diels-Alder reaction and the Cope rearrangement begins with divergent quantitative predictions from semi-empirical and ab initio methods in spite of the fact that they share the same quantum chemical theoretical basis in molecular orbital theory and both are based on the Hartree-Fock SCF approximation procedure. But the development of semi-empirical and ab initio methods take alternative methodological routes and employ different idealizations used to make the modelling of pericyclic reactions more "manageable." Manageability is a pragmatic desideratum to be traded-off against predictive accuracy, but is realized in different ways both within and between semi-empirical and ab initio methods. I will argue below that diagnostic strategies bring these differences to the surface and that diagnostics imposes non-contingent pragmatic constraints on computational modeling. But before I discuss that in more detail, let me say a little about computational modeling in quantum chemistry.

2 Computational Models

By "computational model" I refer to a model whose construction is procedurally based (i.e., based on an approximation method) and whose function is to *render* computable algorithms for analytically intractable equations in order to permit inferences to be drawn about experimentally inaccessible transition structures. The emphasis on "rendering" is intended as a contrast to the mere application of theory. Equations must be significantly altered—for example, by replacing integrals with parameters in semi-empirical methods or by incorporating specific idealizations that comprise tradeoffs between manageability and predictive accuracy. This reprises familiar themes in general philosophy of science which has often recognized the shortcomings of covering law explanation and exclusively theory-driven accounts of model construction, approximation, and idealization. Take for example, Cartwright's (1983) critique of the generic-specific account of laws in philosophy of science. Among other things, approximations are improvements on the fundamental laws which purportedly cover their domain because the approximations incorporate phenomenological correction factors. Morrison and Morgan (1999) and Morrison (1999) propose that models in many sciences, including physics and economics, are partially autonomous of theory and phenomena in terms of their construction and function. Related ideas have been extended to ab initio quantum chemistry where tradeoffs between explanatory force and predictive power are made in computational modeling (Lenhard 2014a). Drawing on Cartwright, Robin Hendry (1998) argues that in order for quantum mechanics to be capable of application within quantum chemistry, one requires the prior specifications of molecular structure and bonding to solve the Schrödinger equation. If these approximations and idealizations are driven by concepts lying outside quantum mechanics, then there is no reason to suppose that confirmation of models in quantum chemistry confer support upon quantum mechanics (Hendry 1998).

Approximation methods (semi-empirical and ab initio) are the procedures used in the construction of computational models. Molecular orbital theory renders wave mechanics into a form capable of being used to construct computational models one can apply to a given target system. In order to do that, approximation methods are required to render the analytically intractable interpreted theoretical formalism into solvable form. These methods incorporate supra-theoretic elements that lie outside not only the wave mechanical formalism, but can also be quite specific to the interests and methodological commitments of practitioners, incorporating methodological directives, heuristics, and tradeoffs. In this respect and in others, computational models in quantum chemistry share significant features with simulations. According to one influential view (Winsberg 2003, 2010), simulations require computational methods to transform analytically intractable equations by breaking continuous differential equations into discrete algebraic form, rendering computable algorithms describing the dynamics of a given target system. Although motivated by theoretical considerations, practitioners transform theoretical structures by incorporating approximations and idealizations drawn from a number of sources independent of theory (Winsberg 2003, 108–109).

Although the motivational, functional, and transformative characteristics of simulations are similar to computational models in quantum chemistry, whether semi-empirical and ab initio computational models constitute simulations *simpliciter* requires a more dedicated treatment than I can provide here. But I would resist the idea that computational models, unlike simulations, produce only numerical predictions. Winsberg argues that simulations are distinctive because they employ the methodological tools of experimental science to run data analyses and to draw inferences from that data (often in the form of realistic images), new knowledge, and are not be construed as merely producing numerical predictions because they "provide information about systems for which previous experimental data is scarce" (Winsberg 2003, 111–112). Sometimes chemists talk of computers as the focus for "experimentation." For example, Dewar and co-workers foresaw using computers "as the chemical instruments in a new kind of 'experimental' technique to be used by chemists on the same basis as infrared or NMR spectroscopy" (Bingham et al. 1975, 1285). These "computational experiments" might be read as going beyond mere number crunching. Instead, the processing of computer algorithms might be thought of as an "experiment" on a digital computer in order to generate "data."[6] Another function that can be attributed to computational models of pericyclic reactions—using both semi-empirical and ab initio methods—is to permit chemists to draw model-based inferences about experimentally inaccessible transition states and the dynamics of pericyclic reactions. These inferences are drawn from computations of bond lengths, bond angles, and other properties. Using transition state theory, chemists can then diagrammatically represent the geometries, energies, and the relative timing of bond breaking and formation in transition structures as the peaks and troughs on a two-dimensional graph representation of potential energy against time.

Aside from what computational models in quantum chemistry might share with simulations, it is important to consider how the use of elements from outside the theoretical formalism in computational model construction results in a somewhat fuzzy distinction between ab initio and semi-empirical methods. Trivially, semi-empirical methods are as much "bottom-up" as "top-down" because parameterization and hence computation is driven in part by the experiment. But computational models using both semi-empirical and ab initio methods facilitate the drawing of inferences about the properties of transition structures that would otherwise be difficult to access experimentally. Furthermore, ab initio methods incorporate elements external to the quantum mechanical formalism because they cannot be performed literally from first principles of quantum mechanics without *any* reference to experimental data. While offering opposing ideas of rigor in quantum chemistry, Jeffery Ramsey (1997, 2000) and Eric Scerri (2004a, 2004b) argue that there is only a difference of degree rather than kind between ab initio and semi-empirical methods. Ramsey (2000, 552 n. 4; 562) claims that while chemists have interpreted the difference as a difference of kind, this is a matter of historical contingency. There are multiple "styles" of "theory production and articulation" and the appropriateness of a method is contextual, depending on the "kinds of problems theorists confront and traditions of theory structure and problem solving within a given discipline" (Ramsey 2000, 550–551). Scerri (2004a) takes the line that while methods are "inextricably linked" (1086), semi-empirical methods tend

---

[6] On the relationship between simulations and experiments in general philosophy of science, see for example Hughes (1999), Winsberg (2010), Parker (2009), and Morrison (2009).

to get better results "only because we have built in a chunk of nature," but the "ultimate theoretical goal is to try to model nature as closely as possible without getting one's theoretical hands too dirty" (1094).

That semi-empirical and ab initio methods are inextricably linked is true not just because in practice one cannot compute properties of molecular structure, bonding, and dynamics without reference to any experimental data. It is also true because modern semi-empirical and ab initio computational models are developments of the Hartree-Fock SCF procedure. This might seem to pose a problem for the veracity of the claim that there was such a thing as a "dichotomy of methods" in computational chemistry. But one does not need a difference in kind to distinguish approximation methods. When confronted with cases where the act of making a distinction necessarily results in a degree of arbitrariness, one can nonetheless seek relatively clear-cut cases. Predictive divergence in the quantum chemistry of pericyclic reactions is such a case. A constitutive methodological difference between approximation procedures is established empirically by their conflicting results. The trouble is, this does little to determine the *epistemological* significance of divergent results among multiple models. Predictive divergence might merely be an artefact of the approximation methods that perhaps highlights the pragmatic limitations on computations in quantum chemistry, but this need not be an epistemological limitation.[7] Given practical constraints on computations in terms of computational power and resources, and different strategies of approximation and idealization used to overcome those constraints, it is little wonder that the results can sometimes conflict. But there remain methodologically and epistemologically significant issues concerning how the choice of approximation method can impact on results and how chemists respond to conflicting results.

3 Robustness and Tradeoffs

One way to think of the dichotomy of methods and its philosophical significance is as a situation in which practitioners sought *robust* results among plural models but were instead confronted with a significant case of predictive divergence. I want to frame the dichotomy of methods in this way so that I can analyze the implications of conflicting results, to motivate an idea of pragmatic tradeoffs later in this section, and to then discuss the methodological and epistemic significance of diagnostic strategies in the section to follow.

Let me briefly outline the idea of robustness analysis. It was originally introduced by Richard Levins (1966) to express a characteristic feature of quantitative modeling in population biology, although it is now recognized that robustness and multiple strategies of idealization (tradeoffs) are not limited to

---

[7] Related concerns arise in distinguishing genuine cases of inconsistency when plural models are used to represent the same target system, from plural but complementary models used to represent different aspects of the same target system (see Morrison 2011).

a single scientific discipline.[8] A "robust theorem" is a proposition expressing a common numerical result entailed by a set of independent models employing different tradeoffs while sharing a common structure:

> "…we attempt to treat the same problem with several alternative models each with different simplifications but with a common biological assumption. Then, if these models, despite their different assumptions, lead to similar results we have what we can call a robust theorem which is relatively free of the details of the model. Hence our truth is the intersection of independent lies.
>
> (LEVINS 1966, 423)

In spite of making different idealizations, we can at least in principle determine what structure common to independent models is responsible for convergent results. Building on Levins' ideas, William Wimsatt (1981) introduced the term "robustness analysis" to describe the determination of robust theorems. Michael Weisberg (2006a, 738; 2013, 159) argues that robust theorems have the form of a conditional hypothesis (they are not "theorems" in the formal sense): "*ceteris paribus*, if [common causal structure] obtains, then [robust property] will obtain." Note that robustness analysis does not confirm robust theorems. Robustness analysis conditionally establishes a second-order relation between robust properties (convergent results) and shared causal structure. Confirmation is a first-order relation between evidence and hypotheses or models. Hence Orzack and Sober (1993, 538) argue that robustness is distinct from observational confirmation. However, the structures specified in the antecedent of robust theorems must receive what Weisberg calls "low-level confirmation": "theorists must have data that demonstrate the adequacy of the underlying modeling framework for representing the model's causal consequences" (Weisberg 2013, 169).

Levins' appeal to the epistemological virtues of robustness analysis underwrites the use of multiple idealized models for studying complex target systems where modelers cannot satisfy various modeling desiderata simultaneously. He opposed the idea (the brute force approach) that mathematical models can render faithful representations of complex systems in population biology because it "would require using perhaps 100 simultaneous partial differential equations with time lags, measuring hundreds of parameters, solving the equations to get numerical predictions, and then measuring these predictions against nature" (Levins 1966, 421). Modelers will therefore need to "trade off" specific desiderata in plural idealizing strategies according to their particular modeling aims.[9] These tradeoffs can occur between precision, generality, and

---

[8] For example, Axel Gelfert (2012) argues that tradeoffs are a feature of modelling in condensed matter physics. Weisberg (2008) uses a form of robustness analysis in his critical account of the structural conception of the chemical bonding, while Weisberg (2004) defends the idea of precision-generality tradeoffs for qualitative models in chemistry. A recent volume dedicated to robustness (Sola et al. 2012) includes contributions addressing a range of scientific disciplines including protein chemistry (Wieber 2012).

[9] Although Levins did not use the term "tradeoff" himself (Weisberg 2006b, 636).

realism resulting in three distinct idealizing strategies: Sacrifice generality to realism and precision (Type I), sacrifice realism to generality and precision (Type II), sacrifice precision to realism and generality (Type III).

The significance of robustness analysis to quantum organic chemistry is that while different communities of quantum chemists defended the methodological legitimacy of their respective approximation methods, they nonetheless expected multiple models to instantiate a robust property in the presence of a shared theoretical structure and the Hartree-Fock SCF approach. The failure to establish robust theorems was the source of the dichotomy of methods controversy and perhaps highlights the difficulty of establishing robust theorems in practice. Jacob Stegenga (2009, 2012) claims that "discordant" results are perhaps ubiquitous in the sciences due to the difficulties in generating evidence from independent "techniques," integrating "multi-modal evidence" which are by definition of different kinds, and because it is often difficult to determine the relevance of that evidence. The present case imposes additional constraints on the epistemological merits of robustness analysis. In spite of the absence of agreed standards of approximation, both ab initio and semi-empirical computational approaches to pericyclic reactions were cast within the well-confirmed modeling framework of molecular orbital theory. Given that the veracity of this framework was accepted by the quantum chemists involved in the controversy (its low-level confirmation), and the presence of a shared basis in the Hartree-Fock SCF approximation procedure, the failure to establish robust theorems seems to indicate that the instantiation of common theoretical or causal structure belonging to a well-confirmed theoretical framework *and* at least some shared mathematical techniques need not imply a robust property.

But this may not be a serious drawback to robustness analysis. After all, robust theorems are subject to a *ceteris paribus* clause. As Weisberg (2013, 159) points out, part of what is involved in robustness analysis is a matter of probing the limits of robustness in a number of ways. What I want to do here is to suggest that one can learn about practice in computational chemistry and the limits of robustness analysis by looking to the tradeoffs themselves in order to determine whether the non-robustness of plural computational models raises serious epistemological difficulties or whether they are contingent upon a lack of knowledge and available computational resources. This will mean looking to the *pragmatic tradeoffs* made in computational modelling which get rather less press than tradeoffs between precision, generality, and realism.

There are good reasons why the pragmatics of idealization strategies in modeling might seem relatively humdrum compared to establishing the logical import of tradeoffs between precision, generality, and realism. Much of the focus on tradeoffs in the philosophical literature has been to challenge or defend their logical necessity. Orzack and Sober (1993) propose that if one attends only to the logical or semantical relations between precision, generality, and realism, they do not trade off. Even if tradeoffs are instantiated in scientific modeling, they are not logically necessary. However, Matthewson and Weisberg (2009) propose an account of tradeoffs between precision and generality that effectively undercuts the idea that they are merely a contingent feature of scientific practice. They argue that theory application requires using equations (model descriptions) to pick out classes of models and investigation of the fit between theory and the world via those models. Precision is a property of model descriptions—a "comparative test" of "the fineness of specification of the model description's parameters" (Matthewson and Weisberg 2009, 178) such that a more precise model description will pick out a proper subset of mathematical models picked out by a less precise model description. Generality is a matter of the relationship between models and the world—the number of logically possible or actual target systems a set of models applies to. The most serious kind of tradeoffs are "strict tradeoffs": one cannot increase precision without decreasing the number of logically possible target systems a set of models apply to. But precision will trade off against generality in a weaker sense when one considers the number of *actual* target systems. Increasing precision will result in modelers' inability to simultaneously *increase* the number of actual target systems a set of models apply to ("increase tradeoffs"). So the logical constraints on tradeoffs between precision and generality may differ depending on whether one is concerned with the number of logically possible target systems and the actual target systems a set of models applies to. But in neither case are tradeoffs contingent.

There is a sense in which all tradeoffs are "pragmatic" because human agents cannot simultaneously increase all theoretical desiderata in practice. But there are differing commitments in the literature to the significance of pragmatics. For Matthewson and Weisberg, tradeoffs are not merely a matter of human limitations. Compare strict and increase tradeoffs to what they call "simple attenuations," where increasing the magnitude of one attribute makes increasing the other difficult but not impossible (Matthewson and Weisberg 2009, 170). More computing power might overcome such attenuations, say, but at considerable cost. Simple attenuations—pragmatic tradeoffs—are therefore passed over quickly.[10] Others attribute a more central role to pragmatics. In response to Orzack and Sober's (op. cit.) attempts to deflate trade-offs *simpliciter*, Odenbaugh (2003, 1504) makes the case that even if one does find that theoretical desiderata do not trade off when one attends to the logical or semantical relations between them, "pragmatic conflicts in model building could still be present." Levins' concern, Odenbaugh stresses, was with the complexity of nature and the cognitive limitations of modelers to comprehend it, which imposes pragmatic but "necessary" constraints on tradeoffs.[11]

Of note here is how Odenbaugh's defense of a pragmatic interpretation of Levins' strategies of idealization concerns the idea that manageable models is a

---

[10] As Weisberg points out, tradeoffs between "theoretically important desiderata in a particular domain" may not be contingent and hence what he calls the "multiple-models idealization" is not something that can be expected to abate with scientific progress (Weisberg 2013, 105). And while the motivations for using multiple models making different tradeoffs can sometimes be pragmatic, often they are not.

[11] See also Odenbaugh (2006).

pragmatic goal of tradeoffs between precision, generality, and realism. On the other hand, Weisberg (2006b, 634) refers briefly to manageability noting that while Levins refers to it (Levins 1966, 422) as a desideratum of modeling in itself to be traded off against other theoretical desiderata (generality, precision, and realism), Levins neither discusses it in any detail nor does he define "manageability" (nor the other theoretical desiderata). There appear to be two uses of the term "manageability": as a pragmatic *goal* of tradeoffs, and as a *desideratum* in itself to be retained or traded off as a *means* to multiple modeling goals. Both interpretations of "manageability" are important, but I want to explore the latter in this chapter as a *pragmatic* desideratum because of its historical importance to quantum chemistry. Park (2009) notes that tradeoffs between the manageability of approximations and idealizations on the one hand, and the predictive accuracy on the other, are a central feature in the history of ab initio computational methods in quantum chemistry. One might add that they are just as important to semi-empirical computational modelling. Although manageability—accuracy trade-offs might appear to be relatively trivial contingent "simple attenuations," in quantum chemistry they can have significant methodological and epistemological implications. Moreover, the strategies adopted to probe the effects of pragmatic trade-offs itself imposes what I consider to be additional pragmatic but non-contingent constraints on computational modeling. Even if improvements in computational resources can mitigate the effects of some aspects of pragmatic tradeoffs between manageability and accuracy, there are features of practice that are nonetheless bound by agents' cognitive abilities.

4 Diagnostics

The methodological and epistemological significance of manageability—accuracy tradeoffs in quantum organic chemistry resides in how the choice of idealizations and approximation procedures determines the computational result. When methods and models lead to divergent results, one requires a means to understand how idealizations and approximation procedures impact on one's quantitative predictions and ascertain the epistemic significance of that divergence. I call the inquiries into the effects of manageable models on predictive accuracy *diagnostics*. One can think of diagnostics as, in a sense, the deconstruction of computational models to locate those aspects of approximation methods and idealizations that entail *non*-robust theorems.

At first blush, one might assume modelers can easily identify the features used to construct models and then figure out why results either agree or do not agree. But this is not necessarily the case. As Paul Humphreys (2004, 2009) has argued, the development of computational processes makes it impossible for cognitive agents to possess knowledge of, and justify, all the epistemologically relevant parts of the computational process. Computational science is "epistemically opaque," or as Humphreys (2009 618, n. 5) has more recently suggested, it may be described as partially epistemic opaque. In "hybrid scenarios" where cognitive agents and computational processes are engaged, one cannot dispense with the cognitive abilities of human agents because computational science "must balance the needs of the computational tools and the human consumers" (Humphreys 2009, 617).

These hybrid scenarios in quantum organic chemistry call for an analysis of pragmatic tradeoffs between the accuracy of computational predictions and the manageability of computational models where "manageability" balances *computational tractability* and *cognitive accessibility*. A manageable model is not merely one that can produce computational results given extant technological resources. It also has to be cognitively accessible. A manageable model is one that allows practitioners to work back from the automated computational result to flesh out computational gray-boxes. The consequences of approximations and idealizations used to render tractable computations given available computational resources must be spelled out. One needs a way to determine how to interpret the computational results in light of the effects idealizations have on the accuracy and hence the reliability of those results. This is the aim of diagnostics. It renders automated computational modeling cognitively accessible by bringing to the surface the ways in which different approximations and idealizations used in ab initio and semi-empirical methods contribute to the prediction of incompatible mechanisms. Diagnostics is built into the practice of computational modeling, which becomes especially prominent in cases of epistemic dissent.

Diagnostics move cognitive agents from a position of partial epistemic opacity with respect to the computational processes that produce numerical results to a position in which models are cognitive accessible in the sense that they are open to critical scrutiny, defense, and amendment. It would therefore be an oversimplification to call manageability "contingent" upon technological resources, including temporal restrictions associated with the existing speed of computations, financial constraints, and so on. Manageability entails a constraint concerning cognitive access to the sources of predictive divergence in order to flesh out the reasons why the methods used to compute the result should or should not be considered reliable and applicable to a given target system. Even if computational tractability can change with technological developments in computing such that one can attempt to increase predictive accuracy as one does in ab initio modelling, practitioners will still need to flesh out the means used to generate the results.

One way of thinking of diagnostics is by comparison to what Ramsey (1992) calls an "expanded epistemology of approximation." Ramsey argues that the practice of approximation comprises various routes to assessing approximation validity which cannot be reduced to the "similarity of outputs of approximation to antecedent theoretical or experimental results" because "the deviation or consilience of two results tells us nothing about the worth or reliability of the procedure which produced the results" (Ramsey 1992, 158). This motivates Ramsey's expanded epistemic criteria of approximation veracity: To rule out consilience as a consequence of a fortuitous cancellation of errors, to ensure that our approximations are justified, and that they are "controllable."

The inadequacies of the similarity view of approximation are particularly salient when one considers the dichotomy of methods debate because the similarity view assumes the availability of some antecedent computational model or experimental data against one might determine that results are robust. There *are* standards available for comparison. For example, comparison of computational results with experimental estimates of activation energies in the Diels-Alder reaction and the Cope rearrangement, and both experimental and theoretical determinations of secondary kinetic isotope effects. But these do not exhaust the strategies quantum chemists employed to diagnose the sources of conflicting results. Nor, as we have seen, are they necessarily unproblematic in themselves. Indeed, if the characterization of computational studies of pericyclic reactions earlier in this chapter is correct, the veracity of computational results is established by a coordination of plural methods, models, and experimental and theoretical approaches to producing independent evidence. One could add that the *failure* to achieve convergent predictions, or predictions which do not match some antecede result, need not invalidate the veracity of the method. Even if semi-empirical predictions do not match experimental estimates of pericyclic reaction activation energies, this does not mean semi-empirical methods are untrustworthy in other applications. Similarly, particular strategies of idealization within ab initio methods are regarded as more or less suitable to the computational study of specific pericyclic reactions, but again this need not undermine their applicability to other contexts.

What Ramsey calls the "controllability" of approximations and idealizations is a particularly important, I argue, because in computational modeling it is dependent on the cognitive accessibility of computational models borne out by diagnostic strategies. The idea of controllability originates with Ronald Laymon (1983, 1987) and Ramsey (1992) adopts it as a desideratum for approximation validity. Although it is not always explicitly defined, perhaps the sense in which an approximation strategy is "controllable" concerns our being able to give an account of the effect of counterfactual initial conditions (idealizations) on the accuracy of predictions such that we are able to seek procedures to "improve" upon those approximations in order to improve the predictions. This idea of improvement is central to Laymon's account of confirmation. If one can relax the counterfactuality of the initial conditions, then the predictions become more accurate. In other words, good theories are "monotonic toward truth": A theory is better confirmed if better approximations lead to better predictions; disconfirmed if it does not lead to better approximations (Laymon 1987, 211). Monotonicity is therefore dependent on controllability.

In computational organic chemistry, an idea of "controllability" might also be developed but as I will elaborate below, it is not one that is oriented toward theory confirmation. Rather, it is a diagnostic strategy used to determine the consequences for predictive accuracy when practitioners employ counterfactual initial conditions in approximation procedures that contribute to computational tractability. An approximation procedure P is controllable when for counterfactual initial conditions $I_{1...n}$, one knows how these specific distortions will affect predictive accuracy E given that there are no actual target systems in which the distortions are realized.[12] For some counterfactual limiting case, a non-actual target system, we trivially know the approximation procedure would have been reliable had the target system realized those distortions. But we want to know what effect (if any) those counterfactual initial conditions would have in actual cases. We want to *know how* we get the result that we do get based on the *knowledge* we possess about the effects our approximations and idealizations will have when computing the magnitudes of the properties of actual target systems. That is what diagnostic strategies are for. It is a matter of figuring out how computationally tractable models will trade off against predictive accuracy in actual cases based on what we know about the idealizations we use in constructing our models.

Pragmatic tradeoffs between manageability and accuracy are coordinated with the theoretical desideratum of "realism"—to be understood as the extent of the counterfactual initial conditions in computational modeling (their "counterfactuality" to borrow a term from Laymon).[13] In making manageability–accuracy tradeoffs, one has to determine the extent to which approximations and idealizations used in order to make equations computational tractable undermines realism by determining their effect on predictive accuracy. That will mean establishing the controllability of approximations and idealizations via diagnostic strategies which render computational models cognitively accessible. Models can be more or less "realistic," but again that requires practitioners can determine the effect of counterfactual initial conditions on predictive accuracy.

The controllability of approximations and idealizations is a central diagnostic strategy in the dichotomy of methods controversy. For example, by the late 1970s when Houk and allies noted "inherent differences in calculational techniques" (Caramella et al. 1977, 4512), it was essential to determine the controllability of the neglect of differential overlap (NDO) approximation used in semi-empirical computations. Recall that this approximation ignores the repulsive effects of two electrons occupying the same atomic orbital, or the repulsive effects of electrons occupying atomic orbitals belonging to different atoms. The latter approximation (neglect of diatomic differential overlap) ignores all multicenter two-electron integrals, which cuts down on the number of two-electron integrals requiring calculation (Lewars 2003, 354). In either

---

[12] I take it that this is similar to Ramsey's sense of controllability (Ramsey 1992, 159). See also Ramsey (1990), in which he draws on chemical kinetics in order to argue that additional justification criteria for theories over and above numerical and causal accuracy, like "range of validity" and intelligibility, are called for when the truth-values of the theory's propositions are not known, the theories suffer computational difficulties, or data is unstable.

[13] Although predictive accuracy can tend to be regarded as falling under the desideratum of "realism," the term is complex because Levins (1966, 1993) appears to use it in more than one sense: as predictive accuracy, or an expression of representational success (Weisberg 2006b, 635). It can also be interpreted as the number of independent variables models take into account (Orzack and Sober 1993). For the purposes of this chapter, when I use the term "predictive-accuracy" I do not wish to conflate Levins' expansive term "realism."

case, while the extent to which the zero differential overlap approximation is employed and hence the extent to which one ignores overlap integrals varies in semi-empirical methods, the aim is to cut down on the number of electron repulsion integrals requiring computation, thereby increasing computational tractability. Dewar's approach parameterized these integrals. The Houk group's diagnosis was that "the 'closed shell' (exchange) interactions appear to be a differentiating factor" such that NDO tended to favor predictions of a two-step (non-concerted) mechanism (Lewars 2003). But in the mid-1990s bias could not be excluded from ab initio methods either. Ab initio calculations take differential overlap into account, "and closed-shell repulsion which results from this overlap, favors synchronous transition states"—in other words, a synchronous, concerted mechanism (Houk et al. 1995 86).

Another diagnostic strategy was to determine the effect of omitting electron correlation energies in ab initio methods. Since electrons repel one another, taking these repulsions into account should ensure a reduction in energy of the transition state for the Diels-Alder reaction. Dewar argued that even by the early 1990s, correlation energy calculations in ab initio methods were only feasible for small molecules, but omitting correlation energies was a pragmatic tradeoff too far. Ab initio methods at the time were hamstrung because if they took electron correlations into account, they were rendered inapplicable to target systems of chemical interest (Dewar, cited by Houk et al. 1995, 86; Dewar and Jie 1992, 538). Houk and colleagues responded that inclusion of electron correlations in semi-empirical methods merely biased results toward alternative mechanisms by lowering the energy and thereby increasing the stability of the intermediates crucial to an alternative two-step reaction mechanism defended by the "anti-abinitioists" (Houk et al. 1995, 86).

These exchanges between rival communities of computational chemists might sound negative but there are many positive aspects to epistemic dissent resulting from diagnostics.[14] Within the controversy of methods, modelers had to determine the controllability of approximations and idealizations. In so doing, they improved knowledge of the target systems, knowledge of the models used to represent them, and knowledge of the methods and idealizations used to construct the models themselves. And diagnostics is also crucial to determining the legitimacy of parameterization in semi-empirical methods. It requires practitioners to make explicit the strategies employed to upgrade the predictive accuracy of semi-empirical computational models. For example, semi-empirical methods could be accused of being ad hoc if selecting parameters and upgrading predictive accuracy is carried out on a case-by-case basis. However, relatively early in the development of the Dewar group's semi-empirical modeling, constraints were proposed on legitimate parameterization approaches. For example, one was not to parameterize one case at a time but rather "the same type of function should be used for each energy term, regardless of the atoms involved" (Bingham et al. 1975, 1287). In other words, semi-empirical

---

[14] On the epistemic value of dissent, see de Cruz and De Smedt (2013).

approximations were supposed to make use of a small number of parameters applied to a diverse type of molecules or molecular properties, rather than merely choosing parameters depending on a specific target system. Semi-empirical modelers had to choose parameters relevant to the calculation of both molecular energies and geometries (bond length, bond angles, and dihedral angles), keeping the number of parameters small so that the calculations were manageable. Usually two parameters were to be used for each pair of atoms. Computational manageability is not merely a function of computational tractability and the controllability of approximations and idealizations. It is also a function of the choice of parameters and the ability to determine the implications of that choice. In semi-empirical computational modeling, neither the choice of parameters nor the reduction in their numbers are ad hoc because the parameters (bond lengths and bond angles) constitute a *minimal core set of parameters* necessary to semi-empirically compute the geometries and energies of organic reaction transition states. These kinds of strategies are a part of diagnostics and it demonstrates that not only must diagnostic inquiries attend to approximations and idealizations but also to the laborious task of parameterization itself.

A crucial feature of diagnostic controllability, which distinguishes the strategy from Laymon's idea of monotonic improvement, lies in its aims and what one can legitimately infer from determining the controllability of approximations, idealizations, and parameterizations. For Laymon, monotonic improvement is an indicator of truth. By introducing better approximations and by de-idealization—in other words increasing "realism"—better predictions result because those predictions are rendered increasingly more accurate by relaxing the counterfactual initial conditions. One issue concerns how Laymon's account of de-idealization focuses on monotonic improvements driven by *theory.* Hendry (1998) argues that in quantum chemistry the improvements (de-idealizations) are driven not by quantum mechanics but by classical chemical concepts. Hence, confirmation cannot be conferred upon quantum mechanics but is local to quantum chemical models instead. Let me put that particular issue aside and note that diagnostics in computational chemistry is not concerned with the truth of theories or the warrant one might have for a model, but is instead a means to determine the *reliability* of approximation methods and the idealizations used in computational modeling, and their *applicability* to specific target systems (organic reactions). This is all the more important when methods and models are in dispute and demonstrates how the practice of quantum organic chemistry cannot be reduce to the testing of theory or even the assessment of computational models.[15] Furthermore, "better" approximations and idealizations do not exhaust the strategies used to

---

[15] The reliability of methods or techniques used to construct models—specifically simulation models—has been discussed by Winsberg (2006). Simulation models can be sanctioned by the techniques or methods used to construct the simulated model to render tractable computations. The case discussed in the present chapter concerns computational models where the reliability and even the applicability of quantitative methods or "techniques" can be contested.

increase predictive accuracy. While the development of semi-empirical methods involves some de-idealization, their methodological feature is not continual improvements gained by relaxing counterfactual initial conditions. Semi-empirical methods can partially circumvent a process of de-idealization by parameterization and iterative upgrades in predictive accuracy by drawing on experimental data.

5 *Applicability and Intelligibility*

Methodological justifications such as those adopted in defense of semi-empirical methods are another aspect of diagnostics and highlight how these methods were taken to pose a serious alternative to ab initio methods in computational studies of pericyclic reactions until at least the mid-1990s. Another constructive epistemic outcome of diagnostics lies in how semi-empirical predictions of stable intermediates rather than single transition states drove the development of higher level ab initio methods in the study of pericyclic reactions such as the Cope rearrangement (Bachrach 2014, 219). And while it is fair to conclude that current computational chemistry favors high level ab initio and density functional approaches to pericyclic reactions, and that they support synchronous concerted mechanisms for the Diels-Alder and Cope rearrangements, diagnostics continues to be an essential feature of practice. As Bachrach notes: "Perhaps the most significant result of [ab initio] computational studies is in clarifying the decision concerning which is the appropriate methodology for investigating pericyclic reactions" (25).

For example, it is thought that one must take multiple electron configurations into account in modeling pericyclic reactions and use multiple configuration self-consistent field approaches. However, the reliability of these procedures has to be considered with regard to their strengths and weaknesses via diagnostics. Methods considered reliable in one context need not be reliable in another. Computation of pericyclic reaction transition state geometries taking all dynamic electron correlations into account preferentially locate aromatic transition states. But it is extremely resource intensive. Single configuration methods have a role to play here, but studies of the Cope rearrangement using similar methods at high levels of approximation can get inconsistent results even though they might be considered amongst the most accurate of computational methods (Bachrach 2014, 221). Pragmatic tradeoffs continue to be made between manageability and accuracy. Computationally tractable methods must be compared to reliable standards at high levels of approximation, but no single ab initio or density functional method can be taken as applicable to *all* systems. The methodological choices computational chemists make depend on the system of interest and cannot be guaranteed by their success in a particular reaction (Bachrach 2014, 227). The relative advantages and disadvantages of computational methods must be continually subject to diagnostic scrutiny and, as a consequence, must be cognitively accessible to practitioners.

Cognitive accessibility imposes non-contingent constraints on computational modelling. And it raises a question concerning the importance of intelligibility in chemistry and the relationship between quantitative and qualitative models. Even before the dichotomy of methods controversy had begun, Woodward and Hoffmann proposed orbital symmetry conservation as a qualitative standard against which future developments of quantitative methods should be judged:

> In our work we have relied on the most basic ideas of molecular orbital theory – the concepts of symmetry, overlap, interaction, bonding, and the nodal structure of wave functions. The lack of numbers in our discussion is not the weakness – it is its greatest strength. Precise numerical values would have to result from some specific sequence of approximations. But an argument from first principles or symmetry, of necessity qualitative, is in fact much stronger than the deceptively authoritative numerical result. For, if the simple argument is true, then *any* approximate method, as well as the now inaccessible exact solution, must obey it.
>
> (HOFFMANN AND WOODWARD 1968, 17; original emphasis)

Elsewhere, Woodward and Hoffmann proposed that their "qualitative arguments" "provide a gauge by which any future theoretical treatment must be measured" (Hoffmann and Woodward 1970, 826). Furthermore, one of Hoffmann's ideas of understanding in chemistry is

> being able to predict qualitatively (this forces you to think before) the result of a calculation *before* that calculation is carried out. If the calculation's result differs from what your understanding gives you, well then it's time to think again, do numerical experiments until you rationalize (that's also to 'understand') the results. Until the explanation is so clear that you could kick yourself in the butt for not having seen why. But don't stop, iterate the process, go on. Understanding will build if you follow this way of analysis.
>
> (HOFFMANN 1998, 3; original emphasis)

Computational organic chemists did not appear to have explicitly endorsed Woodward and Hoffmann's view although Dewar's opponents certainly noted the support ab initio studies provided for the Woodward-Hoffmann rules (Houk et al. 1995). However, there is a connection between diagnostics and intelligibility in chemistry. Perhaps one of the reasons for Woodward and Hoffmann's confidence in orbital symmetry is that numerical results taken in isolation are not enough to guarantee the legitimacy of the methods used to get the result. Hoffmann's point concerning understanding in chemistry, which might be associated with what he elsewhere calls "horizontal understanding" that is "expressed in the concepts, definitions, and symbolic structures at the same level of complexity as the object to be understood" (Hoffmann 2012, 29), is that qualitative models convey understanding of chemical processes that quantitative procedures fail to convey even if quantitative predictions are more precise. One can use qualitative models to anticipate the behavior of target systems and then use those predictions to guide the more precise numerical predictions obtained by approximation to quantum mechanics.

Diagnostics, on the other hand, introduces a requirement to be able to provide an account of how the quantitative result was obtained—to render the results "intelligible" by fleshing out how approximations and idealizations used to render models manageable will impact on quantitative predictions in order to determine their reliability and applicability. The same applies to the procedures used to parameterize semi-empirical models. Even if one had access to unlimited computational resources, diagnostic strategies are an ineliminable feature of computational chemistry. Even if we have accurate predictions, understanding their content requires that we can unpack those results. This unpacking is inevitably a matter of cognitive skills and abilities, and although computational tractability *is* contingent upon technological development, cognitive accessibility is not because it imposes negative constraints on computational modeling. Diagnostics is intimately connected to understanding in chemistry and can arise in the strategies adopted to determine the reliability and applicability of quantitative results as much as it can arise in the qualitative anticipation of chemical phenomena.

Concluding Remarks

I have argued that diagnostics is a methodological feature of computational organic chemistry. When the results of plural computational models fail to produce robust theorems, practitioners attempt to determine the sources of divergent predictions in pragmatic tradeoffs between manageability and accuracy via diagnostic strategies. Diagnostics makes the gray-boxed computational procedures used to render computationally tractable equations cognitively accessible. Diagnostics includes the controllability of the approximation methods and idealizations and also parameterization procedures in semi-empirical modelling. Having controllable approximations and idealizations means that one can give an account of how computational tractability will impact on predictive accuracy by inquiring into precisely what effect (if any) our counterfactual initial conditions might have in actual cases. The goal is to determine the reliability and applicability of methods, not necessarily the warrant we have for models or the truth of theories. What counts as predictive accuracy cannot be determined merely by continual improvements via de-idealization. It is determined by a more complex coordination of methods, models, experimental determinations of activations energies, and both theoretical and experimental studies of kinetic isotope effects. Sometimes, monotonic improvements are bypassed to a significant extent by parameterization and iterative upgrades in accuracy relative to spectroscopic and other data in semi-empirical methods.

Diagnostics is not merely restricted to cases of epistemic dissent in quantum chemistry. It is an ongoing methodological feature of computational organic chemistry because even within a generic computational method, results are not always robust. Diagnostics is central to the investigation of the reliability and applicability of the various ab initio and density functional methods in contemporary computational studies of organic reactions. While practitioners might agree that certain ab initio and density functional methods are the correct option for investigating pericyclic reactions, the choice of specific procedures and idealizations can have non-trivial impacts on the quality and even the relevance of results with respect to specific target systems. That the computational result depends on the choice of quantitative method should therefore not be ignored because pragmatic tradeoffs are constitutive of knowledge of pericyclic reactions. While computational tractability may well be a contingent feature of technological development and available resources, cognitive skills and abilities imposes non-contingent constraints on computational organic chemistry. The cognitive access of computational models resulting from diagnostic strategies is a kind of scientific understanding. And while qualitative models might guide quantitative modeling in their anticipations of chemical phenomena, it is also important to be able deconstruct computational model results to render the means to achieve those results intelligible.

Acknowledgments

I thank Buhm Soon Park and Eric Scerri for their comments on earlier drafts of this chapter.

References

Bachrach, S. (2014). *Computational Organic Chemistry* (2 ed.). Hoboken, NJ: John Wiley & Sons.

Bingham, R.C., Dewar, M.S.J., and Lo, D.H. (1975). "Ground states of molecules. XXV. MINDO/3. An improved version of the MINDO semiempirical SCF-MO method," *Journal of the American Chemical Society,* 97(6), 1285–1293.

Caramella, P., Houk, K.N., and Domelsmith, L.N. (1977). "On the dichotomy between cycloaddition transition states calculated by semiempirical and ab initio techniques," *Journal of the American Chemical Society,* 99(13), 4511–4514.

Cartwright, N. (1983). *How the Laws of Physics Lie.* Oxford: Clarendon Press.

De Cruz, H. and De Smedt, J. (2013). "The value of epistemic disagreement in scientific practice. The case of *Homo floresiensis,*" *Studies in History and Philosophy of Science,* 44, 169–177.

Dewar, M.J.S. (1984). "Multibond reactions cannot normally be synchronous," *Journal of the American Chemical Society,* 106, 209–219.

Dewar, M.J.S. and Jie, C. (1992). "Mechanisms of pericyclic reactions: The role of quantitative theory in the study of reaction mechanisms," *Accounts of Chemical Research,* 25, 537–543.

Gavroglu, K. and Simões, A. (2012). *Neither Physics nor Chemistry—A History of Quantum Chemistry.* Cambridge, MA: MIT Press.

Gelfert, A. (2012). "Strategies of model-building in condensed matter physics: Trade-offs as a demarcation criterion between physics and biology?," *Synthese*, 190(2), 253–272.

Hendry, R.F. (1998). "Models and approximations in quantum chemistry." In *Idealization IX: Idealization in Contemporary Physics–Poznan Studies in the Philosophy of Science and the Humanities 63* (Shanks, N., ed.). Amsterdam: Rodopi, 123–142.

Hoffmann, R. (1998). "Qualitative thinking in an age of modern computational chemistry–or what Lionel Salem knows," *Journal of Molecular Structure (Theochem)*, 424, 1–6.

Hoffmann, R. (2012). "What Might Philosophy of Science Look Like if Chemists Built It?" In *Roald Hoffmann on the Philosophy, Art, and Science of Chemistry* (Kovac, J. and Weisberg, M., eds.). New York: Oxford University Press, 21–38.

Hoffmann, R., and Woodward, R.B. (1965). "Selection rules for concerted cycloaddition reactions," *Journal of the American Chemical Society*, 87(9), 2046–2048.

Hoffmann, R., and Woodward, R.B. (1968). "The conservation of orbital symmetry," *Accounts of Chemical Research*, 1, 17–22.

Hoffmann, R., and Woodward, R.B. (1970). "Orbital symmetry control of chemical reactions," *Science*, 167, 825-831.

Houk, K.N., Gonzalez, J., and Li, Y. (1995). "Pericyclic reaction transition states: Passions and punctilios, 1935–1995," *Accounts of Chemical Research*, 28, 81–91.

Hughes, R.I.G. (1999). "The Ising Model, Computer Simulation, and Universal Physics." In *Models as Mediators* (Morgan, M.S. and Morrison, M., eds.). Cambridge: Cambridge University Press, 97–145.

Humphreys, P. (2004). *Extending Ourselves: Computational Science, Empiricism, and Scientific Method.* Oxford: Oxford University Press.

Humphreys, P. (2009). "The philosophical novelty of computer simulation methods," *Synthese*, 169, 615–626.

Laymon, R. (1983). "Newton's Demonstration of Universal Gravitation and Philosophical Theories of Confirmation." In *Testing Scientific Theories—Minnesota Studies in the Philosophy of Science*, Vol. X (Earman, J., ed.). Minneapolis: University of Minnesota Press, 179–199.

Laymon, R. (1987) "Using Scott's domains to explicate the notions of approximate and idealized data," *Philosophy of Science*, 54, 194–221.

Lenhard, J. (2014a). "Autonomy and automation: Computational modeling, reduction, and explanation in quantum chemistry," *The Monist*, 97(3), 341–360.

Lenhard, J. (2014b). "Disciplines, models, and computers: The path to computational quantum chemistry," *Studies in the History and Philosophy of Science*, 48, 89–96.

Levins, R. (1966) "The strategy of model building in population biology," *American Scientist*, 54(4), 421–431.

Levins, R. (1993). "A response to Orzack and Sober: Formal analysis and the fluidity of science," *The Quarterly Review of Biology*, 68(4), 547–555.

Lewars, E.G. (2003). *Computational Chemistry: Introduction to the Theory and Applications of Molecular and Quantum Mechanics.* Dordrecht: Springer.

Matthewson, J., and Weisberg, M. (2009). "The structure of tradeoffs in model building," *Synthese*, 170, 169–190.

Morrison, M. (1999). "Models as Autonomous Agents." In *Models as Mediators* (Morgan, M.S. and Morrison, M., eds.). Cambridge: Cambridge University Press, 38–65.

Morrison, M. (2009). "Models, measurement, and computer simulation: The changing face of experimentation," *Philosophical Studies*, 143, 33–57.

Morrison, M. (2011). "One phenomenon, many models: Inconsistency and complementarity," *Studies in History and Philosophy of Science*, 42, 342–351.

Morrison, M., and Morgan, M. (1999). "Models as Mediating Instruments." In *Models as Mediators* (Morgan, M.S. and Morrison, M., eds.). Cambridge: Cambridge University Press, 10–37.

Odenbaugh, J. (2003). "Complex systems, trade-offs, and theoretical population biology: Richard Levin's 'The strategy of model building in population biology' revisited," *Philosophy of Science (Proceedings)*, 70(5), 1496–1507.

Odenbaugh, J. (2006). "The strategy of 'The strategy of model building in population biology'," *Biology and Philosophy*, 21, 607–621.

Orzack, S.H., and Sober, E. (1993). "A critical assessment of Levin's 'The strategy of model building in population biology'," *Quarterly Review of Biology*, 68(4), 533–546.

Park, B.S. (2003). "The 'hyperbola' of quantum chemistry: The changing practice and identity of a scientific discipline in the early years of electronic digital computers, 1945–65," *Annals of Science*, 60, 219–247.

Park, B.S. (2009). "Between accuracy and manageability: Computational imperatives in quantum chemistry," *Historical Studies in the Natural Sciences*, 39(1), 32–62.

Parker, W. (2009). "Does it really matter? Computer simulations, experiments, and materiality," *Synthese*, 169, 483–496.

Ramsey, J. (1990). "Beyond numerical and causal accuracy: expanding the set of justification criteria," *Philosophy of Science (Proceedings)*, 1, 485–499.

Ramsey, J. (1992). "Towards an expanded epistemology for approximations," *Philosophy of Science (Proceedings)*, 1, 154–164.

Ramsey, J. (1997). "Between the fundamental and the phenomenological: The challenge of 'semi-empirical' methods," *Philosophy of Science*, 64, 627–653.

Ramsey, J. (2000). "Of parameters and principles: Producing theory in twentieth century physics and chemistry," *Studies in History and Philosophy of Modern Physics*, 31(4), 549–567.

Scerri, E. (2004a). "Principles and parameters in physics and chemistry," *Philosophy of Science (Proceedings)*, 71(5), 1082–1094.

Scerri, E. (2004b). "Just how ab initio is ab initio quantum chemistry?," *Foundations of Chemistry*, 6, 93–116.

Singleton, D.A., Schulmeier, B.E., Hang, C, Thomas, A.A., Leung, S.W., and Merrigan, S.R. (2001). "Isotope effects and the distinction between synchronous, asynchronous, and stepwise Diels-Alder reactions," *Tetrahedron*, 57, 5149–5160.

Sola, L, Trizio E, Nickles, T., and Wimsatt W.C. (eds.). (2012). *Characterizing the Robustness of Science: After the Practice Turn in Philosophy of Science*. Dordrecht: Springer.

Stegenga, J. (2009). "Robustness, discordance, and relevance," *Philosophy of Science*, 76, 650–661.

Stegenga, J. (2012). "*Rerum Concordia Discors*: Robustness and discordant multimodal evidence." In *Characterizing the Robustness of Science: After the Practice Turn in Philosophy of Science* (Sola, L, Trizio E, Nickles,T., and Wimsatt W.C., eds.). Dordrecht: Springer, 207–226.

Townshend, R.E., Ramunni, R., Segal, G., Hehre, W.J., and Salem, L. (1976). "Organic transition states. v. the Diels-Alder reaction." *Journal of the American Chemical Society,* 98(80), 2190–2198.

Weisberg, M. (2004). "Qualitative theory and chemical explanation," *Philosophy of Science,* 71, 1071–1081.

Weisberg, M. (2006a). "Robustness analysis," *Philosophy of Science (Proceedings),* 73(5), 730–742.

Weisberg, M. (2006b). "Forty years of 'The Strategy': Levins on model building and idealization," *Biology and Philosophy,* 21, 623–645.

Weisberg, M. (2008). "Challenges to the structural conception of chemical bonding," *Philosophy of Science,* 75, 932–946.

Weisberg, M. (2013). *Similarity and Simulation: Using Models to Understand the World.* Oxford: Oxford University Press .

Wieber, F. (2012). "Multiple Means of Determination and Multiple Constraints of Construction: Robustness and Strategies for Modeling Macromolecular Objects." In *Characterizing the Robustness of Science: After the Practice Turn in Philosophy of Science* (Sola, L., Trizio E., Nickles, T., and Wimsatt W.C., eds.). Dordrecht: Springer, 267–288.

Wimsatt, W.C. (1981). "Robustness, Reliability, and Overdetermination." In *Scientific Inquiry and the Social Sciences* (Brewer, M. and Collins,B., eds.). San Francisco: Jossey-Bass, 124–163.

Winsberg, E. (2003). "Simulated experiments: Methodology for a virtual world," *Philosophy of Science,* 70(1), 105–125.

Winsberg, E. (2006). "Models of success versus the success of models," *Synthese,* 152, 1–19.

Winsberg, E. (2010). *Science in the Age of Computer Simulation.* Chicago: University of Chicago Press.

Woodward, R.B., and Hoffmann, R. (1965a). "Stereochemistry of electrocyclic reactions," *Journal of the American Chemical Society,* 87(2), 395–397.

Woodward, R.B., and Hoffmann, R. (1965b). "Selection rules for sigmatropic reactions," *Journal of the American Chemical Society,* 87(11), 2511–2513.

Woodward, R.B., and Hoffmann, R. (1969). "The conservation of orbital symmetry," *Angewandte Chemie—International Edition,* 8(11), 781–932.

CHAPTER 15 | Mathematical Chemistry, a New Discipline

GUILLERMO RESTREPO

THE AIM OF THIS chapter is to ponder and discuss the relationship between chemistry and mathematics, taking into account some early research we have performed on the subject (Restrepo and Schummer 2014; Restrepo and Villaveces 2012, 2013; Restrepo 2013). In those works we have discussed some criticism and some support throughout history regarding the relationship. We analyzed the opinions of scholars ranging from Venel and Denis Diderot (eighteenth century) to Pierre Laszlo (twentieth century), all of whom are critical of mathematical chemistry. We also analyzed opinions by Brown and Paul Dirac (nineteenth and twentieth centuries, respectively), who sought a fruitful relationship between mathematics and chemistry. We discussed Kant and his double opinion regarding such a relationship as well. (In summary, Kant initially did not consider chemistry to be a science because of its apparent lack of mathematization, an idea Kant supported in the apparent a priori background of mathematics and in the a posteriori one of chemistry. Kant's revised opinion about the relationship between mathematics and chemistry is totally different, Kant now thinks chemistry contains elements of mathematics.) We have also analyzed Comte's opinions on the necessity of mathematics for chemistry and for the advancement of the latter (Restrepo 2013).

Our work on the philosophy and history of the relationship between mathematics and chemistry is driven by the attention research on the field has gained between the 1960s and the present. A wealth of knowledge in the border between the two sciences has been generated but little attention has been paid to the philosophy and history of the subject (Restrepo and Schummer 2012). Thus in recent years scholars (Balaban 2005, 2013; Basak 2013; Deltete 2012; Gavroglu and Simões 2012; Restrepo and Villaveces 2012, 2013; Restrepo 2013; Schummer 2012; Hosoya 2013; Klein 2013), including the author of the present chapter, have decided to study such a relationship from both a historical and a philosophical viewpoint. One example of the increased interest are the special issues of *Hyle*,[1] which

---

[1] Open access to the papers is available at the website of the journal: http://www.hyle.org/index.html (Accessed 3 July 2013).

were dedicated to mathematical chemistry. Restrepo and Schummer (2012) stated in one of those special issues that mathematical chemistry is a new discipline according to standard criteria of discipline generation. This chapter will further explore that assertion.

1 Defining Mathematical Chemistry

There have been several attempts to define mathematical chemistry as discussed by Klein (2013) and Restrepo (2013). Perhaps the most general definition has been proposed by Klein (2013), who says that mathematical chemistry entails mathematically novel ideas and concepts adapted or developed for use in chemistry. Klein further explains by stating that the "definition distinguishes mathematical chemistry somewhat from simple routine mathematics for chemical problems and even from rather complex mathematics used repeatedly in some standardized manner (perhaps in the form of a 'canned' computer program)" (Klein 2013, 36). Hence, mathematical chemistry is not using mathematics in chemistry but adapting mathematics—or developing new mathematics—for solving chemical questions. (By *adapting* we mean the act of modifying mathematical theories, with mathematical rigor, to face chemical questions.)

Restrepo and Villaveces (2012, 2013) mooted a definition similar to Klein's: Mathematical chemistry is the realization of the mathematical way of thinking in chemistry. The definition combines Comtean ideas on the relationship between chemistry and mathematics (Comte 1893) with Weyl's (1940) assertions on the mathematical way of thinking. Comte stated that "the perfection of chemistry might be secured and hastened by the training of the minds of chemists in the mathematical spirit .... Besides that mathematical study is the necessary foundation of all positive science, it has a special use in chemistry in disciplining the mind to a wise severity in the conduct of analysis" (Comte 1893, 257). Weyl follows the functional thinking of Felix Klein's (1872) Erlangen Programme, which is based on variables, symbols, and functions: The general idea is to address chemical problems by looking for relevant variables, symbolizing them, and finding functions that relate the selected variables. In Restrepo and Villaveces (2012) and Klein (2013), several examples of the mathematical way of thinking in chemistry are discussed, from ancient times to the present. Klein (2013) has particularly shown that mathematical chemistry is not only related to physical chemistry and chemical physics, as many would expect, but it also pervades other subdisciplines of chemistry such as polymer chemistry, organic chemistry, and chemometrics, among others.

We now proceed to show how mathematical chemistry meets the standard criteria of discipline formation.

2 Exploring Disciplinary Criteria of Mathematical Chemistry

We follow the disciplinary criteria discussed by Nye (1993, 4), who states that the construction of a disciplinary identity consists of six elements: (i) genealogy and family descent, including historical mythology of heroic origins and heroic episodes; (ii) core literature defining the group's archetypical language and imagery; (iii) practices and rituals; (iv) physical homeland, including institutions based on citizenship rights and responsibilities; (v) external recognition; and (vi) shared values and unsolved problems.

2.1 Mythology and Genealogy in Mathematical Chemistry

Nye states that genealogies "take their point of origin not in ordinary people but in heroic figures who fought enemies, even villains, and who won their battles by wit and dexterity" (Nye 1993, 22). Once those "heroes" are found, one needs to look for their genealogies, if any. The questions that arise are: Who are the "heroes" of mathematical chemistry, and who are its "enemies"? and Did the "heroes" give place to a school of thought? These are difficult subjects that may ruffle some feathers and that may lead to the foundational myth. One can look for the seeds of mathematical chemistry in the very roots of chemistry by, for example, analyzing Lavoisier's work or by considering Mendeleev's and Brown's contributions.

Lavoisier indeed used the mathematical way of thinking. He took some chemical substances and analyzed them to find their chemical elements; each one of those elements received a symbol; the symbols, by the application of chemical rules of combination (functions), yielded known and unknown substances (Restrepo and Villaveces 2011). Restrepo and Villaveces have shown that Lavoisier's work can be framed in the field of algebra of sets, but such mathematics was not yet developed in Lavoisier's time. Mendeleev, with his work on the periodic table, can likewise be framed in the mathematical way of thinking. He took properties of chemical elements such as atomic weight, valence, acidity, atomic volume, oxidation state, and so on as variables and looked for functions of the form $P = f(A)$, relating the atomic weight A to each one of the properties P (Restrepo unpublished; Restrepo and Villaveces 2012). The regularities of the periodic law, especially the lengths of the periods, inspired Mendeleev to discover the underlying mathematical structure for the chemical elements (Restrepo and Pachón 2007). Twenty years after his seminal work of 1869, Mendeleev reviewed some approaches to explain, on mathematical grounds, the nature of chemical periodicity (Mendeleev 1889). However, these mathematical (numerical) investigations did not stand the test of time.[2] Brown (1864) published a paper on molecular structure that was full of mathematics (Klein unpublished). In his paper Brown describes substances by reducing them to molecules, which are characterized as graphs, which constitutes a very important epistemological shift—namely one of passing from treating substances as macroscopic entities made of chemical elements, to another where substances are made of microscopic assembles (molecules), where

---

[2] One may think that Henry Moseley's work is the culmination of research looking for numerical relationships for the chemical elements. However, what Henry Moseley found was an almost perfect match between Mendeleev's ordering of the elements and the ordering based on the number of protons in the atoms of each element. Henry Moseley's work was indeed a reformulation of the periodic law, but in terms of protons rather than atomic weight.

chemical elements are reduced to the concept of atoms and where the relationships (bonds) among atoms are an important part of the understanding of the substances. This is a clear example of the mathematical way of thinking in chemistry, which Restrepo and Harré (unpublished) have recently explored from a mereological viewpoint. Brown was able to find a link between substances (i.e., those materials stored in bottles reacting with each other) and mathematical objects (i.e. graphs),[3] which for the case of the molecular structure were made of atoms and bonds. Hence, what Brown found were functional relationships of the sort $Substance = f(Molecule)$ and $Molecule = f(Atoms, Bonds)$, the latter expressed in mathematical terms as $Molecule = f(Graph)$. As pointed out by Klein (unpublished), the paper was not regarded as a mathematical one mainly because it lacked numerical mathematics; the mathematics involved was not yet developed (the word "graph" had not yet been coined) and the language was too chemical. Although Lavoisier, Mendeleev, and Brown are clear instances of the mathematical way of thinking and they played important roles in chemistry and in shaping or consolidating chemistry as a discipline, they did not create a school of thought with mathematical grounds. Hence, they cannot be regarded as "heroes" of mathematical chemistry.

However, some other scientists may indeed fill the heroic role. Klein (2013) has recently noted that there was a school of thought with a mathematical foundation after Amedeo Avogadro, Henry Louis Le Chatelier, Rudolph Clausius, van't Hoff, Ostwald, Arrhenius, Gibbs, Maxwell, Ludwig Boltzmann, Nernst, and Haber and Gilbert N. Lewis, who can be considered influential scientists in the now broad field of thermodynamics, where differential equations are of great importance.[4] Electrochemistry, Klein (2013) points out, owes much to the mathematical insight of Gustav Kirchhoff; stereochemistry to van't Hoff and Joseph Achille Le Bel; and crystallography to Auguste Bravais, Arthur Schoenflies, and Fedorov (Klein 2013). For an interesting account of other schools of thought regarding seminal chemistry works with mathematical background, the reader is directed to Klein's survey (2013).

Now, what about mathematics? Are there "heroes" of mathematical chemistry who are mathematicians? Besides the previously mentioned Rudolph Clausius, Maxwell, Arthur Schoenflies and Evgraf Fedorov, other mathematicians critically important to mathematical chemistry include Sylvester (1878a,b) and Cayley (1875), who in the nineteenth century pondered an algebra for chemistry and the enumeration of isomers, respectively. Sylvester introduced the idea of chemico-graphs for representing molecular structures, which both gave rise to the use of the term *graph* in mathematics and strengthened the development of graph theory, now an important branch of discrete mathematics. In the 1960s and 1970s, the use of graph theory in chemistry had a kind of renaissance, and extensive

---

[3] A graph is a couple (V, E) of vertices V and edges E, which in turn are couples of vertices.
[4] Besides differential equations, there are some other mathematical approaches to thermodynamics, as pointed out by Klein (2013), e.g. Weinhold's (1975) development of a geometric Riemannian metric for thermodynamic manifolds, where the empirical laws of equilibrium thermodynamics are brought into correspondence with the mathematical axioms of an abstract metric space.

research was performed by a school of thought that still uses chemical questions to develop both mathematics and chemistry. Klein (unpublished) noted that Sylvester was one of the first scientists in recognize the mathematical flavor of Brown's paper, which we mentioned earlier.

In 1875, Cayley published a paper on the calculation of the number of possible alkane isomers fulfilling the formula C_nH_{2n+2}, and of alkyl radicals fulfilling C_nH_{2n+1}, with n varying from 1 to 13. This paper is the first to consider, and treat, the problem of counting isomers using mathematical insight. Cayley's work on chemical enumeration was further advanced more than 50 years later by the mathematician Pólya with his foundational combinatorial theory of enumeration under group-mediated equivalences. Pólya's approach is the continuation of Cayley's school—where a wealth of mathematics has been generated to solve particular aspects of enumerative chemistry, taking into account the manifold diversity of chemical structures associated to chemical substances.

And what about the enemies of the aforementioned "heroes"? We think the most important battle has been against a misconception of mathematics. As Kemeny (1959) points out, as late as the 1950s there was still a prevalent idea that mathematics is concerned only with the study of numbers and space; we also assert that the layman—and a large part of the scientific community—still think that mathematics has to do with numbers, and that it is not possible to do mathematics without numbers. Kemeny sought to eradicate this misconception in his paper "Mathematics without Numbers" (Kemeny 1959). Mathematics dealing with numbers, in other words, classical mathematics, arose from physical questions characterized by measurements mapped onto numbers. In physics the study of space was also needed and so geometry came into play. But as Kemeny (1959) points out, there is plenty of mathematics not overlaying numbers, such as topology, and graph, group, and order theories—mathematics that arose as the abstract process of particular questions not involving numbers.[5]

A battle caused by not finding mathematical chemistry works of the 1960s and 1970s as part of standard fields such as physical chemistry or quantum chemistry led to the creation of the *International Academy of Mathematical Chemistry*. Randić (2005) summarizes the tone of the issue by claiming that critics of mathematical chemistry "tend to perceive topological indices[6] as frivolous, fraudulent, fortuitous, fiction, fabrication, foreign, fictitious, fallacious, flimsy, folly, and foolish" (Randić 2004, 10). A detailed description of this battle is found in Restrepo and Villaveces (2013). These kinds of battles are normal when a new discipline is blossoming and it is the natural result of creating a new vocabulary and new methods that vary from standard fields of science (Lattuca 2001).

Thus, the heroes of mathematical chemistry have been scientists like Amedeo Avogadro, van't Hoff, Ostwald, Gibbs, Gilbert N. Lewis, Cayley, and Sylvester,

---

[5] For example, topology and graph theory seeds are found in the problem of the Königsberg bridges, a puzzle, making use of the folklore of the eighteenth century inhabitants of Königsberg's and solved by Euler (Biggs et al. 1998, 1–11).

[6] A particular concept of discrete mathematical chemistry.

who created schools of thought, giving place to a scientific genealogy. These heroes and their followers have fought against the misconception of mathematics as being based solely upon numbers.

According to Nye (1993, 4), the school of thought of heroes is consolidated by creating their own language, developed through their own imagery. This is what keeps heroes connected with their followers. Such a common language is created by pointing out to the core literature, to classics. We explore the elements of that language and imagery for mathematical chemistry in the following subsection.

2.2 Language and Imagery in Mathematical Chemistry

It is not enough having a hierarchy of heroes and the complete record of their epic battles to claim a new discipline has arisen. As Nye states (1993, 24), novices must be introduced to a series of classics of the discipline, written by the heroes. Through those classics, novices are trained in the basic vocabulary and methods of the discipline, which are needed both for further discussion and for advancing the disciplinary knowledge.

There are different methods of addressing the question of what comprise the classics of mathematical chemistry. One method is to look for the most-cited papers—which, following Latour (1987), become black boxes, that is, general and basic knowledge for the discipline. If one is willing to follow this approach, then the next question is how to find the right place to look for those highly cited documents. For this particular case, one would assume the answer is to look in the specialized journals on mathematical chemistry (which we will discuss in another section). But there is a problem with this approach: Because those journals are launched when a critical amount of knowledge on the subject is already developed, it is difficult to find the seminal papers that shaped the discipline. These papers are customarily published in journals or books belonging to different subjects, where battles are a given because the methods and vocabulary used in the seminal papers do not entirely match those of the journals where the material is published. Now, even if one assumes that a seminal paper was published in a journal belonging to the discipline the paper is helping to shape, the paper then becomes a black box and it is no longer cited as it passes on to become part of the folklore of the discipline. In short, looking for the classics of mathematical chemistry in mathematical chemistry journals is useless.

Another approach to finding the seminal papers of mathematical chemistry is to scan the documents cited in papers published in specialized journals of mathematical chemistry. Those frequently cited papers are very likely to constitute the core documents for the discipline. However, this approach requires further bibliometric work that we do not attempt to develop here. In any case, the work has to be done, and not only in a static fashion; it has to be redone periodically to explore the dynamics of knowledge establishment within the discipline. If the work is done this way, then it is possible to see the dynamics in which a document becomes a black box. Cayley's seminal paper is easily recognized as a classic in the early days (1960s and 1970s) of the chemical graph renaissance, but it is rarely cited in current papers on the subject, for it has now become a black box—no longer cited, part of the folklore of chemical graph studies.

A static approach that we explore in this chapter is that of studying the most cited publications by scholars recognized as leading figures in the field. The approach is static because the recognition analyzed is static too. In the 1960s scholars were recognized who are different from those recognized now in the 2010s. And again the black boxes show up, for Cayley is no longer found as a leading figure because his contribution has been absorbed by the discipline. Knowing the right place to look for the recognized scholars is another issue arising from our static approach. As we will discuss later, as well as journals and books devoted to mathematical chemistry, there are societies dedicated to the subject; one of those is the International Academy of Mathematical Chemistry,[7] comprising about 100 scientists of various countries. We used Google Scholar to find each of the Academy members, to try to gauge the importance of their works by the use and circulation those works have received. Note that the use of those documents may not only come from the interior of the discipline; it can be used by several others. An example of this situation occurred recently when we were studying the scientific influence of a leading scholar of mathematical chemistry (Restrepo et al. 2013); we found that several cites to his work come from chemical engineering, pharmacology, and other areas than mathematical chemistry. The first 30 scientists of the Academy with more citations to one of their works are shown in Table 15.1.

The subjects of these publications are manifold. They mix traditions, methods, and vocabulary from at least two of the following areas: physical chemistry, chemical physics, quantum chemistry, spectroscopy, organic chemistry, chemoinformatics, chemometrics, new materials, inorganic chemistry, group theory, virtual screening, graph theory, statistics, representation theory, computer science, and network theory.

By analyzing the titles, abstracts, and table of contents of each document shown in Table 15.1, we found the following set of frequent words, constituting the underlying vocabulary for those works: molecular, energy, number, chemical, objects, atoms, network, descriptors, algorithm, theory, density, matrix, electron, graph, index, structure, topological, set, and distribution. A search in Google for different combinations of these words yields an enormous number of scientific documents. For example, the combination "molecular descriptors" yields 187,000 results (28 June 2013). As discussed by Restrepo et al. (2013) there is a community using a common vocabulary.

The common language element for mathematical chemistry is mathematics in their different flavors, such as discrete or continuous mathematics, with their further subdivisions. Current mathematical chemistry language tries to frame chemical knowledge, or chemical questions, in a mathematical

---

[7]This academy was created in 2005; and its website is http://www.iamc-online.org/. (Accessed 3 July 2013).

TABLE 15.1. Thirty Members of the *International Academy of Mathematical Chemistry* with Their Most Frequently Cited Documents.

SCIENTIST	NUMBER OF CITES	DOCUMENT TITLE	JOURNAL/BOOK PUBLISHER	PUBLICATION YEAR
Robert G. Parr	46,606	"Development of the Colle-Salvetti correlation-energy formula into a functional of the electron density"	*Physical Review B*	1988
F. Albert Cotton	19,791	*Advanced Inorganic Chemistry*	John Wiley & Sons	1980
Pierre Hansen	8,828	*Theory of Simple Liquids*	Academic Press	2006
Roald Hoffmann	4,066	"An extended Hückel theory. I. Hydrocarbons"	*The Journal of Chemical Physics*	1963
Rudolph A. Marcus	3,574	"On the theory of oxidation-reduction reactions involving electron transfer. I"	*The Journal of Chemical Physics*	1956
Paul von Rague Schleyer	3,510	"Efficient diffuse function-augmented basis sets for anion calculations. III. The 3-21+G basis set for first-row elements, Li–F"	*Journal of Computational Chemistry*	1983
Dragos M. Cvetković	2,944	*Spectra of Graphs: Theory and Application*	Deutscher Verlag der Wissenschaften	1980
Horst Sachs	2,944	*Spectra of graphs: Theory and application*	Deutscher Verlag der Wissenschaften	1980
Peter Willett	2,891	"Development and validation of a genetic algorithm for flexible docking"	*Journal of Molecular Biology*	1997
Milan Randić	2,369	"Characterization of molecular branching"	*Journal of the American Chemical Society*	1975
Roberto Todeschini	2,346	*Handbook of Molecular Descriptors*	Wiley-VCH	2000
Viviana Consonni	2,346	*Handbook of Molecular Descriptors*	Wiley-VCH	2000

(continued)

TABLE 15.1. Continued.

SCIENTIST	NUMBER OF CITES	DOCUMENT TITLE	JOURNAL/BOOK PUBLISHER	PUBLICATION YEAR
Johann Gasteiger	2,298	"Iterative partial equalization of orbital electronegativity—a rapid access to atomic charges"	Tetrahedron	1980
Adalbert Kerber	2,281	The Representation Theory of the Symmetric Group	Cambridge University Press	1984
Peter John	2,208	Unimolecular Reactions	Wiley-Interscience	1972
Michael E. Fisher	2,099	"Linear magnetic chains with aniso-tropic coupling"	Physical Review	1964
Lemont B. Kier	1,758	Molecular Connectivity in Structure-Activity Analysis	John Wiley	1986
Lowell H. Hall	1,758	Molecular Connectivity in Structure-Activity Analysis	John Wiley	1986
Jiri Čížek*	1,580	"On the use of the cluster expansion and the technique of diagrams in calcu-lations of correlation effects in atoms and molecules"	Pitman Press	1969
Gheorghe Păun	1,456	"Computing with membranes"	Journal of Computer and System Sciences	2000
Douglas M. Hawkins	1,436	Identification of Outliers	Chapman and Hall	1980
Vladimir Batagelj	1,420	Exploratory Social Network Analysis with Pajek	Cambridge University Press	2005
Mel Levy	1,313	"Density-functional theory for frac-tional particle number: derivative dis-continuities of the energy"	Physical Review Letters	1982
Alan R. Katritzky	1,270	Handbook of Heterocyclic Chemistry	Elsevier	1985

Author	Title	Publisher	Year	Citations
Patrick W. Fowler	*An Atlas of Fullerenes*	Clarendon Press	1995	1,198
Haruo Hosoya	"Topological index. A newly proposed quantity characterizing the topological nature of structural isomers of saturated hydrocarbons"	*Bulletin of the Chemical Society of Japan*	1971	1,174
Ivan Gutman	*Mathematical Concepts in Organic Chemistry*	Springer-Verlag	1986	1,083
Ahmed H. Zewail	"Femtochemistry: Atomic-scale dynamics of the chemical bond"	*The Journal of Physical Chemistry A*	2000	1,005
Mati Karelson	"Quantum chemical descriptors in QSAR/QSPR studies"	*Chemical Reviews*	1996	914
Jure Zupan	*Neural Networks for Chemists: An Introduction*	John Wiley & Sons	1993	858

* Book chapter.

context. One of the most widespread elements of language is graph theory and its terms: node, vertex, edge, graph, and so forth. Another element of language is quantum theory with its associated terms like electronic structure, density functional, and Hartree-Fock theories. Many of these elements of language go along with others of theory of computation such as algorithm, model, computational complexity, and so on. All in all, the language and imagery of current mathematical chemistry comes from different mathematical traditions and from several subfields of chemistry.

Nye (1993, 4) claims that there are also other elements, different from language and imagery, which are shared by practitioners of a discipline. These additional elements are practices and rituals. In the following subsection we explore them for the mathematical chemistry case.

2.3 Practices and Rituals in Mathematical Chemistry

Through its practices and rituals, the methods of a discipline can be transferred to novices by the more experienced scholars (Nye 1993, 25). For example, the practices of mathematical chemistry are transferred through oral and written examinations in MSc and PhD works. Writing research proposals is also of importance in these practices, as the novice is trained to use the vocabulary and is aware of the institutions involved in the research (for example for funding purposes). So far, no work has been done in this area, that is, in determining the agencies funding chemomathematical research and the kind of discourse used to attain such support (Restrepo and Schummer 2014).

Currently there is no academic degree or specialty in mathematical chemistry, and so there are no dissertations on the subject, but the leading figures of the past and present have advised students who have been able to bring new knowledge to the field. As seen in Table 15.1, the subjects of the most cited works are varied and the same trend is found in the backgrounds of people working in the community. Some of the disciplines feeding mathematical chemistry with trained scientists are mathematics, chemistry, physics, biology, biochemistry, computer sciences, and the several interdisciplinary fields found in the borders between those disciplines.

The rituals, as Nye call them (1993, 25), include Festschrifts, obituaries, conferences, congresses, colloquiums, seminars, workshops, and informal meetings that reaffirm the legitimacy of the discipline and its practitioners, and reenact traditions. A particular ritual in any scientific community, which is taken as a step forward in becoming an expert in the field, is being selected as a member of editorial boards or an editor-in-chief of a specialized journal of the community. The same ritual occurs for editors of specialized books on the discipline. These rituals are present in mathematical chemistry too, as seen with the editors of books and journals on the discipline, which will be discussed later. An important ritual in mathematical chemistry is being nominated and elected to the *International Academy of Mathematical Chemistry*, which is a clear evidence of recognition by the community of mathematical chemists.

Regarding Festschrifts and obituaries highlighting what we today recognize as the chemomathematical side of the honored scientist, we have several examples: for instance those of Gibbs (Hastings 1909), Ludwig Boltzmann (Meyer 1904), Onsager (Longuet-Higgins and Fisher 1991), and Pólya (Harary 1977). More recently, Trinajstić,[8] Randić,[9] Balaban,[10] Hosoya,[11] Kier,[12] and Klein[13] Festschrifts, to name but a few that have appeared in specialized chemistry journals and which constitute part of the settled rituals of the mathematical chemistry community.

Disciplines, according to Nye (1993, 4), have institutions where key concepts are discussed and where members of the community supporting the discipline gather. These institutions for mathematical chemistry are discussed in the following subsection.

2.4 Mathematical Chemistry Institutions

The institutions that structure the discipline of mathematical chemistry are the collection of activities performed by scientific academies, societies, and editorial boards of specialized journals that all codify membership criteria, relationships, and responsibilities (Nye 1993, 27). These institutions bring collective identity, which is given by a structure of well-differentiated rights and responsibilities, to the community of mathematical chemistry. For example, the *International Academy of Mathematical Chemistry* is a selected group of scientists actively working on mathematical chemistry, as is stated in article four of the statutes of the Academy[14]: "The Academy shall be composed of persons (hereafter 'the Members') chosen from scientists, regardless of their country, who have distinguished themselves by the value of their scientific work and have thus significantly contributed to the advancement of Mathematical Chemistry, and/or who have been pioneers or leaders of particular research directions in the field, and/or who have locally promoted the advancement of Mathematical Chemistry." How new members are elected is stated in article nine: "New Members of the Academy are chosen at the A.G.M. [Annual General Meeting] by Members present at the meeting. New Members shall be nominated at the A.G.M. at least one year before election. Any Member present at the A.G.M. can nominate one person for election by giving an informal presentation in support of his/her candidate. At the next A.G.M., the proposer for

---

[8] Festschrift in honor of Nenad Trinajstić to mark his 65th birthday and to acknowledge his distinguished research in mathematical chemistry. Volume 77 of *Croatica Chemica Acta*, 2004.
[9] A special issue of *Current Computer-Aided Drug Design* honoring Professor Milan Randić on his eightieth birthday. Volume 9, 2013.
[10] Professor Alexandru T. Balaban's 75th Anniversary. *Revue Roumaine de Chimie*, 2006, vol. 51.
[11] The MATH/CHEM/COMP of 2006 was dedicated to Professor Haruo Hosoya on the occasion of his 70th birthday.
[12] Special issue of Current Computer-Aided Drug Design, dedicated to Professor Lemont B. Kier on the occasion of his 80th birthday. Volume 8, 2012.
[13] The Klein Festschrift, a special issue of *Croatica Chemica Acta* on the occasion of his 70th birthday. Volume 86, 2013.
[14] http://www.iamc-online.org/statutes/index.htm (Accessed 3 July 2013).

each candidate should give formal justification, in writing, for the nomination. After as much discussion as is deemed necessary, voting shall take place for each candidate. To be accepted as a new Member, the candidate shall be required to have accrued at least two-thirds of the votes cast."

Other institutions contributing to shape mathematical chemistry are the editorial boards of *MATCH Communications in Mathematical and in Computer Chemistry*, the first journal in the subject (started in 1975); the *Journal of Mathematical Chemistry* (initiated in 1987); and the *Iranian Journal of Mathematical Chemistry*, recently launched (2010). There are also important meetings in the subject that constitute places where members of the community have the opportunity to discuss the advances in research, and where novel ideas and research projects take shape: For example, the *MATH/CHEM/COMP* meetings (traditionally organized in Croatia), the *Indo-US Workshop on Mathematical Chemistry Series*, the *Indo-US Lecture Series on Discrete Mathematical Chemistry*, and more recently the *Mathematical Chemistry Workshop of the Americas*, which have been held in Colombia.

Although a discipline is mainly made by its actors (genealogy) and by the communication between them—attained through meetings and publications— it is of central importance to have external recognition. It is thanks to this recognition that news from the works made inside the discipline start to spread out, to pervade other disciplines and to make the discipline more active. It is *for* external recognition that some scholars become heroes. In the following subsection we discuss some aspects of such recognition for mathematical chemistry.

2.5 External Recognition of Mathematical Chemistry

The exterior of mathematical chemistry is given by chemistry, mathematics, and several other disciplines such as physics, biology, biochemistry, and computer sciences. In (Restrepo and Villaveces 2013) we discussed the recognition in chemistry and in mathematics. In chemistry, Hosoya has summarized such recognition by stating that "conservative [chemistry] professors tried to repel new and strange ideas and intruders from their self-perceived closed territory" (Hosoya 2002, 429). However, Hosoya considers that the situation has changed and now the discipline has gained some acceptance in chemical circles. Randić (2004) has pointed out another possible cause for the rejection, this time particularly referring to the case of topological indices (mathematical representations of the molecular structure), widely used in discrete mathematical chemistry. He states that detractors of these indices disregard them because of their "lack of interpretability." Here it is important to mention that the sought-for interpretability is to lay down the index to a physical framework. As Randić (2004, 10) asks, "Why should [they] have a 'physical interpretation'? He draws attention to the point that there are several other fields where lacking physical interpretability is not taken as a downside of the field; for example, the lack of physical interpretation of a molecular orbital in quantum chemistry does not negate the validity of quantum chemistry. Randić (2004) points out that concepts such as topological index and molecular orbital have a wealth of meaning within the particular field of knowledge where they were created and used. It is too much to ask for meaning and usefulness of those concepts in distant contexts.

Another possible cause for the rejection of mathematical chemistry in chemistry, mentioned by Randić (2004), is the rapid proliferation of the subject, which is exemplified by the growing number of publications. He particularly analyzes the case of chemical graph theory, which gathered only 20 papers in 1973—and about 2,000 in 2000. This might be a result of the rapid coding of the algorithms generated by mathematical chemistry in packages that may be used as "canned" computer programs yielding many papers, which are often devoid of chemical and mathematical insight.

In mathematical circles the recognition of mathematical chemistry has been accepted only slowly (Restrepo and Villaveces 2013). Instances of that recognition are the 1988 and 1996 publications of special issues on mathematical chemistry by the journal *Discrete Applied Mathematics*. The *Workshop on Discrete Mathematical Chemistry*, organized in 1998 by *DIMACS* (Discrete Mathematics & Theoretical Computer Science) is another instance of acceptance, as is the *VII Simposio Nororiental de Matemáticas* organized in Colombia in 2009—at which there was a special session for mathematical chemistry; it was treated at the same level as traditional mathematical areas such as analysis, algebra, combinatorics, and topology. In 2010 the *Xiamen University* (China) organized the *International Conference on Mathematical Chemistry*, which was chaired by mathematicians. This mathematical interest is also evident in the organization of the 2016 meeting of the Academy, which is in charge of the Center for Combinatorics of the *Nankai University* (China).

Industry and some governmental agencies have also recognized the value of mathematical chemistry research. As claimed by Restrepo and Villaveces (2013, 24), "in a scientific world, where results from complex experiments and expensive computations are common, providing advance estimates of those results based on mathematics are noteworthy." Estimating the properties of substances using insights from mathematical chemistry saves time and money for pharmaceutical companies and environmental agencies, which in a matter of weeks need to either give a "green light" to or ban the production of a new chemical based on several biological, biochemical, and environmental tests that are currently assisted by knowledge of mathematical chemistry (Restrepo and Basak 2011). A current breakthrough application of mathematical chemistry with promising industrial use is the design of synthetic routes to particular products by avoiding multistep reactions, different reactors, and attached separation processes; these are called one-pot reactions (Gothard et al. 2012).

In general, the recognition of mathematical chemistry in mathematics has been easier than recognition in chemical circles. Perhaps mathematicians do not fear the results of mathematical chemistry while chemists do. Here it is important to mention that at the bottom of mathematical chemistry is the sought-for simple rules and theories for chemistry, which may ruffle some feathers in scientists who traditionally try to do the same, but through other traditions and methods such as those of quantum chemistry. Perhaps this is the reason for the initial rejection of mathematical chemistry within chemistry. We think the point is not about who develops a theory or discovers a rule in chemistry, but about discovering with the necessary tools and methods, which is what in the end guarantees that those discoveries and theories may result in opening new fields of research. All in all, as claimed by Schummer (2012), chemistry is a science of methodological pluralism where no single theory accounts for the whole science.

As a discipline is a dynamic phenomenon, there must be something pushing for change. That "something," according to Nye (1993, 4) is made by unsolved problems and for shared values between members of the discipline. In the following subsection we discuss them.

2.6 Shared Values and Unsolved Problems in Mathematical Chemistry

We think the most important value of mathematical chemistry is its effort to maintain and enhance the mathematical way of thinking in chemistry. All scientists working in mathematical chemistry, perhaps without knowing it, are looking for fundamental rules or trends underlying chemistry: for example, when studying how the encoding of the molecular structure may also encode features of the substances the characterized molecules represent; or when considering that the molecular structure may be used to estimate properties of substances by ordering their molecules; or when giving insight on the number of possible substances based on the complexity of the molecular structure. There are also examples of mathematization of chemistry where the molecular structure (understood as a collection of atoms related by bonds) is not the central point, such as approaches using category theory (Bernal 2012, Bernal et al. 2015) and topology (Stadler and Stadler 2002, Restrepo et al. 2004, Flamm et al. 2015), to look for patterns such as functional groups or families of chemicals in chemical networks. Or, for example, the study of the network of organic chemistry (Fialkowski et al. 2005), which despite its apparent complexity shows a well-defined topological structure (Grzybowski et al. 2009). We do not want to cover all instances of the mathematical way of thinking in chemistry in the current chapter, but only show that it is present in mathematical chemistry studies and even in some areas where it has not been recognized as such. Klein (2013) gives examples in his survey, such as foundational equilibrium thermodynamics and chemoinformatics. For a recent account of novel results in mathematical chemistry, see the two-volume *Advances in Mathematical Chemistry* (Basak et al. 2014, 2015).

One way of maintaining and enhancing the mathematical way of thinking in chemistry is by advising the research work of novices in different areas of knowledge related to chemistry. As mentioned before, another way of recruiting new minds to the mathematical way of thinking is through the organization of scientific meetings, congresses, conferences, symposia, and workshops.

Perhaps a more direct strategy is the organization of schools of mathematical chemistry where leading scholars teach about their subjects rather than showing their most recent results. These kinds of schools have the advantage of presenting to students the history and reasons to open a particular subject and the questions that motivated its development. Examples of successful schools of this kind are the Summer Schools of Quantum Chemistry organized by Löwdin around 1958, which gave place to the International Winter Institutes at Sanibel Island and Gainesville. Some schools have been organized in the MATH/CHEM/COMP meetings and in the Indo-US Lecture Series on Discrete Mathematical Chemistry. In the end, all these efforts look for something Comte foresaw in the nineteenth century when stating "See how the perfection of chemistry might be secured and hastened by the training of the minds of chemists in the mathematical spirit and astronomical philosophy. Besides that mathematical study is the necessary foundation of all positive science, it has a special use in chemistry in disciplining the mind to a wise severity in the conduct of analysis: and daily observation shows the evil effects of its absence" (Comte 1893, 247).

To make an account of the open problems in mathematical chemistry would be too pretentious. To have an account of all the different subspecialties of mathematical chemistry is quite difficult, as Klein (2013) has pointed out; there are many of those subspecialties with very particular subjects where only the specialists in each subject are aware of their open questions. We think the point is not to give a list of open questions but to show the manifold opportunities for mathematical chemistry, for there are always new results to be explained and predictions to be made through the use of the mathematical way of thinking. However, the specialists need to be aware of the mathematical flavor of their works; and they need to be in touch with the current mathematical knowledge and with mathematicians—otherwise "the cement that keeps the members of the disciplinary group together" (Nye 1993, 30) will not be found and each specialist may then be working isolated from the mathematical chemistry community.

Conclusions and Outlook

By following Nye's (1993) criteria of disciplinary formation, we found that mathematical chemistry is a new discipline with a scientific genealogy, core literature, defined practices, particular institutions, and external recognition and shared values. Philosophers and historians of science have a new field of work that needs further study, and where some of the open questions are (Restrepo and Schummer 2011):

Which are the specific methods, methodologies, and epistemology of mathematical chemistry that distinguish it from both mainstream chemistry and mathematics as well as from other subdisciplines of chemistry and mathematics?

Is mathematical chemistry a theoretical discipline as opposed to experimental? Could there be an experimental mathematical chemistry?

Does mathematical chemistry require specific ontological or metaphysical assumptions or positions regarding the (mathematical) constitution of the world or the reality of mathematical entities?

Are there specific branches of mathematics that are particularly relevant to mathematical chemistry? If so, does that tell something about chemistry in general and mathematical chemistry in particular?

Are there particular links between mathematical chemistry on the one hand, and philosophy of chemistry and philosophy of mathematics on the other?

Does the history of the chemistry–mathematics relationship provide any clues as to what has fostered, and hindered, its cooperative development?

Could mathematical chemistry have developed differently under different historical conditions? Could there be other definitions, other main areas, or even other methodologies and epistemologies of mathematical chemistry?

Although Klein (2013) recently asked, "Where does mathematical chemistry fit?" and he found that mathematical chemistry is central to organic, inorganic, and analytical chemistry—and to biochemistry and physical chemistry—we think further discussion is needed. For example, what is the role of mathematical chemistry in theoretical chemistry? Is mathematical chemistry embedded in theoretical chemistry, or is it merely related to theoretical chemistry while maintaining links to other non-theoretical parts of chemistry?

In 1959 Kemeny claimed that "There is every reason to expect that the various social sciences will serve as incentives for the development of great new branches of mathematics and that someday the theoretical social scientist will have to know more mathematics than the physicist needs to know today" (Kemeny 1959, 578). We think that such a claim is also valid for chemistry, whose complexity requires applying novel mathematical approaches and even developing novel mathematics able to cope with its diversity. As Comte (1893, 247) pointed out, "The perfection of chemistry might be secured and hastened by the training of the minds of chemists in the mathematical spirit." The question is how the mathematical chemistry community, which has already begun to walk the path of training chemical minds in the mathematical way of thinking, is considering, for example, curricula subjects to enhance the process of incorporating more mathematical thinking into chemistry.

References

Balaban, A.T. (2005). "Reflections about mathematical chemistry," *Foundations of Chemistry*, 7, 289–306.

Balaban, A.T. (2013). "Chemical graph theory and the Sherlock Holmes principle," *Hyle*, 19, 107–134.

Basak, S.C. (2013). "Philosophy of mathematical chemistry: A personal perspective," *Hyle*, 19, 3–17.

Basak, S.C. Restrepo, G., & Villaveces, J.L. (eds.). (2014). *Advances in Mathematical Chemistry and Applications*, Volume 1. Sharjah: Bentham.

Basak, S.C. Restrepo, G., & Villaveces, J.L. (eds.). (2015). *Advances in Mathematical Chemistry and Applications*, Volume 2. Sharjah: Bentham.

Bernal, A. (2012). *On Chemical Activity Foundations of a Classificatory Approach to the Study of Chemical Combination*. PhD thesis, Universidad Nacional de Colombia, Bogota.

Bernal, A., Llanos, E., Leal, W., & Restrepo, G. (2015). "Similarity in chemical reaction networks: Categories, concepts and closures." In *Advances in Mathematical Chemistry and Applications*, Volume 2 (S.C. Basak, G. Restrepo, & J.L. Villaveces, eds.). Bentham: Sharjah.

Biggs, N., Lloyd, E., & Wilson, R. (1998). *Graph Theory, 1736–1936*. New York: Oxford University Press.

Brown, A.C. (1864). "On the theory of isomeric compounds," *Transactions of the Royal Society of Edinburgh*, 23, 707–719. (Later reprinted in Brown, A.C. (1865) "On the theory of isomeric compounds," *Journal of the Chemical Society*, 18, 230–245.)

Cayley, A. (1875). "Über die analytischen Figuren, welche in der Mathematik Bäume genannt werden und ihre Anwendung auf die Theorie chemischer Verbindungen," *Berichte der deutschen chemischen Gesellschaft*, 8, 1056–9. (English version: "On the analytical forms called trees, with application to the theory of chemical combinations," *Report of the British Association for the Advancement of Science*, 257–305).

Comte, A. (1893). *The Positive Philosophy of Auguste Comte*, trans. H. Martineau. New York: Calvin Blanchard. (Published originally in French as *Discours sur l'Esprit positif*, 1844.) Available online at https://archive.org/details/positivephilosop01comtuoft

Deltete, R. (2012). "Georg Helm's chemical energetics," *Hyle*, 18, 23–44.

Fialkowski, M., Bishop, K.J.M., Chubukov, V.A., Campbell, C.J., & Grzybowski, B.A. (2005). "Architecture and evolution of organic chemistry," *Angewandte Chemie International Edition*, 44, 7263–7269.

Flamm, C., Stadler, B.M.R., & Stadler, P.F. (2015). "Generalized Topologies: Hypergraphs, Chemical Reactions, and Biological Evolution." In *Advances in Mathematical Chemistry and Applications*, Volume 2 (S.C. Basak; G. Restrepo, & J.L. Villaveces, eds.). Bentham: Sharjah.

Gavroglu, K. & Simões, A. (2012). "From physical chemistry to quantum chemistry: How chemists dealt with mathematics," *Hyle*, 18, 45–69.

Gothard, C.M., Soh, S., Gothard, N.A., Kowalczyk, B., Wei, Y., Baytekin, B. & Grzybowski, B.A. (2012). "Rewiring chemistry: Algorithmic discovery and experimental validation of one-pot reactions in the network of organic chemistry," *Angewandte Chemie International Edition*, 51, 7922–7927.

Grzybowski, B.A., Bishop, K.J.M., Kowalczyk, B., & Wilmer, C.E. (2009). "The 'wired' universe of organic chemistry," *Nature Chemistry*, 1, 31–36.

Harary, F. (1977). "Homage to George Pólya," *Journal of Graph Theory*, 1, 289–290.

Hastings, C.S. (1909). *Biographical Memoir of Josiah Willard Gibbs 1839–1903 (Biographical Memoirs, Part of Volume VI)*. Washington, DC: National Academy of Sciences.

Hosoya, H. (2002). "The topological index Z before and after 1971," *Internet Electronic Journal of Molecular Design*, 1, 428–42.

Hosoya, H. (2013). "What can mathematical chemistry contribute to the development of mathematics?," *Hyle*, 19, 87–105.

Kemeny, J.G. (1959). "Mathematics without numbers," *Daedalus*, 88, 577–591.

Klein, D.J. (2013). "Mathematical chemistry! Is it? And if so, what is it?," *Hyle*, 19, 35–85.

Klein, F. (1872). *Vergleichende Betrachtungen über neuere geometrische Forschungen*. Erlangen: Verlag von Andreas Deichert.

Latour, B. (1987). *Science in Action*. Cambridge, MA: Harvard University Press.

Lattuca, L.R. (2001). *Creating Interdisciplinarity*. Nashville, TN: Vanderbilt University Press.

Longuet-Higgins, C. & Fisher, M.E. (1991). *Lars Onsager 1903–1976, A Biographical Memoir*. Washington, DC: National Academy of Sciences.

Mendeleev, D.I. (1889). "The periodic law of the chemical elements," *Journal of the Chemical Society*, 55, 634–656. URL: http://web.lemoyne.edu/~giunta/mendel.html (Accessed 3 July 2013).

Meyer, S. (1904). *Festschrift Ludwig Boltzmann gewidmet zum sechzigsten geburtstage 20. Februar 1904*. Leipzig: Verlag von Johann Ambrosius Barth.

Nye, M.J. (1993). *From Chemical Philosophy to Theoretical Chemistry: Dynamics of Matter and Dynamics of Disciplines, 1800–1950*. Berkeley: University of California Press.

Randić, M. (2004). "Nenad Trinajstić – pioneer of chemical graph theory," *Croatica Chemica Acta*, 77, 1–15.

Randić, M. (2005). "Academy activities," International Academy of Mathematical Chemistry. URL: http://www.iamc-online.org/who_we_are/index.htm (Accessed 3 July 2013).

Restrepo, G. (2013). "To mathematize, or not to mathematize, chemistry," *Foundations of Chemistry*, 15, 185–197.

Restrepo, G. & Basak, S.C. (2011). "Editorial, hot topic: Applications of graph theory, network theory, and chemotopology to structure-activity relationships and characterization of metabolic processes," *Current Computer-Aided Drug Design*, 7, 81–82.

Restrepo, G., Llanos, E.J., & Silva, A.E. (2013). "Lemont B. Kier: A bibliometric exploration of his scientific production and its use," *Current Computer-Aided Drug Design*, 9, 491–505.

Restrepo, G., Mesa, H., Llanos, E.J., Villaveces, J.L. (2004). "Topological study of the periodic system," *Journal of Chemical Information and Computer Sciences*, 44, 68–75.

Restrepo, G. & Pachón, L. (2007). "Mathematical aspects of the periodic law," *Foundations of Chemistry*, 9, 189–214.

Restrepo, G. & Schummer, J. (2011). "Call for papers: chemistry & mathematics," *Hyle*, 17, Calls section.

Restrepo, G. & Schummer, J. (2012). "Editorial introduction," *Hyle*, 18, 1–2.

Restrepo, G. & Schummer, J. (2014). "Philosophical aspects of mathematical chemistry," *MATCH Communications in Mathematical and in Computer Chemistry*, 72, 589–599.

Restrepo, G. & Villaveces, J.L. (2011). "Chemistry, a lingua philosophica," *Foundations of Chemistry*, 13, 233–249.

Restrepo, G. & Villaveces, J.L. (2012). "Mathematical thinking in chemistry," *Hyle*, 18, 3–22.

Restrepo, G. & Villaveces, J.L. (2013). "Discrete mathematical chemistry: Social aspects of its emergence and reception," *Hyle*, 19, 19–33.

Schummer, J. (2012). "Why mathematical chemistry cannot copy mathematical physics and how to avoid the imminent epistemological pitfalls," *Hyle*, 18, 71–89.

Stadler, B.M.R. & Stadler, P.F. (2002). "Generalized topological spaces in evolutionary theory and combinatorial chemistry," *Journal of Chemical Information and Computer Sciences*, 42, 577–585.

Sylvester, J.J. (1878a). "On an application of the new atomic theory to the graphical representation of the invariants and covariants of binary quantics, with three appendices," *American Journal of Mathematics*, 1, 64–125.

Sylvester, J.J. (1878b). "Chemistry and algebra," *Nature*, 17, 284.

Weinhold, F. (1975). "Metric geometry of equilibrium thermodynamics," *The Journal of Chemical Physics*, 63, 2479–83.

Weyl, H. (1940). "The mathematical way of thinking," *Science*, 92, 437–46.

| # Science, φ-Science, and the Dual Character of Chemistry

REIN VIHALEMM

1 Introduction

A central question in philosophy of chemistry is the status of chemistry as a science: Is chemistry simply a physical science, a science of its own type, or something else? In traditional philosophy of science, physics has been considered the epitome of science, and chemistry was long regarded as a physical science. Recently, however, the "physical" interpretation of chemistry has become unpopular, because it implies in one way or another that chemistry can be reduced to physics—an idea which has come to be seriously questioned. Philosophers of chemistry now emphasize that all sciences need not be similar to physics. They have argued that chemistry is its own type of science, as, for example, biology has been recognized as a science in its own right. This view has been most directly expressed and systematically developed by Joachim Schummer, who observes: "Because it seems hard to decide whether chemistry more resembles physics, biology, technology, or whatever, I propose to handle it as its own type of science" (Schummer 1997, 329–330; cf. Schummer 2006; see also, e.g., van Brakel 1999, 134; 2000, 71–73).

Understanding chemistry as its own type of science emphasizes the experimental nature of chemistry and its contrast to the experimental basis of physics. Drawing on historical and scientometric studies, it has been argued that in contrast to natural history; to biology as an initially descriptive, empirical-inductive science; and to physics—as the epitome of mathematical, hypothetico-deductive science for which experiments(in which measurements are primary) are only tools for testing theories, thus keeping theoretical knowledge connected with "empirical reality"; and, on the other hand, of natural history—"chemistry has always been the laboratory science per se, such that still in the nineteenth century the term 'laboratory' denoted a place for experimental research in which *chemical* operations were performed" (Nye 1993, 50). The chemical laboratory became the model for all the other laboratory sciences when they replaced "thought experiments" by real experiments. Although chemistry is no longer the only experimental science, it is by far the biggest one and historically the model for all others" (Schummer 2004, 397–398). According to Schummer: "In the experimental sciences, experiments are not epistemic tools for checking theories; instead theories are instruments for guiding experiments" (407) and in chemistry "'experiment' for the most part means synthesis and analysis of new substances" (406). In connection with the emergence in the later twentieth century of mathematical chemistry, alongside mathematical physics, special attention has also been paid recently to chemistry's relations to mathematics (e.g., in two special issues of the journal *Hyle*—2012, 2013) as the relation between "the epitome of the experimental laboratory science" and "the deductive science *par excellence*" (Restrepo and Schummer 2013, 1). Although mathematics (which is not limited to numbers and space) has not been alien to chemistry, it has been shown that mathematics alone does not make the study of nature scientific. One main conclusion of recent scholarship is:

> Less than fifty years ago, philosophy of science was totally confined to philosophy of mathematical physics.... Although philosophy of science nowadays still keeps that disciplinary focus to a considerable degree, it has slowly become aware of the rich diversity of scientific methods and traditions, among which mathematical physics is unique in many regards and hardly comparable or transferable to other disciplines, like chemistry, biology, and even experimental physics.... Mathematical chemists should be aware that they cannot take mathematical physics as a model; they have to go their own way instead.
>
> (SCHUMMER 2012, 85–86)

Here I shall argue for the dual character of chemistry. Mathematical physics is certainly an important model in philosophy of science, though it should not be regarded as the only model. Chemistry, as "the epitome of the experimental laboratory science" (and the largest scientific discipline today, thanks to its industrial significance), is not limited to its experimental laboratory component. Even mathematical chemists, being chemists, cannot use mathematics constitutively (in the ways typical of some branches of physics). A central methodological question is: Why is mathematical physics "the" epitome of science, and to what extent is it (rightly) the very *model* of science? When analyzing this issue, special attention should be paid to the most fundamental question in the philosophy of science: What is science? Why has physics become *the* paradigm of science? Why and how does biology differ from that paradigm? Why and how does chemistry differ from that paradigm?

I address this question here by summarizing and further developing some of my previous analyses (Vihalemm 1995; 1999; 2001; 2003a, b; 200 4; 2005; 2007a, b, c; 2011a, b; 2013). These have developed a theoretical concept of science, as an idealized theoretical model of science, which is based on physics, but which includes comparative analysis of chemistry.

I call this model "φ-science." This term designates a physics-like model of exact, quantified, constructive-hypothetico-deductive science. When speaking about science generically, in the contemporary vernacular, two main types of cognition can be distinguished: (i) φ-scientific cognition, and (ii) non–φ-scientific cognition, in the form of natural-historical,[1] descriptive, and classificatory science such as classical geology or classical biology.[2] Modern chemistry has a dual character: It is a hybrid of constructive-hypothetico-deductive inquiry (i.e., φ-science) and classifying-historico-descriptive inquiry (i.e., non-φ-science, or "natural history"). Accordingly, it is incorrect to claim that chemistry *as a science* has a specific character (e.g., that chemical laws of nature or theories are peculiar).[3] This is not, however, to neglect what is distinctive about chemistry. Even regarding chemistry *qua* science, we cannot forget that it is not pure[4] φ-science, but is anchored in its non–φ-scientific origins. This dual character of chemistry provides an important opportunity for clarifying the practical and technological origins of science, and for exploring the ways in which, and the extent to which, the exact-scientific theoretical approach (regarded as "purely scientific"), having certain premises and limits, can be introduced into a technologically oriented, classifying-historico-descriptive inquiry.

---

[1] The term "natural history" is used, e.g., by Stephen Toulmin (see Toulmin 1967). He writes about "the differences between explanatory sciences, such as physics, and descriptive sciences, such as natural history" (40). On the one hand there are "crucial differences" (45), but on the other it is "interesting to consider how far the aims of any particular science are explanatory and how far they are descriptive. Most of the sciences which are of practical importance are, logically speaking, a mixture of natural history and physics" (50). In fact, such distinctions were already used by Immanuel Kant in his theory of natural science (Plaass 1994, van Brakel 2006, van den Berg 2011). Kant's conception is considered below.
[2] Actually, this non–ϕ-scientific, classifying-historico-descriptive type of cognition (and knowledge) can be construed both in a broader and in a narrower sense. Here it is used in the narrow sense—as natural history. In a broader sense, it includes social studies and the humanities as well. However, it is reasonable to regard, at least in some cases, the social sciences and the humanities as different from science in general (i.e,. not only as different from φ-science), although surely they should not be classified as studies of an "inferior type" (for details see Vihalemm 2011a, 86–92).
[3] According to some authors—see, e.g., Christie and Christie (2000)—laws of nature and scientific theories are indeed peculiar in chemistry. It can be argued, however, that laws of nature and scientific theories as categories of philosophy of science cannot vary in their nature from one discipline to another, granting that these disciplines are regarded as sciences (Vihalemm 2003a, 2005).
[4] Bernadette Bensaude-Vincent and Jonathan Simon have published an excellent book with a characteristic title *Chemistry—The Impure Science* (Bensaude-Vincent and Simon 2008). The authors refer to the hybrid nature of chemistry, to "its constant mix of science and technology" (5). The view of chemistry as "impure science," or a science with dual or hybrid nature, not only in the sense of being between science and technology, but also as lying between "the two venerable scientific traditions of physics and natural history" (p. 212), largely coincides with my treatment of the dual character of chemistry. This view of Bensaude-Vincent and Simon (see also Bensaude-Vincent 2008, Simon 2011) will be discussed further below.

2 Why Has Physics Obtained the Status of the Paradigm of Science? A Practical Realist Account of the Question in Historical Context

While it is true that science in general cannot be identified with physics, and thus, e.g., laws and theories of chemistry are not identical in kind to those in physics, it is also true that physics has been accorded the status of the standard for science. Why is this so? What gives physics, its laws and theories, this special status as the very epitome of science as such? Obviously, physics is not regarded as science simply because it is physics, and neither are physical laws and theories regarded as scientific simply because they are physical. It is little help to say that physical theories are scientific because they are mathematically formulated, because we also need to understand philosophically what such mathematical formulation amounts to, and why in physics it has succeeded so very well.

It is essential to recognize the premises and limits of knowledge and research believed to have the status of perfect, exact science (such as physics). Exact scientific cognition is paradoxical insofar as that theoretical knowledge presupposes empirical knowledge, yet this latter also presupposes the former. This is not an allusion to the familiar "theory-ladenness" of observation (which is not paradoxical). Instead, I want to highlight the origin of scientific theory, including the origin of knowledge assumed and required for making scientific observations (a priori as such knowledge seems to be); in other words, how is it at all possible to construct a non-speculative, physical theory?

2.1 In What Sense Is Kant's Notion of "Proper Science" Still Topical?

The paradox of exact scientific knowledge just mentioned is in fact Kant's problem, associated with his famous "Copernican revolution" concerning experience and apodictic knowledge generally: instead of assuming that "all our knowledge of objects must conform to objects," when we "suppose that objects must conform to our knowledge," it will "be possible to have knowledge of objects a priori, determining something in regard to them prior to their being given" (*CPR*: B xvi). Kant further developed this idea specifically with regard to how science as apodictic knowledge of nature is possible (Plaass 1994, van den Berg 2011). Kant's well-known dictum says:

> I assert... that in any special doctrine of nature there can be only as much *proper science* [*eigentliche Wissenschaft*] as there is *mathematics* therein. For, according to the preceding, proper science, and above all proper natural science, requires a pure part lying at the basis of the empirical part, and resting on *a priori* cognition of natural things.
>
> (MFNS 4: 470)

This pure part consists of the pure critical philosophy of nature in general— general metaphysics—as rational knowledge from mere concepts (i.e., the a priori concepts of the understanding of categories), together with mathematics as rational knowledge through the construction of concepts.[5] Kant's *Metaphysical Foundations of Natural Science* (*MFNS*) serves as the explanation and justification of the possibility of mathematical physics providing the special metaphysics which contains mathematics as the tool of a priori specification of empirical concepts.[6] Van den Berg (2011, 25) nicely summarizes why *only* mathematical sciences, exemplified by physics, are proper sciences:

> According to Kant, it is mathematics alone that provides a priori insight of specific quantitative properties of individual physical objects. This is a consequence of the fact that mathematics provides a priori models (individual and concrete representations) of physical objects....[In this way] mathematics provides a priori principles for cognizing physical laws [in cases affording apodictically certain knowledge—*R. V.*].

So, Kant's classical view of the scientific character of knowledge is that scientific knowledge is apodictically certain knowledge achieved by cognizing laws of nature. According to Kant, knowledge of laws of nature is only possible in sciences founded on pure philosophy and employing mathematical construction—as exemplified by physics. Only then can knowledge with empirical content have apodictic certainty, which always requires an a priori basis.

Generally speaking, knowledge which is both *synthetic* and a priori is transcendental;[7] however, Kant also regarded critical metaphysics as a kind of a priori synthetic knowledge. As is shown in Westphal (2004), Kant was not clear

---

[5] "The determination of an intuition a priori in space (figure), the division of time (duration), or even just the knowledge of the universal element in the synthesis of one and same thing in time and space, and the magnitude of an intuition that is thereby generated (number),—all this is the work of reason through construction of concepts, and is called *mathematical*" (*CPR*: A 724/B 752).

[6] It should be mentioned, however, that without reinterpretation (see also the note 7), "Kant's quasimathematical constructive metaphysical procedure is specious. Since that procedure is rooted in transcendental idealism" (Westphal 2004, 176).

[7] "Transcendental" here means the universal and necessary conditions determining in regard to objects of cognition, prior to their being given to a cognitive subject, *how* they as identified objects are possible at all; more concretely, "transcendental" denotes epistemic conditions that make synthetic and a priori knowledge possible for the human mind. Kant's philosophical position is empirical realism combined with transcendental idealism. Nevertheless I agree with Kenneth Westphal who argues in his (2004) and several articles that "transcendental idealism is not, *pace* Kant, required for" the critical tasks of his philosophy, and that "Kant's transcendental idealism is unsupported, false, nor can it fulfill some of the key aims Kant claims it alone can fulfill" (Westphal 2004, 34). He offers a positive reinterpretation of Kant's critical philosophy as unrestricted realism, or as he puts it, "a genuinely transcendental proof of realism *sans phrase*" (35). Westphal shows that Kant's defense of transcendental idealism was based on a "disjunctive syllogism: either empiricism or transcendental idealism is true; empiricism faces insuperable difficulties; therefore transcendental idealism is true. The problem with this disjunctive syllogism is Kant's inadequate effort to examine and defend its major premise" (83). According to Westphal, Kant did not consider the realist alternative. Kant, admittedly, considered and rejected transcendental realism as the opposing (transcendental) view to transcendental idealism, however, the realist alternatives to both empiricism and to transcendental idealism are *not* limited to transcendental realism in the sense of a metaphysical realism which presupposes the theocentric model of knowledge (Allison 2004, 27–38)—in other words, that knowledge represents the

about the fact that his first *Critique* actually already contained impure synthetic a priori knowledge of a kind which the *MFNS* identifies as officially metaphysical. This metaphysical method in turn was untenable, "because that method [was] rooted in Kant's transcendental idealism" (Westphal 2004, 175). According to Kant (*Prol.* 4: 320): "Even though it sounds strange at first, it is nonetheless certain, if I say with respect to the universal laws of nature: *the understanding does not draw its* (a priori) *laws from nature, but prescribes them to it.*"

This a priori prescription of laws to nature should be understood in the context of Kant's "Copernican revolution." In just this connection, Kant refers to scientific experiment:

> Accidental observations, made according to no previously thought-out plan, can never be made to yield a necessary law, which alone reason is concerned to discover. Reason, holding in one hand its principles, according to which alone concordant appearances can be admitted as equivalent to laws, and in the other hand the experiment which it has devised in conformity with these principles, must approach nature in order to be taught by it.

> (CPR: B XIII)

The apparently paradoxical character of exact-scientific cognition (previously noted) will cause no major difficulties if we deal, as in physics, only with experimental-theoretical research which, operating with experimentally substantiated idealizations, *constructs* its object of research, considering nature only through idealized, mathematically projected situations. This point is crucial, and is further examined below. Physics exemplifies experimental exact science in general, and in its purest form, making it possible to study the methodological structure and functions of exact science theoretically. This grounds the concept of φ-science, developed below.

Kant's "Copernican revolution" is doubtless of great importance for understanding science, though it needs to be interpreted—to use Georg Wilhelm Friedrich Hegel's term—through "sublation" (*Aufhebung*), that is, Kant's transcendental idealist account should be (1) "canceled," (2) "preserved," and (3) "transcended." I think it can be done from the position of practical realism (Vihalemm 2011c, 2012, 2013). The central issue is, how do we best understand our cognitive activity, and how we relate to the objects of inquiry through our cognitive activities? For practical realism, the starting point is within practice, thus precluding any transcendental idealist metaphysics with fixed, immutable prior principles. Knowledge, the knower, and the world which is known are all formed in and through practice. In brief, practice is human activity as a social-historical, critically purposeful, normative, constructive, material interference through interaction with nature and society, thus producing and reproducing the human world—culture—within nature. Opposing both contemporaneous idealism and materialism, including Ludwig Feuerbach's, Marx first showed (in his *Theses*

---

world as it is in itself, independently of any cognitive subject. This issue is discussed further below.

on Feuerbach) that the key defect in previous forms of materialism is that they all conceived the world, "the thing [*Gegenstand*], reality, sensuousness...only in the form of the *object* [*Objekt*] or of *intuition* [*Anschauung*], but not as *human sensuous activity, practice*, not subjectively. Hence it happened that the active side, in contradistinction to materialism, was developed by idealism—but only abstractly, since, of course, idealism does not know real, sensuous activity as such." Even Ludwig Feuerbach failed to grasp "human activity itself as *objective* [*gegenständliche*] activity" (Marx 1845, 1st Thesis). Note that when in the first sentence Marx describes human activity "subjectively," he does not mean "mentally," but instead *actively* as agents in practice, as indicated in the second sentence. In his second Thesis Marx stressed: "The question whether objective [*gegenständliche*] truth can be attained by human thinking is not a question of theory but is a *practical* question" (Marx 1845). In the *Economic and Philosophical Manuscripts of 1844* (in his *Critique of Hegel's Philosophy in General*), he wrote: "*nature* ..., taken abstractly, for itself—nature fixed in isolation from man—is *nothing* for man" (Marx 1844).

In practical realist philosophy, the subject and its practical activity, recognized as a legitimate part of objective (material) reality, have objective characteristics as well. The subject is incorporated into reality as its specific component, and consciousness is no longer regarded as its only constituent property: Literally, the subject is incorporated within reality; there are no incorporeal subjects. The impact of practice on reality is brought about not from "outside" but from "within" reality. It is the impact of one form of objective reality upon another—the impact of reality "in the form of activity" on reality "in the form of an object." The traditional model of knowledge acquisition treats subject and object as separate realities in their specific and independent existence, with their independent sets of characteristics. Activity is one subject's properties and, therefore, is external to any object. Thus, the object is also external to the activity, and independent of it; the problem is reduced so to speak to "tracing the contours" of this external object, or "cutting it at its joints." The practice-based approach implies instead that practical activity has a status more fundamental than the status of individual object-things. An individual object is identified as existent only through specifically defined activities within the context in which these objects appear as specific invariants.

2.2 Heidegger's Insights into the Essence of Exact Science

Martin Heidegger was one of the major philosophical thinkers of the twentieth century who illuminated the nature of exact science (see also, e.g., Glazebrook 2000, 2001, 2012; Rouse 2005a, b; Kochan 2011). Among much else, his analysis shows that Kant's reasoning is still vital today (see especially, Heidegger 1977a, b; 2011a, b, c). Here it is appropriate to quote some of Heidegger's more characteristic views on science, since I have myself—proceeding from the perspective of practical realism—reached almost the same conclusions (see, e.g., Vihalemm 2001; 2004). Heidegger writes (1977a, 117; cf. also 1977b, 157):

When we use the word "science" today, it means something essentially different from the *doctrina* and *scientia* of the Middle Ages, and also from the Greek *epistēmē*. Greek science was never exact, precisely because, in keeping with its essence, it could not be exact and did not need to be exact.

Heidegger calls today's science *research*. The latter, being mathematical,[8] is founded on the scientific world picture, on a fixed ground plan or projection of natural events. (Here parallels can be drawn to, for example, Imre Lakatos's research programs or to Thomas Kuhn's paradigms.) According to Heidegger (1977a, 119–120):

> Every event must be seen so as to be fitted into this ground plan of nature. Only within the perspective of this ground plan does an event in nature become visible as such an event. This projected plan of nature finds its guarantee in the fact that physical research, in every one of its questioning steps, is bound in advance to adhere to it. This binding adherence, the rigor of research, has its own character at any given time in keeping with the projected plan. The rigor of mathematical physical science is exactitude. Here all events, if they are to enter at all into representation as events of nature, must be defined beforehand as spatiotemporal magnitudes of motion. Such defining is accomplished through measuring, with the help of number and calculation. But mathematical research into nature is not exact because it calculates with precision; rather it must calculate in this way because its adherence to its object-sphere has the character of exactitude. The humanistic sciences, in contrast, indeed all the sciences concerned with life, must necessarily be inexact just in order to remain rigorous.

Science as project-based exact research is realized by using experiments, but as stressed by Heidegger—and here one must completely agree with him again:

> Physical science does not first become research through experiment; rather, on the contrary, experiment first becomes possible where and only where the knowledge of nature has been transformed into research. Only because modern physics is physics that is essentially mathematical can it be experimental. Because neither medieval *doctrina* nor Greek *episteme* is science in the sense of research, for these it is never a question of experiment.
>
> (GLAZEBROOK 2000, 121)

---

[8] "Mathematical" has to be construed here in a broader sense than referring to some particular discipline of mathematics. In a very general philosophical sense, as indicated by Heidegger, mathematics is a certain kind of a priori knowledge: "*Ta mathemata* means for the Greeks that which man knows in advance in his observation of whatever is and his dealings with things" (Heidegger 1977a, 118; see also 1977b, 170). The mathematical in science is its projective framework. Yet Heidegger's a priori is different from Kant's. As pointed out in (Glazebrook 2000, 50): "What he takes Kant to mean by "a priori," Kant in fact conveys by "pure". [...] Since the pure contains "no admixture of anything empirical" (B3), it is bound to transcendental idealism.... As the a priori carried the epistemic force of certainty for Kant, so the mathematical entails the certainty of givenness in Heidegger's analysis.... He looks not to Kant but to the ancients to raise the issue of epistemic certitude, and he raises that issue not as the question of the a priori, but as the question of the mathematical."

The defining features of experiment as a procedure of exact science consist in the following:

> Experiment begins with the laying down of a law as a basis. To set up an experiment means to represent or conceive [*vorstellen*] the conditions under which a specific series of motions can be made susceptible of being followed in its necessary progression, i.e., of being controlled in advance by calculation. [...] Experiment is that methodology which, in its planning and execution, is supported and guided on the basis of the fundamental law laid down, in order to adduce the facts that either verify and confirm the law or deny it confirmation. [...] The modern research experiment...is a methodology ... related to the verification of law in the framework, and at the service, of an exact plan of nature.
>
> (GLAZEBROOK 2000, 121–122)

Furthermore, a general cultural precondition for the birth of science in the form of physics (mechanics), founded by Johannes Kepler, Galileo, and Newton in the seventeenth century, was the scientific world picture, which is the world picture of the era of technology (see, e.g., Heidegger 1977a, 2011c; Stepin 2005). The scientific world picture provides a basis for treating the world as a modeled reality, and makes it natural to treat the world in this way. The world is understood, literally, as through a certain picture which expresses a construction, or a mechanism, based on a known project. Additionally, the scientific vision of the world means observing phenomena under conditions where these phenomena behave as idealizations: They can be reproduced and described mathematically, since they are subject to universal quantitative laws of nature. Such conditions are determined experimentally. In this sense, science itself determines which aspects of the world it investigates and how. No phenomenon exists for science that could be given for observation independently of the scientific way of treating it. In principle, Galileo and Newton began to connect mathematics with experiment. They began to study, through experiment, things that are subject to mathematics. They posed the problem in a way that was simultaneously experimental and mathematical—and thus, mathematically visible and provable. In the scientific world picture, the world is represented as an object without a (human) subject, as a rationally constructed mechanism operating in accord with the laws of nature.

The answer to the question—considered within the philosophical context—of why physics, rather than any other branch of science, is an exact science with corresponding experimental and theoretical constituents, and as such, a paradigm of science, should now be entirely clear.

Yet some further remarks may clarify how the main points of the present section relate to the position mentioned in the introduction, according to which chemistry can be seen—relying primarily on historical and scientometric studies—as the epitome of specifically *experimental* science and its relation to mathematics and to mathematical physics. Several points presented in this section (and later) may raise such questions as these: *Are* physical experiments exact, and are they exact *because* modern physics is essentially mathematical (as apparently championed by Galileo and Newton)? On the other hand, is it not the case that in the real world exact experiments are not at all possible? Even if, as quoted in the introduction, "mathematical physics is unique in many regards and hardly comparable or transferable to other disciplines," the use of mathematics is hardly limited to physics, and is widespread in chemistry, which is supposed to be "the epitome of the experimental science," is it not?

Regarding such questions please consider the following. It is of course possible simply to abandon a systematic approach by adopting a pluralistic view of science, according to which "science, construed simply as the set of knowledge-claiming practices that are accorded to that title, is a mixed bag" and "should be seen as a [Wittgensteinian] family resemblance concept" (Dupré 1993, 242). However, I think we can do better than that (Dupré's position is criticized in Vihalemm 2007, 229, also Vihalemm 2011: 87–88), though without recourse to a single, unified general account of all the scientific disciplines and practices which have emerged in our inquiries into our complex, inexhaustible world. The present chapter uses practical realism to analyze chemistry via appeal to a theoretical model of science which shows that chemistry has a dual character. This and the following sections highlight some key ideas from philosophical classics to show that the theoretical conception of science I develop has a firm backing in history of philosophy, including historical philosophy of science. The overarching question is how to understand at all theoretical knowledge (including mathematics) and its relationship to the real world. I approach these issues *via* practical realism which, together with an idealized model of exact mathematical science (φ-science), shows that theoretical scientific knowledge is rooted in an experimentally substantiated idealization procedure. (This "Galilean idealization" is explained in section 5). Exact scientific theoretical cognition of the world requires modelling it by using experimentally substantiated idealization procedures only regarding laws of nature.

3 The Relevance of Kant's Legacy to The Philosophy of Chemistry

Kant's concept of proper science (*eigentliche Wissenschaft*) has had enormous though unfortunate and unfavorable impact on the development of philosophy of chemistry. In his most famous works, and in his *Metaphysical Foundations of Natural Science* (1786), Kant treated chemistry as a merely empirical, albeit systematic and experimental art. Thus, chemistry became regarded as a "counter model" to proper science because it lacked a basis in the synthetic a priori, because it lacks metaphysical and mathematical principles:

> Hence, the most complete explanation of given appearances from chemical principles still always leaves behind a certain dissatisfaction, because one can adduce

no *a priori* grounds for such principles, which, as contingent laws, have been learned merely from experience.

(MFNS 4: 469)

This view has prevailed until very recently, with the important difference that physics has been substituted for metaphysics, and chemistry has come to be seen as proper science only to the extent that it is based on physics.[9] Recently, however, Kant's legacy for the philosophy of chemistry has been analyzed anew by Jaap van Brakel (2006), who has shown that Kant's later views on chemistry give reason to reconsider his significance for philosophy of chemistry. In the present context, too, there is good reason to take a fresh look at Kant's legacy, by interpreting it from the viewpoint of practical realism. From this perspective, Kant's quest for identifying both the metaphysical basis of natural science and the apodictically certain knowledge based upon it (represented by laws of nature), should rather be understood, respectively, as the problem of the scientific world picture, based on social-historical practice, and the problem of experimentally substantiated idealization procedures.

As described by van Brakel, Kant's interest in chemistry grew and deepened toward the end of his life and

> in the *Reflections on Physics and Chemistry* of the late 1790s and in the *Opus postumum*…of the same period, his interest in the work of contemporary chemists moved center stage and became an integral part of his philosophical project.…Chemistry presents the philosophy of nature with new problems, "unbeknownst to mathematical physics," viz. to give an account of the variety of substances. Not being able to give a philosophical account of the variety of substances, Kant started to see as a gap in his philosophy.
>
> (VAN BRAKEL 2006, 73)

In fact, a certain shift in Kant's attitude toward chemistry as a science is already evident in the preface to the second edition (in 1787) of the *Critique of Pure Reason*, where Kant mentions Stahl, the founder of the phlogiston theory, alongside Evangelista Torricelli and Galileo, as a scientist who best illustrates the "Copernican revolution":

> When Galileo caused balls, the weights of which he had himself previously determined, to roll down an inclined plane; when Torricelli made the air carry a weight which he had calculated beforehand to equal to that of definite volume of water; or in more recent times, when Stahl changed metals into oxides, and oxides back into metal, by withdrawing something and then restoring it, a light broke upon all students of nature. They learned that reason has insight only into that which it produces after a plan of its own, and that it must not allow itself to be kept, as it were, in nature's leading-strings, but must itself show the way with principles

---

[9] It should be noted that Kant's view was not reductive, i.e., he did not think that, in order to qualify as proper science, chemistry must be reducible to mathematical physics.

of judgment based upon fixed laws, constraining nature to give answer to questions of reason's own determining.

(CPR: B XII-XIII)

Usually, when chemistry is discussed in philosophy of science, this statement by Kant is neglected. Largely this is because the birth of scientific chemistry is not associated with the phlogiston theory, although there is very good reason to do so (see Vihalemm 2004, 2000), but rather with its refutation by Lavoisier in the process of the so-called Chemical Revolution.

Unanimously and quite rightly, Galileo has been regarded as the first exact scientist.[10] The law of free fall he discovered typifies a law of nature. (Below we shall reconsider this example as an instance of experimentally substantiated idealization.) However, what is often ignored but should be emphasized is that Stahl's accomplishments in chemistry are actually comparable to Galileo's law: Speaking in terms of "phlogistication" and "dephlogistication," he discovered (in today's terms) the reversibility of reduction and oxidation reactions (redox reactions, for short). Kant's example is precisely about how it became possible—indeed in *chemistry*—to go beyond mere empirical generalization, to achieve knowledge which is a kind of synthetic a priori knowledge, that is, to achieve apodictically certain knowledge represented by quantified laws of nature. On Kant's view, as previously noted, this counts as a priori prescription of laws to nature, though it accords perfectly with Heidegger's account of the relation between scientific experiment and laws of nature due to a specific kind of mathematical research, in other words, exact science.

Stahl's chemistry does not entirely meet Kant's requirements for a proper science—that is, knowledge founded on a fully developed metaphysical and mathematical foundation, as described by Kant in his *MF N*. However, further immersion into chemistry, as Kant came to give "chemistry a central place in his (later) philosophy" (van Brakel 2006, 81), forced Kant to acknowledge that there was a "gap" in his transcendental philosophy of nature. Kant was "deeply dissatisfied with the *MAdN* [*Metaphysical Foundations of Natural Science*] (or even his whole critical philosophy)" (van Brakel 2006, 81–82); indeed, a new a priori science was missing from Kant's system. I am very sympathetic to the conclusion in van Brakel (2006, 80–81, 82):

> Kant's view of mathematics as the "ideal" model changed, because he realized that he would not be able to account for the experience of the variety of substance on the basis of metaphysics modeled on mathematics: hence his tripartite "pact" between philosophy, mathematics, and natural science.... Given the crucial *philosophical* importance of the problems thrown up by chemistry, perhaps the notion of science had to be rethought such that chemistry could be a proper part of it.

---

[10] Galileo himself, however, could not call himself scientist—he was a mathematician and a natural philosopher—because the term "scientist" was coined only in the 1830s by William Whewell (Ross 1991, 9–10).

Significant here is that Kant already recognized that reconsidering the very notion of "science," so as properly to recognize chemistry as science, also requires reconsidering how and why mathematical physics had been regarded as *the* proper model for science as such. Regarding the relevance of chemistry to philosophy, specifically to philosophy of nature and to philosophy of science, Kant is still right: These important issues still require further attention from us today. However, there is a distinctive, superior alternative to both Kant's quest for a single, integrated philosophy of nature as a "metaphysical foundation" for science, and to the sheer pluralism about "the" sciences (advocated, e.g., by Dupré).

4 The Task of Philosophy of Science

So far I have discussed science mainly within a purely philosophical context, drawing from two great philosophical innovators: Kant and Heidegger, neither of whom is a philosopher of science, although both consider science to be of great philosophical importance. Hence philosophy of science must take a position regarding their views. We saw Kant, following the model of classical science—exemplified by physics and explicated by Kant as proper science as such—tried to refute dogmatic metaphysics and to replace it with a properly critical metaphysics that can present its scientific credentials (to recall the full title of his *Prolegomena*). Heidegger (2011b) also explicated the essence of science, exemplified by physics, as mathematical and (in a distinctive sense) as a kind of a priori, and explicated experimental research "in one concise statement... *Science is the theory of the real*" (Heidegger 1977b, 157). Two centuries after Kant, Heidegger found that philosophy *qua* metaphysics, pretending to be *the* fundamental theory of the real, had reached its end by attaining its utmost possibility in the natural sciences. Therefore, the issue of presenting metaphysics as a science became moot:

> The sciences are now taking over as their own task what philosophy in the course of its history tried to present in certain places, and even there only inadequately, that is, the ontologies of the various regions of beings (nature, history, law, art)....The operational and model-based character of representational-calculative thinking becomes dominant.
>
> (HEIDEGGER 2011b, 314)

We are living in the age of the scientific world picture (in his specific sense of "picture"; see Heidegger 1977a, 128 ff.). This is the world picture of the era of technology; in particular:

> One of the essential phenomena of the modern age is its science. A phenomenon of no less importance is machine technology. We must not, however, misinterpret that technology as the mere application of modern mathematical physical science to praxis. Machine technology is itself an autonomous transformation of praxis, a type of transformation wherein praxis first demands the employment of mathematical physical science. Machine technology remains up to now the most

visible outgrowth of the essence of modern technology, which is identical with the essence of modern metaphysics.

(HEIDEGGER 1977a, 116)

If philosophy is no longer justified—either as metaphysics or as positivist denial of metaphysics (and hence still dependent upon what it rejects!)—then how should the relation between science and philosophy be understood, and especially: How should we view philosophy of science and its potential for studying science? Heidegger emphasizes that the end of philosophy *qua* metaphysics is not at all the end of philosophical thinking, which retains an important task, "a task accessible neither to philosophy as metaphysics nor, even less, to the sciences stemming from philosophy" (Heidegger 2011b, 314); this is thinking as reflection (*Besinnung*) (Heidegger 1977a, 115–116, 137–138; 1977b). This kind of reflection is a certain kind of questioning about the very being of beings (entities, things)—namely, how such beings are intelligible and accessible to us at all. In the case of science, then, this questioning concerns the being of science and of those beings studied by science.

This "ontological difference," as Heidegger calls it, can be interpreted, as I do, from the practical realist viewpoint. Consider Heidegger's claim that metaphysics is Platonism, in one way or another: "All metaphysics, including its opponent, positivism, speaks in the language of Plato" (Heidegger 2011b, 320). Indeed, Heidegger associated the end of philosophy *qua* metaphysics (and anti-metaphysics) with the reversal of metaphysics by Karl Marx: "Throughout the entire history of philosophy, Plato's thinking remains decisive in its sundry forms. Metaphysics is Platonism. Friedrich Nietzsche characterizes his philosophy as reversed Platonism. With the reversal of metaphysics that was already accomplished by Karl Marx, the uttermost possibility of philosophy is attained" (Heidegger 2011b, 313). The single most important aspect in the reversal of metaphysics by Marx—primarily his (1845) mentioned above—is that Marx laid the foundations of practical realism (Vihalemm 2011c, 2012). As mentioned (§2.1, end), knowledge—both the knower (the subject) and the world (the real) which is known; and more specifically, how the world, existing independently of the subject, becomes intelligible for the subject (the real *as* a reality)—are all formed within practice. Heidegger himself refers to practice as *being-in-the-world*, or *Dasein*—literally, being here and now, within one's circumstances and presented by one's circumstances. One crucial question is, How do one's present concerns and activities structure how one regards and deals with those circumstances, what one selects as salient, what one regards as settled and fixed, what one regards as negotiable? These are among the factors which structure one's engagement with and awareness of those beings within one's circumstances which one regards as objects of whatever kinds and potentials.

Practical realism understands the subject-object relation in ways closely analogous to Heidegger's, which Kochan (2011, 89) characterizes as follows:

> On Heidegger's account...the root problem of both realism and antirealism is that they both begin their analysis *one level too late*. They start at the level of subject and object rather than at the more primitive level of *Dasein* [being-in-the-world] and world. Insofar as both realists and antirealists rely upon a theoretical

conception of the world *as* reality [i.e., an ontological construal of the world *qua* objecthood (Kochan 2011, 95)—*R.V.*], they both fail to get at the real in a way which discloses entities in their more fundamental, non-objective state.

Moreover, practical realism has no difficulties with Heidegger's distinction between being and entities, characterized by Rouse as "widely misunderstood" (Rouse 2005a, 2):

> In posing the question of being (of what it means to be, or of the intelligibility of entities *as* entities), Heidegger sought to circumvent unexamined assumptions about knowledge or consciousness, and engage in a more radical philosophical questioning. Drawing upon Greek and medieval philosophy, he spoke of the "being" of an entity as a way of considering its intelligibility as the entity it is. In taking over the term, Heidegger sought to avoid assuming that the intelligibility ("being") of entities is itself an entity (a meaning, an appearance, a concept, or a thought).
>
> (ROUSE 2005a, 3)

In this sense of reflective, philosophical thinking, the task of philosophy of science is a well-formed task.

Consider the title of the well-known book by Alan Chalmers (1999), *What Is This Thing Called Science?* Chalmers stresses:

> There is a sense in which the question that forms the title of this book is misguided, [because] there is no general account of science and scientific method to be had that applies to all sciences at all historical stages in their development. Certainly philosophy does not have the resources to provide such an account. Nevertheless, a characterisation of the various sciences at various stages is a meaningful and important task.
>
> (CHALMERS 1999, 247)

Scientists themselves are typically not very good at handling this task; neither are they particularly interested in handling it. "Although it is true that scientists themselves are the practitioners best able to conduct science and are not in need of advice from philosophers, scientists are not particularly adept at taking a step back from their work and describing and characterizing the nature of that work" (Chalmers 1999, 252). This descriptive and reconstructive task is the task of philosophy of science construed as reflection upon the nature and status of science. Hence we can see in what sense the question "What is this thing called Science?" is not misleading. Chalmers (1999, 248) shows that the key issue is not a universal definition of science, but "debates about what is or is not to count as science, exemplified, for example, in disputes about the status of 'creation science'.... The main aim of those who defend creation science under that name is to imply that it has a character similar to that of acknowledged sciences such as physics." Therefore, it is important to clarify what kind of thing physics is, or has been, which is exactly what Chalmers does in his book, largely by means of historical examples. His account justifies viewing physics as a standard of science.

5 A Theoretical Model of Science—φ-Science—as a Galilean Idealization

I suggest that Chalmers's account can be made more precise by using the concept of φ-science. This concept helps understand in what sense physics represents exact experimental science in general, in its purest form, making it possible to study the methodological structure and functions of exact science theoretically. The main idea is that scientific inquiry into nature focuses on idealized models. I have also applied this principle at the metalevel to studies of science: φ-Science is an idealized model of exact science based on physics. The model has its prerequisites and limits. First, like any model, it need not correspond entirely to the object studied. This means that φ-science does not correspond entirely to the actual phenomena called "science," which is so poorly captured by any single definition. What is important is that this model enables us to determine whether, or to what extent, a particular field of study can be modeled as similar to research widely regarded as science and seen as (nearly) "ideal inquiry," even though this kind of research has its own prerequisites and limits.

As noted, a general cultural prerequisite for the formation of exact science is the emergence of the so-called scientific world picture; φ-Sciences are research domains which best correspond to this picture and provide it with content. However, we must bear in mind Heidegger's (1977a, 129) statement: "World picture, when understood essentially, does not mean a picture of the world but the world conceived and grasped *as* picture." The formation of this picture was originally one of the general cultural prerequisites of science.

The prerequisites and limits of exact science (φ-science) can now be interpreted in terms of Kant's "Copernican revolution." Exact science (constructive-hypothetico-deductive cognition) presupposes a correspondence of its object to its theoretical preconceptions of cognition; in the sense specified earlier, these preconceptions are the practical a priori orientation to and preparations for the research so conducted. Accordingly, the objects of inquiry are limited to those of their features which suit the model provided by such disciplined research. In this connection, consider again Kant's and also Heidegger's discussions of Galileo the revolutionay founder of exact-scientific method of cognition.[11] Galileo is essential to understanding the procedure of experimentally substantiated idealisation (Akhutin 1982, McMullin 1985, Matthews 2005). This procedure is central in φ-science: It is the key to solving the paradox of how exact scientific cognition can be both theoretical and empirical. Galileo's experimental thinking provides an excellent illustration of the principles of practical realism, because experiment is the main form of practice within science. This allows us to show, from the viewpoint of practical realist philosophy of science, how it is possible to overcome (or "sublate," as suggested earlier)

---

[11] On Heidegger's discussion of Galileo see Glazebrook 2000: 82–86 *passim*; Cahoone 1986.

Kant's transcendental-idealistic apriorism; more specifically, it shows how practical realism can interpret constructively Heidegger's understanding of a priori (as mathematical), and his statement "Science is the theory of the real."

Galileo was, first and foremost, a theorist thinking in terms of experiment. The foundation for his experimental way of thinking was laid in his early encounters with mechanical arts. This connectedness to mechanics never ceased. Galileo still had to resist Aristotelean ways of thinking, which accorded mechanical arts no theoretical attention because these construct artificial situations, whereas the theorist's task was to explore nature in its original, natural, uncontrived state. Yet Galileo, already acting in an age of technology, believed that precisely the "mechanical arts" help to discover laws of nature by revealing hidden aspects of natural phenomena. The theorist cannot rely merely on direct observation of nature because in direct observation natural phenomena do not manifest themselves in "pure form," with all their characteristic aspects. The task of the theorist is to obtain knowledge which is definite, general, and necessary by its nature; hence knowledge is not gained by observing various individual phenomena, but instead presumes as its object a definite, stable, self-identical, general, and necessary phenomenon as the causal-structural possibility of all relevant individual phenomena. Such an object must be constructed in a special way. Here the constructing experience of "mechanical arts" becomes helpful, but these arts must be theoretically understood. One must know how to perform thought experiments, and how to construct idealized objects, because only such objects are free from all sorts of extraneous influences, are identical to themselves, and are absolutely precise. However, Galileo's idealizations have fixed limits, and are not speculative. This is crucial. Galilean thought experiment is an idealization of real experiments constructed in a specific way. The theorist must find certain aspects of real phenomena, or a particular combination of these phenomena, for creating conditions in which they reveal their theoretical characteristics, they so to speak "idealize themselves" by closely approximating an idealized object. This means that the theoretical world of a science, with all its thought experiments and idealized objects, is not at all independent of real experiments and objects: to the contrary, it is shaped by these experiments and objects, and represents them. The theoretical world is constructed by a disciplinary practice in a way which makes possible a transition from theoretical objects—idealizations—to the real world. In modern scientific theories it is often difficult to trace the connections between theory and experiment (see Stepin 2005), although these connections do exist, whereas for Galileo we see clearly that a theoretical view defines certain schemes for experiment: that theory is experimental, and experiment is theoretical. Experiment "serves not only to connect the real and the ideal, for the entire theory participates in the experiment—in its design and execution—so it is the experiment that is the theory itself, now become the instrument for investigating the world" (Akhutin 1982, 298).

This point deserves both emphasis and clarification, to understand both Galileo's scientific innovation and my idealized model, φ-science. Mathematics is applied in many ways in many sciences, both physical and social. However, there is a very specific way in which idealizations, especially mathematical idealizations, are connected with experiments such as Galileo's kinematics and in φ-science. In φ-science as in Galileo's kinematics mathematics is not simply applied; Galileo's kinematics and φ-science are *essentially* mathematical insofar as the "world" of Galileo's kinematics, and the "world" of φ-science, is so constituted that its entities are to be mathematically defined; they are constituted *as* objects of scientific knowledge only insofar as they are mathematically construed, such as "mass" and "force" in classical mechanics, or other kind of forces in other branches of physics, which connect mathematical, φ-scientific theory with experiment. Put otherwise, φ-scientific theory represents its mathematically defined idealizations as really existing objects which in specified experimental situations have mathematical characteristics and accordingly behave as idealized objects. Galileo demonstrated this perfectly in his famous two *Dialogues* (Galilei 1952, 1967).

Consider how Galileo discovered the law of free fall. In accordance with principles discussed above, Galileo began with the conception that the "real" free fall of bodies is not manifest in usual conditions, in the "natural state." In order to understand the phenomenon studied, one must necessarily take into account other, unusual conditions; in fact, one should take into account *all* possible conditions. In other words, one must explore which conditions influence the phenomenon, in order to obtain it in pure form (free fall, for instance), or to specify relevant ceteris paribus conditions. Relying on various observations and experiments, Galileo saw that the velocity of freely falling bodies depends on the resistance of the medium. Free fall, properly speaking, only occurs in idealized conditions—in a vacuum. Thus Galileo's concept of free fall is a theoretical concept expressing an idealized object, and the law of free fall governs the behavior of such idealized objects. Consequently, it expresses theoretical knowledge *without* relying upon speculative reasoning. Galileo reached the conclusion that the velocity of freely falling bodies does not depend on their weight by making experiments which showed that, as the resistance of the medium is decreased, the difference between the velocities of bodies with different weight also decreases. It is thus logical to conclude that if we reach the ideal situation in which resistance is eliminated, the difference between the velocities of different bodies disappears altogether: in a vacuum, all bodies fall at the same rate. In order to demonstrate this law, Galileo constructed an ingenious experiment which "embodied" idealized objects and conditions, by reducing the issue to isochronic oscillations of pendulums and showing that, as the resistance of the medium is decreased (in Galileo's construction, this can be approximated by reducing the amplitude of oscillations or by using longer strings), oscillations of pendulums become more and more isochronic (Galilei 1952, 167 ff.). We see that the process of idealization turns out to be a transition to the limit: a tendency in the change of empirical factors is extrapolated to infinity.[12] Such

---

[12] "The final move is a transition from probability to reality, which is related to a sort of jump across infinity, to a transition towards a vacuum. This transition is identical to that which, in differential calculus, is called extremal transition. Experimental idealisation is a transition from

idealized objects—obtained from real objects by such a limit process—are mathematically definable; these objects themselves have become mathematical objects. Instead of the real body, Galileo obtained a point; instead of the real medium, he obtained the geometrical space; he thus reduced the physical question to a geometrical one. The mathematical construction is a natural continuation of the experimental construction: We can make the transition from the one to the other, and can model one on the other. This is why Galileo could say

> Philosophy is written in this grand book, the universe, which stands continually open to our gaze. But the book cannot be understood unless one first learns to comprehend the language and read the letters in which it is composed. It is written in the language of mathematics, and its characters are triangles, circles, and other geometric figures without which it is humanly impossible to understand a single word of it; without these, one wanders about in a dark labyrinth.

(GALILEO 1957, 237–238)

Galileo's use of such mathematical idealizations, drawn directly from Archimedes, is especially clear in his account of the rate and path of the free fall of a stone along a tower, in which the stone appears to us to fall straight down, parallel to the tower, but were it viewed from the earth's axis of rotation, would be seen to inscribe part of one of Archimedes' spirals moving at a uniform rate. His demonstration is too intricate to abbreviate here, but speaks brilliantly for itself.[13]

Now in an important sense, φ-science too is a Galilean idealization, an instrument for investigating an important phenomenon of the modern age: its science. The concept of φ-science enables us to elucidate the basic presumptions and limits of physics serving as the standard of science. It helps us assess the possibility (or, respectively, the impossibility) of applying the kind of inquiry typical of physics to other disciplines. Science understood as φ-science means a specific way of seeing and understanding the world: it means modeling the world from the viewpoint of making it subject to mathematically exact laws of nature, thus enabling scientific explanations and predictions. What cannot be captured by such objective laws simply has nothing to do with φ-science; it is not "seen" by it, and needs some other form of inquiry and cognition. It is not an aim of φ-science to provide a truthful picture of the world in all its diversity, but to discover specific kinds of laws. When φ-science explores some phenomenon, it aims to discover those aspects of the phenomenon subject to quantified, mathematically de-

---

an infinite sequence of real situations to a situation that is infinitely different from them, yet also a mental continuation of them is the external transition from real experiment to its mental continuation. Then a return journey has to be made. ... Consequently, the worlds of real terrestrial events and ideal mathematical objects are bound together by the experiment in the extremely idealized transition, although they differ infinitely (a mathematical object cannot exist and existing objects cannot have mathematical definiteness)" (Akhutin 1982, 296).

[13] His demonstration is roughly in the middle of the second day of his *Dialogue Concerning the Two Chief World Systems* (Galileo 1967, 164–167).

fined laws, to determine how exactly, and to what extent, the phenomenon is subject to those laws, and to establish what is possible or impossible according to these laws. Therefore, let us consider the central concept of φ-science—a law of nature.

From the viewpoint of logic, laws of nature have traditionally been understood as a certain type of propositional form expressed in terms of predicate calculus as follows:

$\forall x(P(x) \rightarrow Q(x))$ or, for short: $\forall x F(x)$

The inductive empiricist interpretation of these sentences is

$F(a)$ & $F(b)$ & $F(c)$....

This formula expresses an infinite conjunction: a is F (a has the quality or characteristic F) and b is F and c is F etc., where a, b, c, and so on are random objects from their respective total set. However, unless we take into account *all* these objects (and there's an indefinite number, potentially an infinity of them), we have no right to formulate an apodictic law concerning the objects of this set.

A law is something altogether different. If $\forall x F(x)$ expresses a law—an apodictic proposition—then x expresses only a variable: The truth value of the corresponding proposition does not depend on the concrete values of this variable. If, in contrast, $F(x)$ is a propositional function expressing a concept, then the outcome of substituting concrete values for x is either a true or false proposition. But a law says nothing about the concrete objects, that is, the elements belonging to the set. It only concerns the predicate F which is the same for all objects (in the whole "infinite conjunction"). Therefore, it is more exact to express the logical form of a law of nature in the second-order predicate logic as follows (Gryaznov 1982, 130):

$\exists F \forall x F(x)$.

The sentence reads: "There exists a predicate F, such that in case of any x, x is F. Here, the characteristic F does not depend on the concrete value of x, that is, it does not depend on the individual object that is substituted for x.

From the logical point of view, the scientist's task in discovering laws of nature is to find a predicate—of general and apodictic nature—for expressing a law. From the perspective of actual scientific activity, finding such a predicate means constructing an idealized object. For example, the law of free fall means in fact the constructing of the idealization of the "freely falling body." This construction has nothing to do with generalizing from individual cases; instead it fabricates the general case, thus obtaining of the phenomenon in its "pure form." In the condition of free fall, any object moves with the same acceleration and with velocity $v = gt$, and it covers the distance $s = gt^2/2$.

Thus, the concept of φ-science enables us to explain how and in what sense it is possible to meet Kant's requirements for "proper science," to show the possibility of apodictically certain knowledge, without positing any metaphysical foundations for natural science, based on the untenable transcendental idealist philosophy of nature.

6 Two Main Types of Cognition and the Dual Character of Chemistry

As suggested in the introduction, when speaking about science in general, two main types of cognition can be distinguished: (i) φ-scientific cognition, having a constructive-hypothetico-deductive character; and (ii) non-φ-scientific (natural-historical) cognition, having a classifying-historico-descriptive character. Using the theoretical model of φ-science, it can be shown that modern chemistry has a dual character. On the one hand, chemistry has features characteristic of φ-scientific cognition, or of "purely scientific" theoretical inquiry. From this point of view, it is a science in the same philosophical sense as mathematical physics. On the other hand, chemistry remains involved with its non-scientific (i.e., non-φ-scientific) origin, retaining some aspects characteristic to natural history, and some aspects characteristic to a technological discipline.

In previous work (e.g., Vihalemm 2001, 2007), I have argued that the dual nature of chemistry arises from the fact that since chemistry (and not only in its pre-scientific stage of evolution as technology) investigates particular kinds of substances, or stuffs, and their transformations, its primary task is not to discover laws of nature (unlike pure φ-science, which can be *defined* by the concept of laws of nature). Chemistry as non-φ-science aims to identify and classify substances and their modes of transformation, whereas chemistry as φ-science investigates *what* in the substances, what in their properties and transformations, can be modeled as obeying laws of nature. Regarding its primary tasks, chemistry belongs to natural history and is radically different from the constructive-hypothetico-deductive inquiry characteristic of φ-science. This has led to difficulties in considering chemistry merely as a physical science and has required seeking an alternative to the paradigm of science modeled on mathematical physics. We saw that Kant's conception of proper science also—and even his Critical philosophy as a whole—ran into difficulties due to chemistry, which revealed an irreparable "gap" in Kant's transcendental philosophy of nature. The problem was how to account for the experience of the variety of substances, while presuming a metaphysics modeled on mathematics to create a comprehensive a priori science. In this connection, as we saw, van Brakel too found that "perhaps the notion of science had to be rethought such that chemistry could be a proper part of it." However, the problem instead is that the Kantian notion of proper science, embodied in mathematical physics, had to be rethought, freeing it from the need for any transcendental idealist philosophy of nature as its metaphysical foundation, by reinterpreting this concept of proper science as a theoretical, idealized model of science within the framework of practical realist philosophy of science, using instead the concept of φ-science.

Due to its dual nature chemistry gives us an especially important opportunity for demonstrating how the conception of φ-science works. (Indeed, this conception has largely been developed on the basis of these characteristics of chemistry.) As pointed out here and elsewhere,[14] philosophical analysis of chemistry precludes our simply identifying exact science with physics. Instead, it suggests that we should find out and philosophically explain why physics obtained the status of the paradigm of science—as natural as this status may seem to be!—and to reveal both the premises and also the limits of this particular type of cognitive practice. One central feature of this explanation is that in actual historical practice—in this culture of the technological era and of the scientific world picture—physics has (after Galileo and Newton) obtained the status of the paradigm of science so securely and, so to speak, naturally, that it blocks any apparent need for critical reflection, for any reconsideration of why "physics" has become more or less synonymous with "real science." It should be realized that theories, laws, concepts, and so on are not scientific simply because they are physical. It is obvious, for instance, that being a mathematically formulated physical theory or law is not itself the reason why this physical theory or law has gained the status of an embodiment of science as such. The fact that chemistry is not a purely physical science, in the sense of not matching exactly the paradigm of proper science, makes its history a very clear example of the introduction of such a paradigm of exact into a field which was originally non-exact science. This paradigm of proper science concerns constructing scientific concepts and theories and formulating scientific laws—provided we bear in mind that the terms "scientific" and "physical" are not synonyms.

For instance, Mendeleev's periodic law, although not a mathematically formulated law of physics (in the typical sense of the term "mathematical"), is a real law of nature; it is exact in the same philosophical sense as are the laws of physics (Vihalemm 2003a; 2005; 2011b, 101–103). Mendeleev's periodic law has been one of the most characteristic and also one of the most puzzling examples of chemical laws and theories. This law seems to be essentially different in its nature from the exact laws of classical physics. However, the periodic system of chemical elements was established by constructing an idealized system of idealized elements. Reference to the theoretical concept of a chemical element is a fundamental idealization substantiated by experimental chemistry—namely, a definite position in the periodical system based on the periodic law.

Comparing chemistry with physics, the philosophy and methodology of science can learn from the actual history of science what the premises and limits of a science as exact, ideal science are. Biology is not a good example of this kind of ideal exact science because the resistance of the material is too strong, so to speak. Biology is clearly regarded as an altogether different type of cogni-

---

[14] Including my presentation at the 14th Congress of Logic, Methodology and Philosophy of Science (Nancy, France, July 19–26, 2011) and the paper based on it, "Philosophy of Chemistry against Standard Scientific Realism and Anti-Realism," submitted for the special issue—as Proceedings of the Congress—of *Philosophia Scientiae*, 19(1), March 2015.

tive practice, although it is traditionally regarded as "science," though of its own type. However, modern biology too has to some degree become a discipline with a dual nature: molecular biology and genetics, for instance, apply φ-scientific models. Yet life cannot be constructed from scratch, investigating living systems requires a classifying-historico-descriptive approach. Curiously, physics too seems to be acquiring a somewhat dual character. The emergence of physical theories concerning self-organization (as developed by Ilya Prigogine and others) indicates that physics itself *qua* φ-science has certain premises, actual aims, and definite limits (see Vihalemm 2007b; 2001: 195-196, 198). The physical study of self-organization is not purely φ-scientific, that is, constructive-hypothetico-deductive, research any more. It is rather a combination of classifying-historico-descriptive inquiry and constructive-hypothetico-deductive research, both characteristic of chemistry. In principle, a self-organizing system cannot be "constructed" because its organization and behavior cannot be prescribed and created by an external source. It emerges autonomously in certain conditions. The research task is to investigate in what kinds of systems and under what kinds of conditions self-organization will emerge.

Several authors (Schummer 1997, van Brakel 2000, Bensaude-Vincent 2008, Simon 2012) have suggested that chemistry has greater affinity to technology than to mathematical physics. This is why chemistry has been regarded as an "impure" science. In this connection Bernadette Bensaude-Vincent and Jonathan Simon (2008) refer to the hybrid nature of chemistry. As noted at the outset, this view coincides with the notion of the dual nature of chemistry. In my view, what is especially valuable in Bensaude-Vincent's and Simon's philosophical analysis of chemistry—which I regard as a practical realist approach—is their insistence on its importance to philosophy of science in general:[15]

> It should, by now, be clear... that the popular image of chemistry as a superficial empirical science obliged to seek its philosophical foundations in other more fundamental science is quite inaccurate, if not philosophically defamatory. Whether this vision of chemistry is the deliberate construction of philosophers of science with a predilection for physics, or just results from the lack of attention paid to chemists' concepts and methods, it does great disservice to philosophy, depriving it of an interesting practice-based approach.
>
> (BENSAUDE-VINCENT AND SIMON 2008, 209)

In Conclusion

To conclude, I think that it is precisely the dual character of chemistry that gives an interesting opportunity for exploring the notion of φ-science, because it shows clearly the difference between physics, as it exists in actual fact, and

---

[15] See also reviews and the authors' responses to them ("Book Symposium") under the informative heading: "Ask not what philosophy can do for chemistry, but what chemistry can do for philosophy" (Chang et al 2010).

physics-like science as the theoretical model of science. Chemistry is not just a physical science whose theoretical foundations are given by physics; it cannot be reduced to physics. Yet neither is chemistry so different from physics—from the perspective of the theoretical model of science—that one should start to protest against the "domination" of the image of science as physics, and aim at a pluralistic understanding of science. φ-Science as a theoretical model of science cannot give—and is not intended to give—a complete description of science, if we take into account all areas of knowledge and inquiry normally included in the general category of "science." The latter may be incommensurable with φ-science, but this does not exclude co-operation between φ-scientific and non-φ-scientific approach (cf. Vihalemm 1999, 86–88; 2007a, 232).

I also wish to remark briefly on the evolution of conceptual systems of chemistry (for details see: Vihalemm 2001, 190–196; 2007c). Having a dual nature, chemistry as φ-science, on the one hand, constructs its object of research as an idealized, mathematically specified object emerging within an experimental (i.e., artificial) situation. On the other hand, it applies non-φ-scientific treatment of nature which corresponds to the principles of *poiēsis*: instead of producing substances by "anti-natural" manufacturing, it obtains them in the natural way as they are found and produced in ecological systems. This contradiction between artificial and natural leads us to appreciate that nature itself is most skillful at constructing substances. It seems plausible that the conceptual structure of the development of chemistry—corresponding to the Aristotelian "four causes"—from theories of chemical composition, to theories of chemical structure, to theories of chemical kinetics, and finally, to theories of chemical self-organization, characterizes a general tendency in the development of natural sciences. These sciences move, so to speak, from the artificial to the natural world, from the constructed and organized phenomena toward the self-organizing and behaving phenomena.

The classical view of science and scientific objectivity is paradoxical. It is "subjective objectivity." The objective scientific world picture is constructed by the subject (the scientist), according to special criteria of what counts as science, and by prescribing specific conditions of cognition. In the scientific world picture, the world is regarded as a subject-free object described from outside, as if the describer did not belong to it. Such a view is illusory, of course, if it is understood as the objectively true picture of the world (nature) itself. φ-Scientific cognition has certain premises and limits, and is effective within the framework of these premises and limits, but it cannot claim to have the status of ideal cognition and knowledge in general.

The natural world of complex processes is not the same as the predictable and invariant world of φ-science. Knowledge about the natural world "in its naturality" cannot be obtained within the framework of the constructive-hypothetico-deductive approach, but also requires classifying-historico-descriptive type of inquiry.

Acknowledgments

The research was supported by the ETF7946 grant and SF0180110s08 grant. I thank Tiiu Hallap for improving the English of this chapter. I give special thanks to Kenneth Westphal, who carefully read the manuscript and suggested many stylistic corrections as well as substantial clarifications, particularly on Kant's transcendental idealism.

References

Akhutin, A. V. (1982). "Galileo's Experiment," *Scientia*, 117, 287–300.

Allison, H. E. (2004). *Kant's Transcendental Idealism*. New Haven, CT: Yale University Press.

Bensaude-Vincent, B. (2008). Chemistry beyond the "Positivism vs. Realism" Debate. In *Stuff: The Nature of Chemical Substances* (Ruthenberg K., van Brakel, J., eds.). Würzburg: Köningshausen & Neumann, 45–53.

Bensaude-Vincent, B. and Simon, J. (2008). *Chemistry—The Impure Science*. London: Imperial College Press.

Cahoone, L. E. (1986). "The interpretation of Galilean science: Cassirer contrasted with Husserl and Heidegger," *Studies in History and Philosophy of Science*, 17, 1–21.

Chalmers, A. F. (1999). *What Is This Thing Called Science?* (3rd ed.). Buckingham: Open University Press.

Chang, H., Nordman, A., Bensaude-Vincent, B., Simon, J. (2010). "Ask not what philosophy can do for chemistry, but what chemistry can do for philosophy," *Metascience*, 19, 373–383.

Chemistry and Mathematics, Part 1. (2012). *Hyle*, 18 (1).

Chemistry and Mathematics, Part 2. (2013). *Hyle*, 19 (1).

Christie, J. R. and Christie, M. (2003). Chemical laws and theories: a response to Vihalemm," *Foundations of Chemistry*, 5 (2), 165–177.

Christie, M. and Christie, J. (2000). "'Laws' and 'Theories' in Chemistry Do Not Obey the Rules." In *Of Minds and Molecules: New Philosophical Perspectives on Chemistry* (N. Bhushan and S. Rosenfeld, eds.). New York: Oxford University Press, 34–50.

Galilei, G. (1952). "Dialogues Concerning the Two New Sciences." In *Great Books of the Western World, 28* (R. M. Hutchins, ed.). Chicago: Encyclopaedia Britannica, Inc., 123–260.

Galilei, G. (1957). "The Assayer." In *Discoveries and Opinions of Galileo* (Selections translated by Stillman Drake). New York: Doubleday & Co., 231–280.

Galilei, G. (1967). *Dialogue Concerning the Two Chief World Systems—Ptolemaic & Copernican*. (Stillman Drake, trans.), foreword by Albert Einstein (2nd ed.). Berkeley: University of California Press.

Glazebrook, T. (2000). *Heidegger's Philosophy of Science*. New York: Fordham University Press.

Glazebrook, T. (2001). "Heidegger and scientific realism," *Continental Philosophy Review*, 34, 361–401.

Glazebrook, T. (ed.) (2012). *Heidegger on Science*. New York: State University of New York Press.

Gryaznov, B. S. (1982). *Logic, Rationality, Creativity*. Moscow: Nauka [in Russian: Грязнов, Б. С. Логика, рациональность, творчество. Москва: Наука].

Heidegger, M. (1977a). "The Age of the World Picture." In *The Question Concerning Technology and Other Essays*. (William Lovitt, trans.). New York: Harper and Row, 115–154.

Heidegger, M. (1977b). "Science and Reflection." In *The Question Concerning Technology and Other Essays*. Trans. William Lovitt. New York: Harper and Row, 155–182.

Heidegger, M. (2011a). "Modern Science, Metaphysics, and Mathematics." In *Martin Heidegger. Basic Writings from* Being and Time *(1927) to* Task of Thinking *(1964)* (David Farrell Krell, ed.). With a foreword by Taylor Carman. London: Routledge Classics, 183–212.

Heidegger, M. (2011b). "The End of Philosophy and the Task of Thinking." In *Martin Heidegger. Basic Writings from* Being and Time *(1927) to* Task of Thinking *(1964)* (David Farrell Krell, ed.). With a foreword by Taylor Carman. London: Routledge Classics, 2011, 311–325.

Heidegger, M. (2011c). "The Question Concerning Technology." In *Martin Heidegger. Basic Writings from* Being and Time *(1927) to* Task of Thinking *(1964)* (David Farrell Krell, ed.). With a foreword by Taylor Carman. London: Routledge Classics, 217–238.

Kant, I. (1902–). *Kants Gesammelte Schriften*, 29 vols., Königlich Preußische (now Deutsche) Akademie der Wissenschaften, G. Reimer (now De Gruyter), Berlin. Cited by volume: page numbers, following initials of the English title.

Kant, I. (1929). *Kant's Critique of Pure Reason* (N. K. Smith, trans.). New York: St. Martin's. Designated herein as "*CPR*," and cited by "A" or "B" for its two editions.

Kant, I. (1995–).*The Cambridge Edition of the Works of Immanuel Kant in Translation* (P. Guyer and A. Wood, eds.). Cambridge: Cambridge University Press.

Kant, I. (2002a). *Theoretical Philosophy after 1781* (H. Allison and P. Heath, eds.). Cambridge: Cambridge University Press.

Kant, I. (2002b). *Prolegomena to Any Future Metaphysics that will be able to come forward as Science* (1783) (G. Hatfield, trans.). Designated herein as "*Prol.*"

Kant, I. (2002c). *Metaphysical Foundations of Natural Science* (1786) (M. Friedman, trans.). Designated herein as "*MFNS.*"

Kochan, J. (2011). "Getting real with Rouse and Heidegger," *Perspectives on Science*, 19 (1), 81–115.

Marx, K. (1844). *Economic and Philosophical Manuscripts of 1844*. http://www.marxists .org/archive/marx/works/1844/manuscripts/hegel.htm

Marx, K. (1845). "Theses on Feuerbach." *http://chss.montclair.edu/English/furr/gned/ marxtonf45.pdf*

Matthews, M. R. (2005). "Idealisation and Galileo's Pendulum Discoveries: Historical, Philosophical and Pedagogical Considerations." In *The Pendulum: Scientific, Historical, Philosophical and Educational Perspectives* (M. R. Matthews, C. F. Gauld and A. Stinner, eds.). Dordercht: Springer, 209–235.

McMullin, E. (1983). "Galilean Idealization," *Studies in History and Philosophy of Science*, 16, 247–273.

Nye, M. J. (1993). *From Chemical Philosophy to Theoretical Chemistry*. Berkeley: University of California Press.

Plaass, P. (1994). *Kant's Theory of Natural Science*. Translation, Analytic Introduction, and Commentary by Alfred E. and Maria G. Miller. With an Introductory Essay by Carl Friedrich von Weizsäcker. Dordrecht: Kluwer Academic Publishers.

Restrepo, G. & Schummer, J. (2013). Editorial. "Mathematics and Chemistry, Part II," *Hyle*, 19(1), 1–2.

Ross, S. (1991). "*Scientist:* The Story of a Word." In *Nineteenth-Century Attitudes. Men of Science*. Dordrecht: Kluwer Academic Publishers, 1–39.

Rouse, J. (2005a). "Heidegger on Science and Naturalism." *Division I Faculty Publications*. Paper 36. http://wesscholar.wesleyan.edu/div1facpubs/36

Rouse, J. (2005b). "Heidegger's Philosophy of Science." In *A Companion to Heidegger* (H. L. Dreyfus, M. A. Wrathall, eds). Oxford: Basil Blackwell, 173–189.

Schummer, J. (1997). "Towards a philosophy of chemistry," *Journal for General Philosophy of Science*, 28, 307–336.

Schummer, J. (2004). "Why Do Chemists Perform Experiments?" In *Chemistry in the Philosophical Melting Pot. (Dia-Logos: Studies in Philosophy and Social Sciences, Vol. 5)* (D. Sobczyńska, P. Zeidler, and E. Zielonacka-Lis, eds.). Frankfurt am Main: Peter Lang Europäischer Verlag der Wissenschaften, 395–410.

Schummer, J. (2006). "The Philosophy of Chemistry: From Infancy Toward Maturity." In *Philosophy of Chemistry: Synthesis of a New Discipline. (Boston Studies in the Philosophy of Science 242)* (D. Baird, E. Scerri and L. MacIntyre, eds.). Dordrecht: Springer, 19–39.

Schummer, J. (2012). "Why mathematical chemistry cannot copy mathematical physics and how to avoid the imminent epistemological pitfalls," *Hyle*, 18(1), 71–89.

Simon, J. (2012). "The production of purity as the production of knowledge," *Foundations of Chemistry*, 14(1), 83–96.

Stepin, V. (2005). *Theoretical Knowledge*. Dordrecht: Springer.

Toulmin, S. (1967). *The Philosophy of Science. An Introduction*, London: Hutchinson.

van Brakel, J. (1999). "On the neglect of the philosophy of chemistry," *Foundations of Chemistry*, 1(2), 111–174.

van Brakel, J. (2000). *Philosophy of Chemistry: Between the Manifest and the Scientific Image*. Leuven: Leuven University Press.

van Brakel, J. (2006). "Kant's Legacy for the Philosophy of Chemistry." In *Philosophy of Chemistry: Synthesis of a New Discipline (Boston Studies in the Philosophy of Science. Vol. 242)* (D. Baird, E. Scerri, and McIntyre, eds.). Dordrecht: Springer, 69–91.

van den Berg, H. (2011). "Kant's conception of proper science," *Synthese*, 183, 7–26.

Vihalemm, R. (1995). "Some comments on a naturalistic approach to the philosophy of science," *Studia Philosophica*, II(38), 9–18.

Vihalemm, R. (1999). "Can Chemistry Be Handled as Its Own Type of Science?" In *Ars Mutandi—Issues in Philosophy and History of Chemistry* (N. Psarros and K. Gavroglu, eds.). Leipzig: Leipziger Universitätsverlag, 83–88.

Vihalemm, R. (2000). "The Kuhn-loss thesis and the case of phlogiston theory," *Science Studies*, 13(1), 68–78.

Vihalemm, R. (2001). "Chemistry as an Interesting Subject for the Philosophy of Science." In *Estonian Studies in the History and Philosophy of Science (Boston*

Studies in the Philosophy of Science, Vol. 219) (R. Vihalemm, ed.). Dordrecht: Kluwer, 185–200.

Vihalemm, R. (2003a). "Are laws of nature and scientific theories peculiar in chemistry? Scrutinizing Mendeleev's discovery," *Foundations of Chemistry,* 5(1), 7–22.

Vihalemm, R. (2003b). "Natural Kinds, Explanation, and Essentialism in Chemistry." In *Chemical Explanation: Characteristics, Development, Autonomy (Annals of the New York Academy of Sciences, Vol. 988)* (Joseph E. Earley, Sr. (ed.). New York: The New York Academy of Sciences, 59–70.

Vihalemm, R. (2004). "The Problem of the Unity of Science and Chemistry." In *Chemistry in the Philosophical Melting Pot. (Dia-Logos: Studies in Philosophy and Social Sciences, Vol. 5)* (D. Sobczyńska, P. Zeidler and E. Zielonacka-Lis, eds.). Frankfurt am Main: Peter Lang Europäischer Verlag der Wissenschaften, 39–58.

Vihalemm, R. (2005). "Chemistry and a theoretical model of science: On the occasion of a recent debate with the Christies," *Foundations of Chemistry,* 7(2), 171–182.

Vihalemm, R. (2007a). "Philosophy of chemistry and the image of science," *Foundations of Science,* 12(3), 223–234.

Vihalemm, R. (2007b). "A. Whitehead's metaphysical ontology and I. Prigogine's scientific ontology: From a point of view of a theoretical conception of science'," *Problemos,* 71, 78–90.

Vihalemm, R. (2007c). "On the Conceptual Systems of Chemistry." In *The History of Science in the Philosphical Context. In Memory of V. I. Kuznetsov (1915-2005)* (A. A. Pechenkin, ed.). St.-Petersburg: RHGA, 49–70. [In Russian: Вихалемм, Р. К вопросу о концептуальных системах химии. В кн.: А. А. Печенкин (ред.). *История науки в философском контексте. Посвящается памяти Владимира Ивановича Кузнецова (1915-2005).* СПб: РХГА, стр. 49–70.]

Vihalemm, R. (2011a). "A Monistic or a Pluralistic View of Science: Why Bother?" In *An den Grenzen der Wissenschaft. Die "Annalen der Naturphilosophie" und das natur- und kulturphilosophische Programm ihrer Herausgeber Wilhelm Ostwald und Rudolf Goldscheid. Die Vorträge der Konferenz, veranstaltet von der Sächsischen Akademie der Wissenschaften zu Leipzig und dem Institut für Philosophie der Universität Leipzig im November 2008* (P. Stekeler-Weithofer, H. Kaden, & N. Psarros, eds.). (Abhandlungen der Sächsischen Akademie der Wissenschaften zu Leipzig. Philologisch-historische Klasse. Band 82, Heft 1: 79-93). Stuttgart/Leipzig: Sächsische Akademie der Wissenschaften zu Leipzig. In Kommission bei S. Hirzel.

Vihalemm, R. (2011b). "The autonomy of chemistry: Old and new problems," *Foundations of Chemistry,* 13, 97–107.

Vihalemm, R. (2011c). "Towards a practical realist philosophy of science," *Baltic Journal of European Studies,* 1(9), 46–60.

Vihalemm, R. (2012). "Practical realism: Against standard scientific realism and anti-realism," *Studia Philosophica Estonica* 5.2, 7–22.

Vihalemm, R. (2013). "What Is a Scientific Concept? Some Considerations Concerning Chemistry in Practical Realist Philosophy of Science." In *The Philosophy of Chemistry: Practices, Methodologies and Concepts* (J.-P. Llored, ed.). Cambridge: Cambridge Scholars Publishing, 364–384.

Westphal, K. R. (2004). *Kant's Transcendental Proof of Realism.* Cambridge: Cambridge University Press.

INDEX

Figures, notes, and tables are indicated by f, n, and t following the page number.

American Chemical Society, 9, 10n8
Ammonia, 195
Analytical ideal, 46n42
Analytical method in physics, 268
Angular momentum. *See* Spin
Anisotropic processes, 224
Annales de Chimie, 40n11, 43
Anodic oxidations, 163
Anomalous configurations, 136–137,
    137*f*, 140–142, 140*f*
Anomalous monism, 111
Ante rem realism, 210
Anthracene, 170–171
Anti-gauche interconversion, 162
Anti-realism
    dual character of chemistry and,
        365–366
    natural kinds and, 258
    property coherence in substances, 214
    reaction mechanisms and, 156
    scientific realism and, 234,
        234–235nn1, 236–238, 240
Anti-reductionism
    reductionism and, 126–127, 126n2,
        135–137, 141–142
    supervenient laws and, 112, 119, 121
Approximation models and procedures,
        87–88, 215, 241, 295, 306,
        308, 310–315, 317, 319–325,
        326–327. *See also* Ab initio
        methods; Hartree-Fock method;
        Semi-empirical modeling and
        approximations
A priori explanations and requirements
    Chemical Revolution and, 37, 46, 53
    dual character of chemistry
        and, 355–356, 359n8, 361–363
    natural kinds and, 257–258, 262
    property coherence in substances
        and, 207
    supervenient laws and, 119
Archimedes, 19, 370
Aristotelian elements and worldview,
        15–17, 21–23, 30, 44–46, 109, 151,
        179–180, 222–223, 268, 375
Aristotle, 15, 44, 108, 151, 179–180,
        179n2, 200, 222, 268, 368, 375

Armstrong, David, 115n13, 203,
    210, 220
Aromatic compounds, 164–167,
    294–295
Arrhenius, Svante, 66, 335
Artificial fiber metaphor, 76
*Aspects de la Chimie Quantique
    Contemporaine* (Hoffman), 69
Astronomical realism, 234–235n1
Astronomy, 38, 234, 245
Asymptotic expansions, 215–216
Atomic chemistry, 236, 246
Atomicity, 107–108
Atomic number
    contingent chemistry and, 145
    natural kinds and, 261, 267
    reductionism and, 129–130, 136, 139
    supervenient laws and, 119–121
Atomic population, 87, 89
Atomic theory of matter, 120–121, 151
Atomic weight
    contingent chemistry and, 145
    mathematical chemistry and, 334,
        334n2
    natural kinds and, 261
    reductionism and, 129–130, 129*f*
    scientific realism and, 236
    supervenient laws and, 120–121.
        See also Periodic law
Atomism
    causality and, 193
    corpuscular chemistry and, 15–36
    natural kinds and, 256n1, 266
    philosophy of mind and, 107–108
    quantum chemistry and, 73
    scientific realism, 235–237, 247
Atoms
    causality and, 190–192, 194
    corpuscular chemistry and, 17–18
    divergence and, 310–311, 322–324
    mathematical chemistry and, 334–335,
        334n2, 338
    natural kinds and, 253–254, 256n1,
        258, 260–262, 264, 266–267,
        269, 271
    philosophy of mind and, 85–92,
        89n12, 94, 98–100, 107–108

scientific realism and, 236
    The Theoretician's Dilemma
        and, 282–285
Potassium nitrate, 21, 21n5
Potential energy, 205–206, 308–309
Potential energy surfaces (PESs), 261
Powerful particular. *See* Causal agency
    and agents
Practical realism, 357–358, 361–362,
    365–368, 374
Practices in mathematical chemistry,
    342–343
Pragmatic Maxim, 201
Pragmatic tradeoffs, 306, 313, 315–320,
    316n9, 318n10, 322–323, 325,
    327–328
Pratten, Stephen, 224
Precipitation, 26–28, 95–96
Precision of methods, 316–319
Predicates defined, 200–201
Predictive accuracy, 9, 309, 319–323,
    322n13, 325, 327
Predictive divergence, 315
Predictive powers, 313
Preservative realism vs. conservationist
    pluralism, 245–249
Pressure
    corpuscular chemistry and, 18–19, 21,
        23–24, 27n10, 30, 32
    natural kinds and, 270
Priest, G., 298n13, 300n14
Priestley, Joseph, 43n28, 48, 48n50,
    243, 244
Prigogine, Ilya, 217, 219
Primacy of Physics Constraint, 300
Primary alkyl halides, 164
Primary levels, 118, 201
Primas, Hans, 73, 216
Primitive affections, 18, 18n2
Principe, Lawrence, 20n4, 34n17
Principia (Newton), 31
Principle of Naturalistic Closure, 300
Principle of relativity (Galilean), 211
Principle of unity, 202–204
Principles of bodies. *See* Elements
Principles of Chemical Revolution, 40,
    44–45, 46, 46n42, 49, 50

Prism models, 239
Probability
    causality and, 181, 195
    reaction mechanisms and, 159, 170–171
    transition to reality from, 369–370n12
Processes, 201–202
Promiscuous ontology, 291–297
Promiscuous realism, 262
Proper science, 355–358, 361–364, 362n9
Properties of theories, 285, 296
Property coherence in substances,
    199–228
    biochemical dissipative structures
        and, 218
    causality and, 192, 196
    causes and determinants of, 222–228
    chemical systems properties and,
        204–208
    closures and, 213–219
    contextual determination and, 226–227
    contingencies in explanations and,
        146–149
    event and agent causality and, 221–222
    intrinsic properties and, 220–221
    limits to agency and, 223–224
    molecular properties and, 204–205
    non-agentive determination and,
        225–226
    ordinary things and, 227–228
    overview, 199–204
    Peircean determinants and, 224–225
    properties and relations of, 200–201
    proposed principle of unity for
        bundles and, 202–204
    reaction mechanisms and, 169, 172
    relative onticity and, 207–208
    spectroscopic properties and, 205
    structuralist revival and, 219–220
    structures and, 219–222
    substance-attribute approach
        alternatives and, 201–202
    substances, defined, 200
    substrate-bundle debates and, 202
    supervenient laws and, 112, 116,
        121–122
    symmetries and, 210–213
    wholes and parts as, 208–210

Property dependence and
supervenience, 113
Property instances, 199, 202, 224–225
Protein chemistry, 316n8
Protein synthesis, 162–163
Protons, 194, 212, 219
Proton transfer, 195
Proust, Joseph, 49, 250
Proximate causes, 223–224
Psillos, Stathis, 221–222, 247, 282,
284n2
Ptolemy, 39, 169, 245
Publications on mathematical
chemistry, 337–338, 339–341t, 345
Purdue University, 150
Purity and purification, 92–94, 256, 270
Purvis, M.K., 186–187
Putnam, Hilary, 98, 107, 126, 201, 254,
263, 266–268, 286n5
Pyle, Andrew, 39n10

QTAIM (Quantum Theory of Atoms in
Molecules), 90
Qualitative approach, 203, 307–308,
308n1, 316n8, 326–328
Quantification and quantitative analysis,
287–288
Chemical Revolution and, 40–41,
41n17, 43, 48, 48n54, 53, 53n77
contingencies in explanations and,
151–152
methods and models, 101, 315–316,
324n15, 326, 328
quantum chemistry and, 66
Quantifiers of theories, 290
Quantization, 204–205
Quantum calculations, 248
Quantum chemistry
contingencies in explanations and,
145–146
diagnostics and, 306–308, 310, 311n4,
313–315, 317, 319, 324
experiments through computers in,
68–70, 314
old problems (re)considered in,
70–74
overview, 74–77

philosophy of mind and, 96–110
revisiting history of, 60–64
role of theory in chemistry and,
64–68
scientific realism and, 243, 248–249
as (sub)-discipline, 60–77
supervenient laws and, 121
The Theoretician's Dilemma
and, 298
Quantum chromodynamics, 235, 250
Quantum Field Theory, 107–109,
247, 248
Quantum mechanics
contingencies in explanations and,
145, 152
diagnostics and, 313
philosophy of mind and, 107
property coherence in substances
and, 213–214
quantum chemistry as (sub)-discipline,
60–61, 63, 67–68, 71, 73, 75–76
reaction mechanisms and, 154,
160, 172
reductionism and, 125–129,
133–137, 141
scientific realism and, 237, 239, 241,
248, 249
supervenient laws and, 115, 121–122
Quantum numbers, 133, 134f, 135, 138
Quantum organic chemistry, 317
Quantum physics, 96, 103, 106–107,
128, 207
Quantum theory, 213–214, 280–281,
296, 298, 300, 302, 342
Quantum Theory Group, 70
Quantum Theory of Atoms in
Molecules (QTAIM), 90
Quarks, 201, 235
Quick, R., 226
Quine, Willard V., 208–209, 247,
287–288, 290

Radiation, 133, 243
Radioactivity, 132
Rainforest realism, 298n13, 299–302
Ramsey, F.P., 283–285, 284n2, 287–288
Ramsey, J., 73, 314, 320–321, 322n12

Ramsey sentences, 282–284, 284n2,
287–288
Randić, M., 336, 343–345, 343n9
Rates of chemical reactions. *See*
Kinetics; Reaction mechanisms
Rationality, 50, 219
Reactants
causality and, 181, 191, 195
diagnostics and, 307–308
natural kinds and, 265
property coherence in substances
and, 206–207, 217
reaction mechanisms and, 156–162,
165, 168–169, 171–172
The Theoretician's Dilemma and, 294
Reaction coordinate diagram, 159
Reaction mechanisms, 154–172
causality and, 185, 190–197
cause and effect and, 168–169
contingencies in explanations
and, 146–147
defined, 156–157
explanatory power of chemical
concepts and, 164–166
falsification and, 169–171
historical background of, 154–155
modeling of, 307–312
natural kinds and, 260, 264–265
organization of, 166–168
overview, 172
philosophical considerations for,
155–172
philosophical views of, 161–163
property coherence in substances
and, 199, 205–207, 215, 216–218
scientific realism and, 155–156,
241, 249
of S_N2 reaction, 157–160
The Theoretician's Dilemma and,
294–296
Reactions, organization of, 166–168
Reactivity, 122
Realism
diagnostics and, 316–317, 319,
322n13, 324
dual character of chemistry
and, 355–361, 356n7, 365–366

natural kinds and, 257, 259
pluralism vs., 245–249
scientific realism, 234–235n1
The Theoretician's Dilemma and,
288n7, 292, 293, 298n13,
299–302
Reasons, 223–224, 226
Reciprocal causality, 224
Reciprocal determination, 224
Redescription, 119–121, 123
Reductionism, 125–142
anomalous configurations
and, 136–137, 137f
anti-reductionism and, 135–136
causality and, 179–182
contingencies in explanations and,
144, 149, 152
dual character of chemistry and,
362n9
electronic configuration and, 137–141,
138–140f
Madelung Rule and, 134–135, 136f
overview, 141–142
periodic system and, 128–132,
129–131f
philosophy of mind and, 83–87,
89–91, 97–107
physics and, 132–133, 132f, 134f
property coherence in substances
and, 201
quantum chemistry as (sub)-discipline,
62, 70–74, 71n4, 76
question of, 125–128
reaction mechanisms and, 161
scientific realism and, 248–250,
249n12
supervenient laws and, 112, 114–115,
114n9, 117, 121, 123
The Theoretician's Dilemma and,
281, 283, 299
Reflections on Physics and Chemistry
(Kant), 362
Regimentation of theories, 287–288
Regiochemistry, 166
Regularity analysis, 180–181, 197
Regularity between events, 185
Reid, Thomas, 183

natural kinds and, 266
overview, 111–113
periodic law as example of, 119–122
property coherence in substances, 215–216
scientific explanations and, 122–123
Supervenience view of truth-making, 288–289, 289t
Suppe, F., 285
Surface science, 63
Sylvester, J.J., 335, 336
Symmetry allowed reactions, 307
Symmetry concept, 210–213
Symmetry forbidden reactions, 307
Symmetry-group, 220
Synthesis. *See also* Reaction mechanisms
  Chemical Revolution and, 33
  contingencies in chemical explanation and, 145–146
  corpuscular chemistry and, 33
  diagnostics and, 308
  dual character and, 353, 356n5
  natural kinds and, 255–257, 269
  philosophy of mind and, 93
  quantum chemistry as (sub)-discipline, 65
  reaction mechanisms and, 160, 162–163, 167–168, 171–172
  scientific realism and, 237, 242
Synthetic chemistry, 242, 307
Systems
  causal power and, 221–222
  closures and, 213–219
  coherence and, 209–213, 210n4
  contextual determination of, 226
  limits to agency in, 223–224
  of ordinary things, 227–228
  properties of, 204–208

Tabery, J.J., 163
Table of elements, 235, 235n2. *See also* Periodic table and system
Table salt, 92, 150
Taxonomic systems. *See* Linguistic framework

Technology, 374
Tellurium, 120
Temperature
  natural kinds and, 258
  property coherence in substances and, 204–205
  reaction mechanisms and, 169
Tenaillon, O., 226
Tennis balls as ordinary entities, 202–203, 227
Terminology. *See also* Linguistic framework
  natural kinds and, 257
  reaction mechanisms and, 166
  The Theoretician's Dilemma and, 279–280, 282–283, 285n4, 286, 292, 300n15
Tetrahedral structures, 146–147, 158, 240
Tetramethylsilane (TMS), 95
Textbooks, idealized accounts in, 30, 121, 135, 141, 148, 241, 280, 293
TGro (grounding view of theories), 6, 281, 288–291, 289t, 293–294, 297–300
Thagard, Paul, 39n10
Themata, 247
Theoretical chemistry, 69, 348
Theoretical desiderata, 318–319, 318n10, 321–322, 322n13
Theoretical dimension, 16–17
Theoretical entities, 180, 202
Theoretical frameworks, 62–69, 72–77
Theoretical reductionism, 128
The Theoretician's Dilemma, 279–302. *See also* Hempel, C.G.
  chemical objects and, 297–299
  contingencies in chemical explanation and, 146
  elements as example of, 291–293
  ontological reduction and, 299–300
  ontology and, 287–291, 297–302
  orbital explanation as example of, 293–297
  overview, 279–282
  quantifiers and, 287–288

Truth
  diagnostics and, 313, 324
  natural kinds and, 267
  scientific realism and, 247–248,
    247n11
  The Theoretician's Dilemma and,
    284–285
Truth-making, 6, 281, 287–291, 288n7,
  289n8, 289t, 293–294, 297–300
Truth-value logic, 208n3, 322n12, 371
TSup (supervenience view of truth-
  making), 288–289, 289t

Ultimate causes of corpuscular
    chemistry, 17–19, 18n2, 22–23,
    223–224
Underdetermination
  scientific realism and, 242,
    244–245, 247
  supervenient laws and, 117
Unidirectional causality, 224
Unimolecular reactions, 161–162
Uniqueness of composition, 208–209
Unity principle for properties, 202–204
UNIVAC (universal automatic
    computer), 70n3
Universalism, 210
Universal laws, 146
Universals, 202, 220, 226. See also
    Attributes
University Computing Laboratory, 70
University of Cambridge, 70
University of Illinois, 144
Unobservable entities, 234–238,
    239, 285
Unrestricted composition, 208–209
Unsolved problems in mathematical
    chemistry, 346–347
Urea, 146

Valence
  causality and, 188, 193–196
  diagnostics and, 310
  philosophy of mind and, 85–92
  quantum chemistry as
    (sub)-discipline, 67

scientific realism and, 239–240
supervenient laws and, 119–122
The Theoretician's Dilemma and,
    294, 297–298
van Brakel, Jaap, 3n1, 73, 266–269, 292,
    362, 363, 372
Vander Laan, David, 210
van der Vet, P., 296
van Fraassen, Bas, 234, 234–235n1, 237
Van't Hoff, J.H., 64–66, 71n4, 335–336
Varela, Francisco, 103
Variation Principle, 89–90
Varzi, A.C., 261
Vector field, 91
Velocity of free-falling objects, 369–370
Vemulapalli, G.K, 90
Venel, Gabriel-François, 40–43, 40n13,
    53, 332
Vermeeren, H., 3n1
Vihalemm, Rein, 8, 352
Villaveces, J.L., 333–334, 336, 345
Virulence, 225–226
Vital air as oxygen, 44n34, 52n75
Voltage switch as mechanism, 161

Water
  natural kinds and, 254, 266, 269, 271
  reaction mechanisms and, 171
Watt, James, 48, 48n50
Wave mechanics, 313
Wave-properties, 214
Weight
  Chemical Revolution and, 41, 44
  contingencies in explanations
    and, 145
  corpuscular chemistry and, 17–19, 21,
    23, 29–30, 32
Weighting coefficients, 85–89,
    89nn12–13
Weinberg, Steven, 155, 156
Weisberg, Michael, 172, 308n1, 316–319,
    316n8, 318n10
Weiss, Paul, 200
Weitz, J., 226
Westphal, Kenneth, 356–357, 356n7
Weyl, Hermann, 211, 333